Milestones in Drug Therapy
MDT

Drugs for Relapse Prevention of Alcoholism

Edited by R. Spanagel and K.F. Mann

Birkhäuser Verlag
Basel · Boston · Berlin

Editors

Rainer Spanagel
Universität Heidelberg
Abteilung für Psychopharmakologie
J5, Zentralinstitut für Seelische Gesundheit
68159 Mannheim
Germany

Karl F. Mann
Klinik für Abhängiges Verhalten und Suchtmedizin
J5, Zentralinstitut für Seelische Gesundheit
68159 Mannheim
Germany

Library of Congress Cataloging-in-Publication Data

Bibliographic information published by Die Deutsche Bibliothek
Die Deutsche Bibliothek lists this publication in the Deutsche Nationalbibliografie; detailed bibliographic data is available in the internet at http://dnb.ddb.de

ISBN 3-7643-0214-3 Birkhäuser Verlag, Basel - Boston - Berlin

Part of Springer Science+Business Media
Printed on acid-free paper produced from chlorine-free pulp. TFC ∞

Printed in Germany
ISBN-10: 3-7643-0214-3
ISBN-13: 978-3-7643-0214-6

9 8 7 6 5 4 3 2 1

www. birkhauser.ch

Contents

List of contributors

Ludovico Abenavoli, Institute of Internal Medicine, Catholic University, Gemelli Hospital, L.go A. Gemelli 8, 00168 Rome, Italy
Giovanni Addolorato, Institute of Internal Medicine, Catholic University, Gemelli Hospital, L.go A. Gemelli 8, 00168 Rome, Italy; e-mail: g.addolorato@rm.unicatt.it
Roberta Agabio, Department of Neuroscience, University of Cagliari, Viale Diaz 182, 09126 Cagliari, Italy; e-mail: agabio@unica.it
Daniel Bachteler, Department of Psychopharmacology, University of Heidelberg, Central Institute of Mental Health (CIMH), J5, 68159 Mannheim, Germany; e-mail: dbach@zi-mannheim.de
Fabio Caputo, G. Fontana, Center for the Study and Treatment of Alcohol Addiction, University of Bologna, Viale Risorgimento 2, 40136 Bologna, Italy
Mauro A.M. Carai, Department of Neuroscience, University of Cagliari, Viale Diaz 182, 09126 Cagliari, Italy; e-mail: macarai@unica.it
Giancarlo Colombo, C.N.R. Institute of Neuroscience, Section of Cagliari, Viale Diaz 182, 09126 Cagliari, Italy; e-mail: colomb@unica.it
Michael S. Cowen, Howard Florey Institute, University of Melbourne, VIC 3010, Australia; e-mail: m.cowen@hfi.unimelb.edu.au
Giosuè deLorenzi, Institute of Internal Medicine, Catholic University, Gemelli Hospital, L.go A. Gemelli 8, 00168 Rome, Italy
Philippe De Witte, UCL – Biologie du Comportement, Place Croix du Sud, 1 bte 10, 1348 Louvain-la-Neuve, Belgium; e-mail: dewitte@bani.ucl.ac.be
Griffith Edwards, National Addiction Centre, Institute of Psychiatry, 4 Windsor Walk, London SE 5AF, UK; e-mail: p.davis@iop.kcl.ac.uk
Anna Ferrulli, Institute of Internal Medicine, Catholic University, Gemelli Hospital, L.go A. Gemelli 8, 00168 Rome, Italy
Douglas R. Funk, Department of Neuroscience, Center for Addition and Mental Health, Toronto, Ontario, Canada M5S 2S1
Giovanni Gasbarrini, Institute of Internal Medicine, Catholic University, Gemelli Hospital, L.go A. Gemelli 8, 00168 Rome, Italy; e-mail: ggasbarrini@rm.unicatt.it
Gian Luigi Gessa, Department of Neuroscience, University of Cagliari, Cagliari; and C.N.R. Institute of Neuroscience, Viale Diaz 182, 09126 Cagliari, Italy; e-mail: lgessa@unica.it
Markus Heilig, Laboratory of Clinical and Translational Studies, National Institute of Alcohol Abuse and Alcoholism, National Institiute of Health, 10 Center Drive (10/1E5330), Bethesda, MD, 20892-1108, USA; e-mail:

markus.heilig@mail.nih.gov

Petri Hyytiä, Department of Mental Health and Alcohol Research, National Public Health Institute, POB 33, FI-00251 Helsinki, Finland; e-mail: petri.hyytia@ktl.fi

Bankole A. Johnson, South Texas Addiction Research & Technology Center, Department of Psychiatry, The University of Texas Health Science Center at San Antonio, 3939 Medical Drive, Suite 100, San Antonio, TX 78229-3900, USA; e-mail: bjohnson@uthscsa.edu, and Departments of Psychiatric Medicine and Neuroscience, University of Virginia, P.O. Box 800623, Charlottesville, VA 22908-0623, USA; e-mail: bankolejohnson@virginia.edu

Henry R. Kranzler, Department of Psychiatry, University of Connecticut Health Center, Farmington, CT 06030-2103, USA; e-mail: kranzler@psychiatry.uchc.edu

Dieter Ladewig, Department of Substance Use Disorders, Psychiatric University Hospital, Wilhelm Klein-Strasse 27, 4025 Basel, Switzerland; e-mail: dieter.ladewig@pukbasel.ch

Anh Dzung Lê, Department of Neuroscience, Center for Addiction and Mental Health, Toronto, Ontario, Canada M5S 2S1; and Department of Pharmacology and Psychiatry, University of Toronto, Toronto, Ontario, Canada M5S 1A8; e-mail: anh_le@camh.net

Lorenzo Leggio, Institute of Internal Medicine, Catholic University, Gemelli Hospital, L.go A. Gemelli 8, 00168 Rome, Italy

Carla Lobina, Department of Neuroscience, University of Cagliari, Viale Diaz 182, 09126 Cagliari, Italy

Karl F. Mann, Department of Addictive Behaviour and Addiction Medicine, University of Heidelberg, Central Institute of Mental Health, J 5, 68159 Mannheim, Germany; e-mail: sucht@zi-mannheim.de

Charles P. O'Brien, University of Pennsylvania, Department of Psychiatry/Philadelphia Veterans Affairs Medical Center, Treatment Research Center, 3900 Chestnut Street, Philadelphia, PA 19104-6178, USA; e-mail: obrien@mail.trc.upenn.edu

David W. Oslin, University of Pennsylvania, Department of Psychiatry/Philadelphia Veterans Affairs Medical Center, Treatment Research Center, 3900 Chestnut Street, Philadelphia, PA 19104-6178, USA; e-mail: oslin@mail.med.upenn.edu

Helen M. Pettinati, University of Pennsylvania, Department of Psychiatry/Philadelphia Veterans Affairs Medical Center, Treatment Research Center, 3900 Chestnut Street, Philadelphia, PA 19104-6178, USA; e-mail: pettinati_h@mail.trc.upenn.edu

John David Sinclair, Department of Mental Health and Alcohol Research, National Public Health Institute (KTL), POB 33, Helsinki 00251, Finland; e-mail: david.sinclair@ktl.fi

Rainer Spanagel, Department of Psychopharmacology, Central Institute of Mental Health (CIMH), University of Heidelberg, J5, 68159 Mannheim,

Germany; e-mail: spanagel@zi-mannheim.de

Howard Tennen, Department of Community Medicine and Health Care, University of Connecticut Health Center, Farmington, CT 06030-2103, USA; e-mail: tennen@nso1.vchc.edu

Todd E. Thiele, Department of Psychology, University of North Carolina, Davie Hall, CB 3270, Chapel Hill, NC 27599-3270, USA

Ulrich von Bardeleben, Psychiatrische Klinik Meiringen, 3860 Meiringen, Switzerland; e-mail: ulrich.vanbardelen@pm-klinik.ch

Friedbert Weiss, Department of Neuropharmacology, The Scripps Research Institute (CVN 15), 10550 North Torrey Pines Road, La Jolla, CA 92037, USA; e-mail: bweiss@scripps.edu

Gerhard A. Wiesbeck, Department of Substance Use Disorders, Psychiatric University Hospital, Wilhelm-Klein-Strasse 27, 4025 Basel, Switzerland; e-mail: gerhard.wiesbeck@pukbasel.ch

Preface

Alcoholism is a pathological behavioural syndrome, characterised by compulsive alcohol use, craving and relapses, even recurring after many years of abstinence. It is suggested that chronic alcohol abuse leads to persistent changes within several neurochemical pathways in the brain and furthermore that an imprinted drug and addiction memory may scarcely be extinguished. Hence, the question arises as to whether there ought to be a reasonable hope that pharmacological drugs will be developed that interfere with an addiction memory, and as a result, finally lead to a cure?

In this book, leading preclinical and clinical experts in the field of alcohol relapse prevention strive to furnish an answer to this question. None of the researchers or clinicians believes in a magic bullet that will be of help to all alcoholic patients in overcoming this disease. However, there is now convincing evidence demonstrating that specific subpopulations of alcoholic patients experience satisfactory benefit from currently available treatments. Today we have two medications for relapse prevention on the market – acamprosate and naltrexone. Although, currently, only a minority of alcoholic patients benefit from these medications, the approval of these compounds may be considered a hallmark in the field of psychopharmacology, even comparable to the era when the first antidepressant compounds were introduced.

In recent years we have been witnessing an enormous growth in the science and knowledge regarding the field of relapse prevention. The combination of elaborated animal models, molecular and genetic approaches along with clinical studies, have revealed novel drug targets as well as more effective pharmacological treatments. In contrast to developments in other areas of psychiatric diseases, convincing neurobiological concepts have evolved and there is a better understanding of how alcohol affects our brain. A driving sociological force behind these rapid developments has been a very serious need in our society, demanding better treatments for alcoholic patients. With the advent of alcoholism having publicly become accepted as a disease of the brain, many stigmas associated with this behavioural syndrome were torn down. Thus, patients are now more willing to accept the notion and the act of taking medication, especially as there are already very promising pharmacotherapeutic drugs on the horizon. However, the pharmaceutical industry is still reluctant to conquer this enormous market, simply because their market analysis is performed based on a profitable short-term run, thus misjudging the situation.

We hope that this work will help in speeding up the integration of anti-relapse treatment therapies into our health-care systems. The book is divided up into blocks of preclinical and clinical research, based on specific compound meeting. The man was conducted to a seat, and then was prayed for by some of the ladies present. The prayers which were so earnestly uttered that night by these women undoubtedly greatly impressed the man...A few days after he was converted to Christ."

Such a story will probably seem somewhat mawkish to the objectively minded modern professional, and today's drunk will not be prayed over in the A and E Department. And yet salvation was a well-articulated model of relapse prevention. It saw the tendency to relapse as residing in the person, and the remedy was only a holistic approach that brought about total change in that person's sense of being. Only that kind of life change could be expected to extirpate the problem. Relapses could be dealt with by more prayer, but true conversion was expected to result not in a transient amelioration of the drinking behaviour but in permanent cure. The Temperance group, the church circle and new and sober friendships, were on hand to provide the long-term after-care.

Temperance workers knew that reform could easily be followed by relapse, and they were not unrealistic in their expectations. But there were plenty of life-long reformed drunkards to give their testimony from the platform. A great Temperance orator, John B. Gough, who was known in his time as "The Temperance Apostle", had a harrowing personal history of alcoholic debauchery to give him credibility as a speaker [11].

The Temperance Movement thus created as a recognisable and popular image the person who is likely today to be designated "the recovering alcoholic". The movement faded but reference to the importance of spirituality in recovery lives on [12]. Whatever the researcher's or professional's personal stance on formal religion, it can hardly be doubted that what the religious approach can contribute to relapse prevention sets good questions. "Getting really saved" is what P.E. Turner [13] deemed to be the heart of therapy, but the kind of personal change-experience identified here is not limited to religious experience alone [14].

Treatment makes its entry as means to prevent the drunken relapse

Treatment when it arrived on this scene did not supplant punishment or salvation as methods for curing relapse. It coexisted with these older remedies and at times absorbed elements from them. But in the 18th and early 19th century, doctors began to appear who claimed that drunkenness was a treatable medical condition [15]. That was revolutionary thinking. These doctors defined the target condition as "drunkenness" and the differentiation between drunkenness and dependence was at that time unknown.

Within the armamentarium of treatment there developed over time two main elements. One was the psychological, culminating in the psychologically based relapse prevention approaches as epitomised today in the formulations offered by Marlatt and his colleagues [1]. The other was physical, with its fruition seen in the sorts of drug treatment that are the subject of this book. We will consider each of these developments in turn. To an extent they mesh with the contrasting underlying formulations as to the nature of the disorder as learnt habit, or alternatively as disease of the brain.

Relapse prevention as a psychological undertaking

It was the concept of excessive drinking as learnt habit that, at the turn of the 18th–19th centuries, legitimised the presence of the medical doctor in this treatment arena. A potent background influence was the ferment of ideas resulting from the European enlightenment, with its emphasis on rationality [16]. Benjamin Rush [17] and Thomas Trotter [18] contributed to emerging views on the nature of drunkenness and its appropriate treatment. Trotter's thinking on these matters was particularly well worked out and we will here concentrate on the ideas elaborated in his 1804 "Essay on Drunkenness" [18].

Trotter asserted that intoxication was an event that lay within the mental sphere, or as he put it "the habit of drunkenness is a disease of the mind": the word "disease" has to be understood as carrying a meaning close to "disorder" rather than anything more specific. He said that it would be useless to treat the physical complications of excessive drinking until "the evil genius of the habit has been subdued". Craving was a manifestation of the habit: -

> "The cravings of appetite for the poisonous draft are to the intemperate drinker as much the inclinations of nature for the time, as a draft of cold water to a traveller panting of thirst in the desert."

The "duty of treatment" was to break "the chain of habit" and the follow-through from theoretical exposition to therapy was close, with the need to identify relapse precipitants explicit. The case analysis should lead to identification of "the particular cause, time and place of his love of the bottle" and "something proposed that will effectually wean his affection from it". The word "relapse" was in Trotter's vocabulary.

Trotter probably derived his theoretical ideas from David Hume, the Edinburgh philosopher who was a powerful enlightenment thinker. There is in Trotter's writing a feeling of the radical break with all that had gone before. The invitation was to understand the behaviour rather than bluntly punish or preach against it. At or around that enlightened time the psychological treatment of drinking problems, as we today know it, was born.

In the 19th century many writers showed the persistence of interest in development of a psychological approach to treatment of the drinking habit [19], although few if any of these authors matched Trotter's clarity of vision. The application in the 1950s of Pavlovian conditioning to the treatment of alcohol dependence [20] represented a formal attempt to apply psychological theory to treatment of the drink habit, but the theoretical analysis was not highly developed. But in the 1950s and against a background of the strong general emergence of behaviour therapy as a theory-based treatment for psychological disorders, reports began to be published on the application of learning theory to the treatment of drinking problems [21–23].

Important further contributions to this field of thought were made by Litman [24]. A book published in 1985 by Marlatt and Gordon [1], then brought together work of a scope and quality to give relapse prevention both a firm theoretical underpinning and an important place in the repertoire of relevant treatments. Indeed, the phrase "relapse prevention" from 1985 onwards for many people became synonymous with the ideas developed by Marlatt and his colleagues. Several reviews have reached positive conclusions on the effectiveness of this approach [25] although one review has reached more guarded conclusions [26]. Relapse prevention as conceived by Marlatt today constitutes a broad array of psychological treatments, but pharmacological approaches are not in the listing [25].

Relapse prevention as physical treatment

The view became general among 19th century physicians that inebriety led to exhaustion of the nerve cells, with this brain disorder then perpetuating the problem [27]. The exhaustion could be relieved by abstinence, rest, and the general provisions of a sanatorium regime. A number of physical treatments could, however, be added to the interventions and were, by implication, agents of relapse prevention.

Norman Kerr's influential textbook on the treatment of "Inebriety or Narcomania" was first published in 1888 [27]. He was in 1884 the inaugural president of the society that later evolved into the present-day Society for the Study of Addiction, and he was a founding editor of the journal which is today published as Addiction [28].

Kerr provided a detailed critique of the then current physical treatments for alcohol dependence. He insisted that inebriety was "a complex disease", and although not dismissive of the psychological components in treatment, he put more emphasis on the physical treatments than Trotter had done. His was the era of disease theories, and disease required somatic treatments. He warned against nostrums and was dismissive of vegetarianism as a cure for inebriety. Various tonics might however be beneficial, including "Gentian and dilute nitro-hydrochloric acid with cardamoms", and "orange and quinine" had a place. Strychnine might cautiously be prescribed. There was a place for cod liver oil as a specific "nerve food". Kerr was dismissive of the then fashionable Turkish bath as a 7-day cure – "Alas for the inebriate this is but an oriental dream".

Kerr's faith in drug treatments to restore the health of his patients' exhausted brain cells was in accord with the therapeutic beliefs of the period. Over the following decades and with no availability of controlled trials, unbridled therapeutic enthusiasm led in this sector of practice to a variety of physical remedies such as to constitute an "era of anything goes" [2].

The availability in the 1940s of disulfiram (antabuse) for the first time put on offer a pharmacotherapy specifically directed at preventing alcoholic relapse [29]. It is difficult to imagine a drug technology more sharply intended to sever Trotter's "chain": If the patient drank, after taking this drug, rather than their experiencing the intended pleasure, they would sustain an unpleasant toxic reaction. After 60 or so years of clinical experience, disulfiram as relapse prevention technology does not, however, appear to have won widespread acceptance. Toxic reactions can occur and compliance will be poor unless dosing is supervised and some doctors will be uneasy about the ethicality of risking, in the name of treatment, a potentially dangerous alcohol disulfiram reaction [30].

In 1980 Robin Murray published in the lineal descendant of Kerr's journal a prescient editorial entitled "Why Are The Drug Companies so Disinterested in Alcoholism?" [31]. Murray at the time argued,

> "...there is no available drug which acts in a fundamental way to hinder the development and progression of the alcohol dependence syndrome...Surely the time is now right for one or more of the pharmaceutical companies to mount a major research programme aimed at developing drugs which interfere with the biochemical mechanisms involving tolerance and dependence on alcohol."

That editorial stimulated the convening in 1983 of an international conference at the Institute of Psychiatry, London, under the title "Pharmaceutical treatments for alcoholism". The Proceedings volume was published in 1984 [32].

The 1983 meeting and the publication that followed, represented an intentional attempt by the academic community to provoke drug firm interest in the treatment of drinking problems. Whether those moves had any tangible impact is an open question. The first controlled report on the use of Acamprosate in the treatment of alcoholic dependence appeared in 1985 [33], while the first major reports on the use of Naltrexone for this purpose came in 1992 [34, 35]. The present volume speaks to the burgeoning of interest, which has since occurred in alcohol-related pharmacotherapies, and in the study of underlying brain mechanisms.

Alcoholics anonymous as agent of relapse prevention

Alcoholics Anonymous is a self-help group that traces its origin to an event in 1933 [36, 37]. A precariously sober stockbroker, "Bill W", called on a recently relapsed surgeon, "Dr. Bob", and helped him to pull out of a relapse and find sobriety. It could be argued that AA from its inception has been an organisation dedicated to relapse prevention.

How AA participation may achieve the relapse prevention aim was described thus by Edwards [38]: -

> "Many aspects of AA actively seem designed to teach the necessary cognitive and coping skills for achievement and maintenance of the drink-free life. Indeed, an AA meeting can sometimes seem almost to take on the guise of a cognitive-behavioural workshop. Don't get tired, don't get angry, learn to laugh at yourself, see the other person's point of view. Look out for the tricks your thinking can play on you, the planned relapse, the planned excuses, the self-justifying resentment. Have something else to do, don't get bored. Carry some chocolate. Count your blessings. Don't go to that party if you are going to find the temptation to drink too great."

AA defines abstinence as the cure. It is an organisation that not only seeks to prevent relapse but has a capacity to deal with relapse and invites its members to learn from the relapse experience. The programme contains within it a spiritual element but the extent to which the individual member aligns with that perspective is optional.

A recent monograph [39] gives an account of AA and other self-help or mutual help groups operating internationally in the alcohol problems field. AA and similar organisations point to the fact that relapse prevention is not a concept which should be limited to what professionals do for or to their patients. It is a concept which should also embrace what people do for themselves to prevent relapse.

Conclusions and directions for the future

On the basis of this brief survey of the past, let us now identify some messages for the future of the relapse prevention endeavour.

Relapse prevention is a useful concept. From the material laid out in this chapter, it is clear that relapse prevention has been the intention of many types of actor and activity, long before the phrase passed into present usage. To an extent the modern phrasing is no more than a restatement of the old and evident truth – the business of treatment is to prevent and deal with relapse. The phrase contains within it an invitation both to the laboratory scientist and the research psychologist better to understand the mechanisms of relapse, and better and more intelligently to interdict those mechanisms. "Relapse prevention" is a phrase usefully to invite the focusing of minds even if it is really only treatment in new wrappings.

Relapse prevention is a multidisciplinary endeavour. Powerful animal models can now assist the laboratory scientist who wishes to investigate pharmacological approaches to relapse prevention, and the new anti-craving drugs are likely to have a significant impact on clinical practice. Psychological methods of relapse prevention have become increasingly sophisticated and broad-based over recent years. In terms of effect size, maintenance of intended impact, cost-effectiveness, treatment compliance, and ability to reach a satisfactory percentage of the population in need, there are still many important unanswered research questions relating to the value of either pharmacological or psychological approaches to relapse prevention. Rather than viewing these two types of intervention as rivals, what should increasingly be built into theory, research and practice, should be concepts that integrate brain science and psychology.

Learning from other substances. Treatment of the use of illicit drugs and of tobacco raises many of the same questions as occur with alcohol. There are, for instance, similar histories to be traced out on competing models of understanding and the competition between physical and psychological approaches to treatment. For present purposes let us note the potential fruitfulness of a comparative historical analysis of the relapse prevention concept across substances, but we will not enter that territory. In both instances the same questions on how most effectively to combine drug and psychological treatments to prevent relapse are being debated as in the alcohol sector.

Is the aim of relapse prevention inevitably total abstinence? Over much of history, and whether punishment, preaching or treatment was the vogue, the prescribed interventions have had abstinence as their goal. With the development of modern psychological ideas on drinking as learnt habit, more flexible approaches have developed with the therapy often aimed perhaps at the unlearning of excess, rather than the learning of life-long abstinence [15]. There is no reason why pharmacotherapy should not similarly embrace this more flexible goal. The appropriateness of different goals becomes here an empirical research question.

One size is unlikely to fit all. History seems often to show the response to the drinker as having been a matter of one size fits all. Every drunkard should be beaten, every drunken sinner prayed over, every inebriate put into a Turkish bath, all alcoholics given antabuse, all presenting clients enrolled in a psychological relapse prevention program. A lesson must surely be that the help given to troubled drinkers has to be differentiated by personal need and circumstance. That awareness becomes increasingly important with the mounting evidence on the clinical importance of co-morbidity [15].

Not forgetting the whole person. Much of the relevant history is a chronicle of well-intentioned authorities doing things to people with drinking problems, with the recipients conceived as passive objects. Punishment, preaching, drugs and psychological treatments, they are sequential scenes in the one long play. The lesson from history is that at the end of the day relapse prevention is, in fact, about the individual's personal choice made within a unique personal, social and cultural context, and it is the outcome of what George Vaillant has termed the "natural healing process" [40].

References

1 Marlatt GA, Gordon JR (eds) (1985) *Relapse preventive maintenance strategies in the treatment of addictive behaviours*. Guilford, New York

2 Lewin L (1931) *Phantastica: Narcotic and stimulating drugs*. Kegan Paul/Trench and Trubner, London [Translation of the second German edition by PHA Wirth.]

3 Pittman DJ, Gordon CW (eds) (1958) *Revolving door: A study of the chronic police case inebriate*. Monographs of the Yale Center of Alcohol Studies No.2. Free Press, Glencoe ILL

4 Pittman DJ (1969) Existing and proposed treatment facilities in the USA. In: T Cook, D Gath, C Hensman (eds): *The drunkenness offence*. Pergamon, Oxford, 115–126

5 Babor T, Caetano R, Casswell S, Edwards G, Giesbrecht N, Graham K, Grube J, Grunewald P, Hill L, Holder H et al. (2003) *Alcohol: No ordinary commodity: Research and public policy*. Oxford University Press, Oxford

6 Hunt GM, Azrin NH (1973) A community-reinforcement approach to alcoholism. *Behav Res Therapy* 11: 91–104

7 Edwards G (2004) *Matters of substance*. Allen Lane, London

8 Edwards G (2004) *Alcohol, the world's favorite drug*. Thomas Dunne Books/St Martin's/Griffin, New York

9 Maxwell MA (1950) The Washingtonian Movement. *Q J Stud Alcohol* 11: 411–415

10 Daniels WH (1878) *The Temperance reform and its greater reformers, an illustrated history*. Nelson and Phillips, New York

11 Gough JB (1881) *Sunshine and shadow or gleanings from my life work*. AD Worthington, Hartford Connecticut

12 Cook CCH (2004) Addiction and spirituality. *Addiction* 99: 539–551

13 Turner PE (1934) Illustrative cases from a women's inebriate home and some underlying causations. In: CC Weeks (ed.): *The Twentieth International Congress on Alcoholism: volume of proceedings*. [No publisher given], London, 343–345

14 Edwards G, Oppenheimer E, Taylor C (1992) Hearing the noise in the system. Exploration of textual analysis as a method for studying change in drinking behaviour. *Br J Addict* 87: 73–81

15 Edwards G, Marshall J, Cook CCH (2003) *The treatment of drinking problems*, 4th edn. Cambridge University Press, Cambridge

16 Porter K (1985) The drinking man's disease: the pre-history of alcoholism in Georgian Britain. *Br J Addict* 80: 385–396

17 Rush B (1785) *An enquiry into the effects of spirituous liquors on the human body*. Thomas and Andrews, Boston Mass

18 Trotter T (1804) *An essay medical, philosophical and chemical on drunkenness and the effects on the human body*. TN Longman and O Rees, London

19 American Association for the Study and Care of Inebriety (1893) *The disease of inebriety from alcohol, opium and other narcotic drugs*. John Wright, Bristol

20 Lemere F (1987) Aversion treatment of alcoholism: some reminiscences. *Br J Addict* 82: 257–258

21 Conger JJ (1956) Reinforcement theory and the dynamics of alcoholism. *Q J Stud Alcohol* 17: 296–305

22 Kingham RJ (1958) Alcoholism and the reinforcement theory of learning. *Q J Stud Alcohol* 19: 320–330

23 Franks CM (1958) Alcohol, alcoholism and conditioning: a review of the literature and some illustrative considerations. *J Mental Sci* 104: 14–33

24 Litman GK (1986) Alcoholism survival: the prevention of relapse. In: WR Miller, N Heather (eds): *Treating addictive behaviours, processes of change*. Plenum, New York, 391–406

25 Parkes GA, Anderson BK, Marlatt GA (2001) Relapse prevention therapy. In: N Heather, TJ Peters, T Stockwell (eds): *International handbook of alcohol dependence and problems*. John Wiley, Chichester, 565–592

26 Miller WR, Wilbourne PL (2002) Mesa grande: a methodological analysis of clinical trails of treatment for alcohol use disorders. *Addiction* 97: 265–277

27 Kerr N (1888) *Inebriety or narcomania. Its etiology, pathology, treatment and jurisprudence*, 3rd edn. HK Lewin, London

28 Berridge V (1990) Special issue: The Society for the Study of Addiction 1884–1988. *Br J Addict* 85: 983–1087

29 Hald J, Jacobsen E (1948) A drug sensitising the organism to ethyl alcohol. *Lancet* 255: 1001–1004

30 Fuller RF, Gordis E (2004) Does disulfirum have a role in alcoholism treatment today. *Addiction* 99: 21–24
31 Murray R (1980) Why are the drug companies so disinterested in alcoholism? *Br J Addict* 75: 113–115
32 Edwards G, Littleton J (eds) (1984) *Pharmacological treatments for alcoholism.* Croom Helm, London
33 Lhuintre JP, Daoust M, Moore ND, Chretien P, Saligaut C, Tran G, Bosimare F, Hillemand B (1985) Ability of calcium-bis-acetyl-homotaurine, a GABA agonist, to prevent relapse in weaned alcoholics. *Lancet* 1: 1014–1016
34 Volpicelli JR, Alterman AI, Hayashida M, O'Brien CP (1992) Naltrexone in the treatment of alcohol dependence. *Arch Gen Psychiatry* 49: 876–880
35 O'Malley SS, Jaffe AI, Chang G, Schottenfield RS, Meyer RE, Rounsaville B (1992) Naltrexone and coping skills therapy for alcohol dependence: a controlled study. *Arch Gen Psychiatry* 39: 881–887
36 Alcoholics Anonymous (1976) *Alcoholics Anonymous: the glory of how many thousands of men and women have recovered from alcoholism*, 3rd edn. AA General Service Office, York NJ
37 Katz E (1991) *Not-God: a history of Alcoholics Anonymous.* Hazeldon, Center City MN
38 Edwards G (1996) Alcoholics Anonymous a mirror held up to nature. In: G Edwards, C Dare (eds): *Psychotherapy, psychological treatments and the addictions.* Cambridge University Press, Cambridge, 220–239
39 Humphreys K (2004) *Circles of recovery: self-help organisations for addiction.* IRMA Monograph. Cambridge University Press, Cambridge
40 Vaillant GE (1995) *Natural history of alcoholism revisited.* Harvard University Press, Cambridge MA

Drugs for Relapse Prevention of Alcoholism
Edited by R. Spanagel and K. Mann

How to measure relapse in animals

Rainer Spanagel

Department of Psychopharmacology, Central Institute of Mental Health, University of Heidelberg, J 5, 68159 Mannheim, Germany

Introduction

In this chapter, two animal models – the reinstatement model [1] and the alcohol deprivation effect model [2] – will be described in detail and the limitations of these models in mimicking craving and relapse as observed in human alcoholics will also be discussed. It would be careless, however, if other models would not be considered to measure these phenomena. For example, the conditioned place preference paradigm [3], second order schedules [4], or the escalation paradigm [5] could easily be adapted for this purpose. Although these and other paradigms are widely used in the drug abuse field to further our understanding on craving and relapse, there has only been little effort to adapt these paradigms to alcohol research. Moreover, the application of these models to alcohol research remains a difficult issue. Thus, alcohol administration to rodents does not usually produce a conditioned place preference [3, 6]. However, some specific experimental conditions have been described in order to obtain alcohol-induced conditioned place preference. One paradoxical experimental condition is the use of DBA/2 mice [7]. They exhibit conditioned place preference to alcohol, despite the fact that they avoid alcohol under free drinking conditions. In contrast, C57BL6/J mice, which show a high preference for alcohol in a two bottle free choice paradigm, exhibit conditioned place aversion following alcohol treatment [7]. Therefore, alcohol-induced conditioned place preference cannot be considered as a reliable and reproducible measure of alcohol craving. It is desirable to see that preclinical researchers put more effort into more elaborate animal models of alcohol craving and relapse in the future; however, currently we have to rely on the two aforementioned models to further our understanding on addictive behaviour.

The reinstatement model

The reinstatement model is currently being used in many laboratories to investigate mechanisms underlying relapse behaviour [1]. However, it should be noted that the reinstatement test is performed under drug-free conditions. In contrast, a typical relapse or lapse in alcoholic patients is defined as compulsive alcohol consumption following a period of abstinence; a relapse can therefore not happen under drug-free conditions. Thus, having this definition in mind, it remains unclear to the author, how the reinstatement model was put forward by many researchers as a model of relapse. This is per definition wrong but does not diminish the value of this model in measuring drug-seeking behaviour, which without a doubt, is one behavioural dimension of craving. In conclusion, the reinstatement model should be considered as a model of craving rather than relapse, even if it will be a long way to change the literature in this respect. Importantly, it is necessary to use semantics in a proper fashion, as it provides the basis for our common and highly specialized scientific language.

In the following, the reinstatement procedure is briefly described: an animal (rat or mouse) is trained to self-administer a drug and is then subjected to extinction – that is, the animal is tested under conditions of non-reinforcement until operant responding appears to be extinguished. When the animal reaches some criterion of unresponsiveness, various stimuli are presented. A stimulus is said to reinstate the drug-seeking behaviour if it causes renewed responding, i.e., lever pressing, without any further response-contingent drug reward. At least three conditions can reinstate responding: (i) drug priming – that is the injection of a small dose of the drug, (ii) stress, and (iii) conditioned stimuli. The data derived from studies using the reinstatement model suggest that the neuronal events mediating drug-, stress-, and cue-induced reinstatement of drug-seeking are not identical [1].

Although reinstatement of intravenous self-administration of psychostimulants and opioids has been established for many years, only a few attempts have been undertaken to transfer this paradigm into the alcohol field. In 1995, the first alcohol reinstatement study in rats was reported by Chiamulera and co-workers [8]. In this study, rats acquired operant responding for alcohol over several months. After stable lever pressing was obtained between subsequent sessions, the rats were tested in extinction, meaning that animals received water instead of alcohol following lever pressing. After eight to ten extinction sessions, re-exposure to a small quantity of ethanol was able to reinstate previously extinguished alcohol-seeking behaviour. These results are consistent with the widely reported description of the "first-drink" phenomenon: ingestion of a small quantity of alcohol may induce in abstinent alcoholics a strong subjective state of craving and, then, relapse to drug-taking behaviour [9]. The "priming effect" due to alcohol preload may be evident even after years of abstinence from the drug [10] and a strong craving for alcohol and higher alcohol intake in social drinkers following alcohol priming was described [11]. Only very recently has the alcohol reinstatement paradigm been followed up by other research groups, and it could be demonstrated that intermittent footshock stress can also reinstate previously extinguished responding for alcohol [12]. Furthermore, it has been shown that alcohol-associated olfactory cues and other cues can reinstate extinguished alcohol-seeking behaviour [13].

Foot-shock stress and response-contingent presentation of an alcohol-associated light cue, acting as a conditioned stimulus, also augment reinstated extinguished responding [14]. Thus, addictive effects of these stimuli on responding are observed, supporting the idea of different neuronal systems mediating stress- and cue-induced reinstatement [1, 14].

The reinstatement model of drug-seeking behaviour is now a well-established model in rats, and it is only recently that it has become possible to transfer this model to mice [15]. However, some methodological transfer problems from rats to mice ought to be considered: operant tasks have to be achieved by the subject during the course of a reinstatement experiment and usually rats acquire goal-directed behaviour more easily than mice. The first goal to be achieved by the subject in the reinstatement procedure is selective responding on a reinforced lever. This is usually an easy task for a rat but certainly a more difficult one for mice, due to higher motor activity. Another confounding variable in mice is that lever pressing is reinforcing *per se*. The same problem might also relate to nose-poking. Thus, whether a reinforcer follows a lever press/nose-poke or not, it might not influence subsequent behaviour of the mouse. Despite the problematic factors of high motor activity and reinforcing effects of lever pressing/nose poking, mice tend to successfully acquire selective responding under a simple fixed ratio (FR1) for the drug. However, on the control lever (i.e., the non-reinforced lever), a higher number of responses is observed in mice. The next task to be achieved, is extinction of lever responding. Thus, lever responding is without any further consequence and the individual does not receive the drug anymore. Rats, on average, show a short burst of responding, with responding gradually declining over the days. Following 10 days, rats usually only show few spontaneous responses. Extinction in mice appears to be more difficult and thus more extinction days are needed as compared to rats and again much higher rates of spontaneous responding are observed. Regarding the third task, reinstatement of drug-seeking behaviour has to be elicited. A combination of a light cue and alcohol priming has been shown to reinstate alcohol-seeking behaviour in mice [15]. However, only a few preliminary studies, using the reinstatement paradigm with mice, have been reported and it will take more effort to work out the optimal conditions for reinstatement of drug-seeking behaviour in mice. This effort, however, will pay off as in the future, this paradigm will most likely be of frequent use in studying conditional knock-out mice so as to precisely pin down the genes and brain sites involved in alcohol craving.

In conclusion, reinstatement of alcohol-seeking behaviour can be used to study the neurobiological and molecular basis of craving, since there appears to be a good correspondence between the events that induce craving in laboratory animals and those that provoke it in humans. Furthermore, acamprosate and naltrexone are known to reduce craving in alcoholic patients and can also reduce or even block cue-induced reinstatement of alcohol-seeking behaviour [13, 16, 17]. Nevertheless, the usefulness of the reinstatement model in mimicking craving in humans experiences some limitations. First, the phenomenon of craving is complex. Thus, although the operational definition of craving as incentive motivation to drink alcohol has the advantage of making the phenomenon of craving measurable, such a definition neglects the fact that craving also has a subjective dimension that is difficult, if at all possible, to assess in laboratory animals [18]. Second, it appears that extinction of alcohol-seeking behaviour usually only plays a minor role in alcoholic patients trying to achieve and maintain abstinence. With the exception of patients undergoing focused extinction therapy, alcoholics generally try to avoid exposure to external alcohol cues during abstinence. In most cases, alcoholics stay abstinent for a while but then may experience craving and subsequent relapse, if they are re-exposed to external cues (e.g., the sight of a bar or smell of alcohol), and in particular, if they are in a vulnerable internal state. Consequently, the animal reinstatement procedure may not accurately reflect the situation of abstinent alcoholics experiencing craving.

The alcohol deprivation effect model

The alcohol deprivation effect model is an animal model to study compulsive alcohol drinking and relapse-like drinking behaviour [2]. Alcohol-experienced animals show a transient increase in alcohol consumption and alcohol preference after a period of forced abstinence (alcohol deprivation), which is termed the "alcohol deprivation effect". It can be seen in long-term alcohol-drinking rats, both under home cage drinking and under operant self-administration conditions. The effect is observed in monkeys [19] and man as well [20]. Interestingly, the alcohol deprivation effect is prolonged and enhanced in alcohol-preferring P- and HAD-rat lines after repeated deprivation phases [21, 22] and it changes its characteristics with repeated deprivation phases [23]. Thus, the alcohol deprivation effect in long-term alcohol self-administering rats, which had experienced repeated deprivation phases, shows some particularly interesting features: during an alcohol deprivation effect, these animals consume large amounts of highly concentrated alcohol solutions and even at unusual times. Pronounced changes in the diurnal rhythm of drinking activity were observed in long-term alcohol-drinking rats, which had repeated deprivation phases [23]. Tested in a fully automated electronic drinkometer device, age-matched control animals showed normal drinking activity. Thus, drinking activity during the active night phase was high, whereas drinking activity during the inactive light phase was very low, reaching zero for some hours. In contrast, in long-term alcohol-drinking rats during the alcohol deprivation effect, the pattern of drinking activity completely changed. In particular, during the inactive phase, most of the animals still showed high drinking activity. Moreover, some animals were found that even demonstrated level drinking, i.e., which implies that the alcohol deprivation effect can be associated with alterations in circadian rhythmicity. This is a finding which probably translates to humans. Thus, a new hypothesis suggests that chronic alcohol intake can

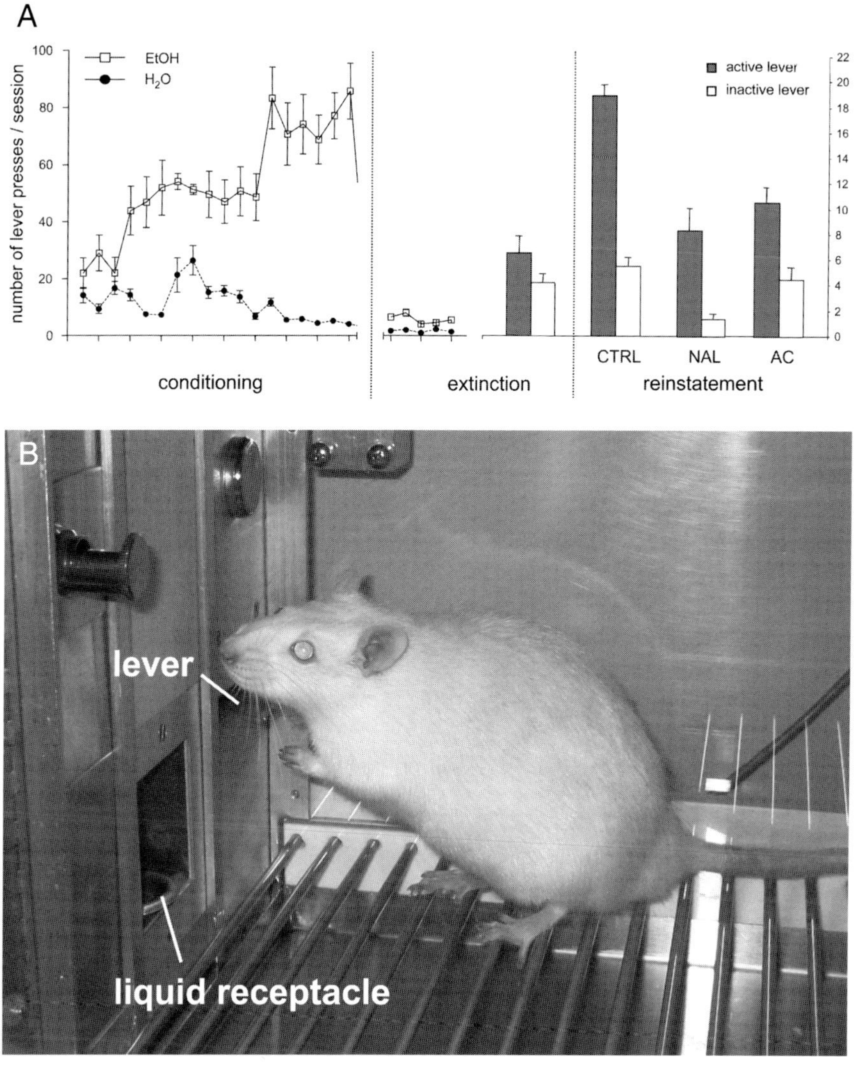

Figure 1. (A) Cue-induced reinstatement of ethanol-seeking. During a phase of conditioning rats acquire stable lever pressing for ethanol in the presence of a distinct set of cues. After extinction, the animals are again exposed to the respective cues, formerly associated with ethanol, leading to renewed responding on the ethanol lever in the absence of the primary reinforcer (CTRL). Treatment with naltrexone (NAL; 0.25 mg/kg) or acamprosate (AC; 200 mg/kg) leads to a significant reduction of ethanol-seeking behavior. (Katner et al. (1999) *Neuropsychopharmacology* 20: 471–479; Bachteler et al. (2005) *Neuropsychopharmacology; in press*). (B) Reinstatement paradigm. Each lever press is followed by a 25–30 µl drop of ethanol as primary reinforcer.

influence the expression and function of clock genes and can thereby alter circadian rhythmicity and it has further been shown that clock gene activity can influence drinking behaviour in mice and man as well [24].

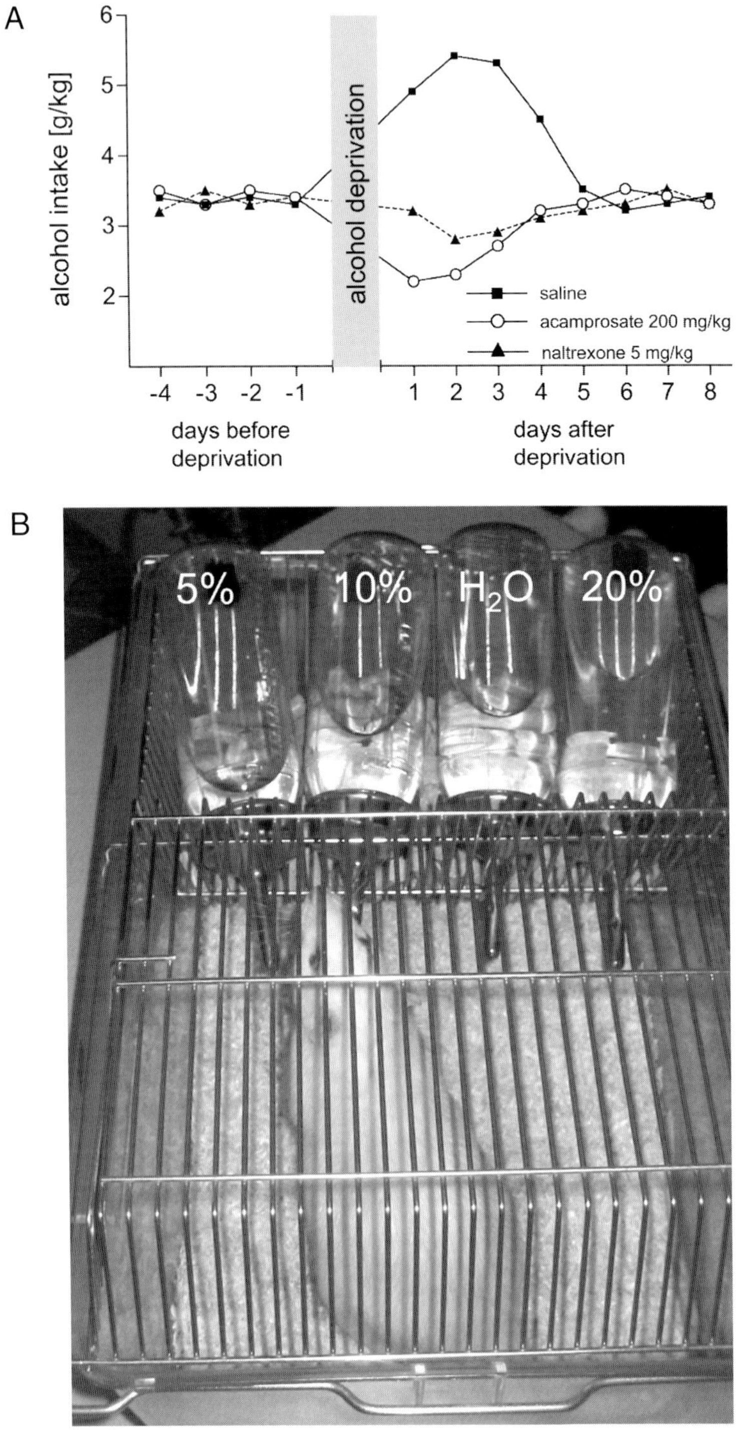
A
alcohol intake [g/kg]
6
5
4
3
2
alcohol deprivation
saline
acamprosate 200 mg/kg
naltrexone 5 mg/kg
-4 -3 -2 -1
1 2 3 4 5 6 7 8
days before deprivation
days after deprivation
B
5%
10%
H_2O
20%

As already mentioned, the alcohol deprivation effect is observed across several species, including monkeys and human social drinkers. Salimov and Salimova [25] were the first who described this effect in mice. In this study, and in subsequent ones of the same group [26], many mice failed to exhibit an alcohol deprivation effect. However, it is important to note that only a short deprivation period of 3 days was introduced in these studies, which might not lead to an effect in most of the mice. The experience in our lab over the years has shown that a deprivation period of at least 2–3 weeks is needed, that would then lead to an alcohol deprivation effect in most mice. However, there will always be a number of mice which will consistently fail to show such an effect. We are now in the course of a selective breeding program, where we are breeding two lines of mice: one which shows a consistent alcohol deprivation effect *vs* one which consistently fails to show such an effect. These two lines will certainly help in identifying the genetic factors underlying the alcohol deprivation effect, and thereby alcohol relapse drinking behaviour. So far, only a few reports on transgenic mice in the context of the alcohol deprivation effect model and the involvement of a particular gene are available [27, 28]. However, the information from these studies has so far failed to have great influence on better treatment strategies regarding relapse behaviour. It is hoped that it has become clear to the reader that with the use of conditional knockout mice, we are about to enter a new research area and it is strongly believed that in the future many labs working on relapse will follow this line of research and will produce helpful information to guide us in new treatment strategies on relapse behaviour.

In summary, the alcohol deprivation effect is a useful model to study compulsive alcohol drinking behaviour and relapse-like drinking behaviour. However, it is important to note that measuring an alcohol deprivation effect only assesses a behavioural outcome and does not provide insight about a subjective state associated with compulsive drinking behaviour. Nevertheless, the fact that the clinically effective anti-relapse drugs acamprosate and naltrexone also reduce or even abolish the alcohol deprivation effect [29] lends predictive value to this animal model for the development of new and better drugs for the treatment of alcoholism. Furthermore, the alcohol deprivation effect model could also be applied as a high through-put pharmacological screen in the pharmaceutical industry in identifying novel compounds which interfere with alcohol relapse. In fact, the alcohol deprivation effect model is currently used as a high through-put screen test to study relapse drinking in a N-ethyl-N-nitros urea

Figure 2. (A) Effects of acamprosate and naltrexone on the alcohol-deprivation effect. Rats had unlimited, voluntary access to water and three different alcohol solutions (5, 10, 20%) for eight months before alcohol was completely withdrawn for two weeks (alcohol deprivation). Intermittent treatment with acamprosate (200 mg/kg) and naltrexone (5 mg/kg) reduced alcohol intake compared to control animals after renewed access to ethanol. (B) Home cage drinking. Rats are exposed to the choice of water and three different alcohol solutions (5, 10, 20%) in their home cage. After few weeks, a stable ethanol-intake and -preference develops. Both intake and preference patterns change after an imposed phase of alcohol deprivation.

(ENU) mutagenesis program. ENU is a chemical mutagen that has been shown to produce a large number of mutations. The most important requirement for an effective ENU screen is a simple and reliable behavioural high through-put test. Finally, the successful application of the alcohol deprivation effect model in an ENU mutagenesis program [30] demonstrates that this test could also be used as a high through-put test for medication development.

References

1 Shaham Y, Shalev U, Lin L, de Wit H, Stewart J (2003) The reinstatement model of drug relapse: history, methodoloy and major findings. *Psychopharmacology* 168: 3–20
2 Spanagel R (2003) Alcohol addiction research: From animal models to clinics. *Best Pract Res Cl GA* 17: 507–518
3 Tzschentke T (2005) Measuring reward with conditioned place preference: an up-date of the literature. *Addict Biol*; *in press*
4 Everitt BJ, Robbins TW (2000) Second-order schedules of drug reinforcement in rats and monkeys: measurement of reinforcing efficacy and drug-seeking behaviour. *Psychopharmacology* 153: 17–30
5 Ahmed SH, Koob GF (1998) Transition from moderate to excessive drug intake: change in hedonic set point. *Science* 282: 298–300
6 van der Kooy D, O'Shaughnessy M, Mucha RF, Kalant H (1983) Motivational properties of ethanol in naive rats as studied by place conditioning. *Pharmacol Biochem Behav* 19: 441–445
7 Cunningham CL, Niehus DR, Malott DH, Prather LK (1992) Genetic differences in the rewarding and activating effects of morphine and ethanol. *Psychopharmacology* 107: 385–393
8 Chiamulera C, Valerio E, Tessari M (1995) Resumption of ethanol-seeking behaviour in rats. *Behav Pharmacol* 6: 32–39
9 Ludwig AM, Wikler A, Stark LH (1974) The first drink: psychobiological aspects of craving. *Arch Gen Psychiatry* 30: 539–547
10 Besancon F (1993) Time to alcohol dependence after abstinence and first drink. *Addiction* 88: 1647–1650
11 de Wit H, Chutuape MA (1993) Increased ethanol choice in social drinkers following ethanol preload. *Behav Pharmacol* 4: 29–36
12 Lê AD, Quan B, Juzytch W et al. (1998) Reinstatement of alcohol-seeking by priming injections of alcohol and exposure to stress in rats. *Psychopharmacology* 135: 169–174
13 Katner SN, Magalong JG, Weiss F (1999) Reinstatement of alcohol-seeking behavior by drug-associated discriminative stimuli after prolonged extinction in the rat. *Neuropsychopharmacology* 20: 471–479
14 Liu X, Weiss F (2002) Addictive effect of stress and drug cues on reinstatement of ethanol seeking: exacerbation by history of dependence and role of concurrent activation of corticotropin-releasing factor and opioid mechanisms. *J Neurosci* 22: 7856–7861
15 Spanagel R, Sanchis-Segura C (2003) The use of transgenic mice to study addictive behavior. *Clin Neurosci Res* 3: 325–331
16 Bachteler D, Economidou D, Danysz W, Ciccocioppo R, Spanagel R (2005) The effects of acamprosate and neramexane on cue-induced reinstatement of ethanol-seeking behaviour. *Neuropsychopharmacology*; *in press*
17 Lê AD, Poulos CX, Harding S, Watchus J, Juzytsch W, Shaham Y (1999) Effects of naltrexone and fluoxetine on alcohol self-administration and reinstatement of alcohol seeking induced by priming injections of alcohol and exposure to stress. *Neuropsychopharmacology* 21: 435–444
18 Spanagel R, Hölter SM (2000) Controversial research areas: a critical review. In: G Zernig, A Saria, A Kurz (eds): *Handbook of Alcoholism: Clinical and Theoretical Approaches*. CRC Press, Boca Raton, Fl, 401–416
19 Sinclair JD (1997) The alcohol-deprivation effect in monkeys. *Psychonom Sci* 25: 21–25
20 Burish TG, Maisto SA, Cooper AM, Sobell MB (1981) Effects of voluntary short-term abstinence

from alcohol on subsequent drinking patterns of college students. *J Stud Alcohol* 42: 1013–1020
21 Rodd-Henricks ZA, McKinzie DL, Murphy JM et al. (2000) The expression of an alcohol deprivation effect in the high-alcohol-drinking replicate rat lines is dependent on repeated deprivations. *Alcohol Clin Exp Res* 24: 747–753
22 Rodd-Henricks ZA, McKinzie DL, Shaikh SR et al. (2000) Alcohol deprivation effect is prolonged in the alcohol preferring (P) rat after repeated deprivations. *Alcohol Clin Exp Res* 24: 8–16
23 Spanagel R, Hölter SM (1999) Long-term alcohol self-administration with repeated alcohol deprivation phases: an animal model of alcoholism? *Alcohol Alcoholism* 34: 231–243
24 Spanagel R, Pendyala G, Abarca C et al. (2004) The circadian clock gene period2 alters the glutamatergic system and thereby modulates alcohol consumption. *Nature Med*; *in press*
25 Salimov RM, Salimova NB (1993) The alcohol-deprivation effect in hybrid mice. *Drug Alcohol Depend* 32: 187–191
26 Salimov RM, Salimova NB, Shvets LN, Maisky AI (2000) Haloperidol administered subchronically reduces the alcohol-deprivation effect in mice. *Alcohol* 20: 61–68
27 Cowen MS, Schroff KC, Gass P, Sprengel R, Spanagel R (2003) Neurobehavioral effects of alcohol in AMPA receptor subunit (GluR1) deficient mice. *Neuropharmacology* 45: 325–333
28 Cowen M, Schumann G, Yagi, Spanagel R (2003) Role of Fyn tyrosine kinase in ethanol consumption by mice. *Alcohol Clin Exp Res* 27: 1213–1219
29 Spanagel R, Hölter SM (2000) Pharmacological validation of a new animal model of alcoholism. *J Neural Transm* 107: 669–680
30 Spanagel R, Sanchis-Segura C (2005) Basic considerations for random ENU mutagenesis in combination with an alcohol screen test. *Addict Biol*; *in press*

Drugs for Relapse Prevention of Alcoholism
Edited by R. Spanagel and K. Mann

How to measure relapse in humans

Henry R. Kranzler[1] and Howard Tennen[2]

Departments of Psychiatry[1] and Community Medicine and Health Care[2], University of Connecticut Health Center, 263 Farmington Avenue, Farmington CT 06030, USA

Introduction

Although the measurement of alcohol consumption and alcohol-related problems has received considerable attention in reviews of treatment research [1–5], to our knowledge these reviews have not focused on issues specific to pharmacotherapy trials. Methods for the conduct of clinical trials testing medications to treat alcohol dependence have undergone substantial change over the past two decades [6–8] and the issues specific to measurement of treatment outcome in these studies warrant consideration. Outcome measures should fit the practical requirements and hypothesized effects of the intervention being examined [9]. Because medications often have specific effects on drinking behavior, pharmacotherapy trials may impose unique requirements on the process of outcome measurement. Furthermore, to the extent that different medications have different hypothesized mechanisms of action and practical constraints, it is unlikely that uniformity of assessment can be obtained across studies [9]. In this chapter, we review developments in the methods used to collect data for the evaluation of outcomes and the specific measures used to evaluate the success of the medications.

Before focusing specifically on pharmacotherapy trials, however, we consider some of the issues that are of general relevance to the measurement of drinking behavior in treatment populations. This will provide the background for a selective review of pharmacotherapy studies that have been published over the past 30 years, a description of recent developments in this field, and speculation on what the future may hold for studies examining the effects of medications on alcohol treatment outcomes. The examples we selected to illustrate trends over time are drawn from randomized, controlled trials, since these represent the "gold standard" for the evaluation of treatment efficacy. However, the most rigorous methods have generally been used in more recent trials, and earlier studies often utilized methods no longer considered acceptable for use in efficacy trials. We focus this review on medications or groups of medications that have received the greatest research interest: lithium, the alcohol-sensitizing medications disulfiram and calcium carbimide, acamprosate and the opioid antagonists naltrexone and nalmefene.

Outcome measures in alcohol treatment trials

Finney et al. [3] examined 404 treatment outcome studies published between 1968 and 1998 that included two or more treatment/control conditions. In general, the number of treatment outcome studies published annually has increased in a linear fashion during this period. All of the studies that specified the method of data collection (96%) indicated that the data were obtained by self-report. The mean number of outcome variables assessed in these studies was 6.8 (median = 5). The correlation of outcome measures with year of publication gave no evidence of a significant increase in the number of outcome variables over time.

Table 1 shows the outcome measures that were used with greater frequency over the 31-year span, along with outcome indicators that have fallen out of favor. The measures that were more widely adopted over time include time-based abstinence (e.g., days of abstinence), time-based alcohol consumption (e.g., drinks per day), physiological markers [e.g., γ-glutamyl transpeptidase (GGT)], dependence symptoms, and drinking-related problems, all of which were more likely to be assessed in later years. In contrast, "improvement" in drinking behavior, measures of global functioning or improvement in global functioning, and measures of occupational functioning or participants' financial situation were less likely to be examined in more recent studies. Noteworthy for their lack of association with date of publication were categorical (i.e., complete) abstinence and psychological functioning.

Finney et al. [3] also examined the popularity of various outcome indicators in studies published between 1990 and 1998. This period is particularly relevant to the focus of this chapter. As a consequence of support from the US government (i.e., the National Institute on Alcohol Abuse and Alcoholism (NIAAA)) and the pharmaceutical industry in Europe, pharmacotherapy trials began to increase substantially in number and sophistication beginning in the 1990s. During this 9-year period, categorical abstinence and quantity-based

Table 1. Significant correlations between percentage of studies measuring class of outcome variable and year of publication (Adapted from [3])

Class of outcome variable	Aggregate r (n = 31)
Alcohol consumption, quantity-based	$.81^{**}$
Physiological marker	$.73^{**}$
Alcohol consumption, time-based	$.72^{**}$
Dependence symptoms	$.71^{**}$
Abstinence, time-based	$.59^{**}$
Drinking-related problems	$.44^{*}$
Rating of improvement in drinking	$-.44^{*}$
Global functioning or improvement	$-.45^{*}$
Occupational/financial situation	$-.46^{*}$

* $p < 0.05$, ** $p < 0.001$

alcohol consumption were most commonly used as outcome measures (used in 54% of published trials), followed by psychological functioning (38%), time-based abstinence (34%), and time-based alcohol consumption (33%).

These findings led Finney and colleagues to the conclusion that the variety of outcome measures in alcohol treatment trials makes it difficult to compare results across studies. The large number of outcomes assessed also raises the question of how to correct for multiple comparisons [3]. This is a particularly thorny issue since the average power to detect a medium effect size in these studies at $p < 0.05$ was only 0.54, with only 23% of studies providing adequate power (i.e., 0.80) to detect such an effect [10].

Methods used to collect drinking data in alcohol treatment trials

A critical issue in the selection and use of outcome measures in pharmacotherapy and other treatment trials, is the feasibility of collecting complete data from each research participant at baseline (i.e., reflecting pre-treatment) and at various follow-up points during and after treatment [9].

Self-report measures

Alcohol treatment researchers have traditionally used self-report assessments to measure alcohol consumption and its consequences. Although these measures are generally considered valid, poorly designed questions, being under the influence of alcohol during the assessment, and fallible memory contribute to a lack of reliability and accuracy of these measures [11]. The risk of inaccurate information, which may be augmented by contingencies that favor misrepresentation, may be particularly pronounced during treatment or afterward, when patients may minimize the extent of their drinking to prevent embarrassment to themselves, to the researchers or to the clinicians treating them.

In response to the commonly held assumption that self-report data have questionable validity owing to patients' inclination to lie or forget, Babor et al. [12] used data from Project MATCH, a large, multi-center study of client-treatment matching, to examine client characteristics associated with discrepancies between subject self-report and other sources of information, including biochemical markers and independent reports from collateral informants. Most participants in this trial appeared not to be lying, sociopathic, or otherwise behaving in an intentionally deceptive manner. Rather, demographic factors, alcohol severity, cognitive functioning, and social stability measures were more closely related to discrepancies among the data sources, than were measures of personality or psychiatric symptoms. Individuals who presented discrepant results had more severe drinking problems, more previous treatments, higher levels of pretreatment drinking, and more cognitive impairment, all of which could contribute to inaccurate recall. In sum, it appears that research participants in clinical trials generally do not minimize their drinking on self-report measures and, when they do, it may be less a result of intentional deception on their part, and more a function of memory deficits associated with a more chronic and severe course of alcohol involvement.

A decade ago, Babor et al. [9] concluded that although biochemical measures and collateral informants' reports may increase an investigator's confidence in treatment outcome results, they add little beyond self-report data. Nonetheless, credibility is a major consideration in the reporting of treatment outcome data, and the added value of alternative measures may be worth their cost and inconvenience. An important consideration, when obtaining data from multiple sources, is how to best integrate the findings in meaningful ways. Moreover, resources devoted to collecting these alternative sources of outcome data might be better invested in procedures aimed at increasing the validity of self-report information, such as providing recall cues and emphasizing the importance of accurate information or, as we describe below, collecting data on a more frequent (e.g., daily) basis.

Methods for the collection of self-reported drinking data

Three approaches to the collection of self-reported drinking data are relevant to clinical trials [5]. Quantity-Frequency methods (QF), which are retrospective estimates of average or usual drinking, do not lend themselves to assessing variation in drinking intensity over time. Given some evidence that medications such as the opioid antagonists have their greatest effects on the intensity of drinking [13, 14], this is an important limitation to the use of QF methods for measurement of drinking outcomes in pharmacotherapy trials. Daily drinking estimation measures (DDE) are retrospective estimates of daily drinking, using methods such as the Timeline Followback method or Form 90. Over the past decade, DDE methods, particularly the Timeline Followback method (TLFB; [15]), have become the preferred method to measure drinking behavior in alcohol treatment trials. The Form 90, a DDE method similar to the TLFB, was developed for use in Project MATCH. Concurrent recall methods (CR) are reports provided in close temporal proximity to the drinking occurrence, using methods such as self-monitoring via paper diaries, palmtop computers, or interactive voice response (IVR) technology. CR methods have not been widely used in alcohol treatment research. However, prospective daily monitoring, a CR approach that has been examined in this context [16–18], appears to offer a number of potential advantages over DDE methods, as described below.

Timeline followback method (TLFB)

The TLFB requires the interviewer to use a calendar of events identified by the respondent to establish anchor points such as holidays, anniversaries, major national events, etc., which serve to facilitate the respondent's retrospective estimates of drinking on a day-to-day basis. The TLFB has been shown to be reliable and valid when administered as an in-person interview [19, 20], when administered by telephone or computer [21] and when used with Swedish-, Spanish-, or English-speaking alcohol abusers [22].

Recent studies of the TLFB have focused on comparisons of aggregate and more fine-grained comparisons with other data collection methods. Carney et al. [23] evaluated the overall correspondence and day-to-day agreement between daily reports and reports obtained using the TLFB [15]. Daily booklets were completed for 28 days by a group of problem drinkers and for 30 days using a palmtop computer by a group of moderate drinkers. Although aggregate levels of drinking were highly correlated across the two methods, the TLFB systematically underestimated drinking relative to daily assessment. There were also large differences among participants in day-to-day correspondence, suggesting that the TLFB may be less useful for measuring patterns (rather than levels) of alcohol consumption.

Prospective daily monitoring

Designs that employ prospective daily monitoring have a number of advantages over traditional cross-sectional designs: (a) rapidly changing processes can be measured closer to their "real time" occurrences; (b) retrospection biases are minimized; (c) powerful within-person analyses that eliminate sources of confounding can be applied to individuals' time series; and (d) temporal sequencing of variables can be established more confidently.

One approach to collecting daily data involves IVR technology, a relatively new and increasingly popular data collection method that uses the telephone to administer survey questions. In using IVR, the respondent answers each question by pressing the keys on the telephone keypad or provides a spoken response, and the responses are entered automatically in a database. IVR enjoys a number of advantages over traditional face-to-face interviews, including interview consistency, access to difficult-to-reach populations, including alcoholics [24], immediate data availability and accessibility, and convenience both for respondents and researchers. Searles and colleagues have demonstrated the feasibility, reliability and validity of daily IVR drinking reports in community and student samples [25–27].

Searles et al. [27] compared daily reports using IVR technology for 366 days with TLFB assessments conducted in person at 13-week intervals in a group of drinkers, some of whom met criteria for an alcohol use disorder. Aggregate correlations for amount consumed, and frequency of drinking days and heavy drinking days were modest, with substantial variability among individuals. Compared to participants without an alcohol use disorder, those with such a diagnosis underreported their drinking.

We [17] recently examined the potential utility of IVR for measuring daily drinking behavior and related mood measures, as well as medication compliance in a pilot trial of naltrexone for problem drinkers. In this study, nine participants provided daily data via IVR during the 12-week treatment period. IVR was well received by study participants, who demonstrated a high degree of adherence to the call requirements. When compared across subjects, we found convergence on both the number of drinking days and mean daily consumption between daily IVR and TLFB, which is in agreement with findings reported by Bardone et al. [26] However, when compared using within-person observations and multi-level modeling (analytic methods of the kind highlighted by Stout [4] for maximizing the statistical power of treatment trials), there was evidence that subjects under-reported alcohol use on the TLFB. On average, participants reported drinking a mean of 0.58 more drinks/day, when assessed using daily IVR compared to the biweekly TLFB, and 0.78 drinks more drinks/day compared to the 12-week TLFB. These differences are consistent with the findings of Searles et al. [27, 28], who found that traditional QF indicators of alcohol use yielded lower levels of alcohol use compared to daily IVR.

In concluding that individuals under-reported their alcohol use on the TLFB, we assume that close to real time drinking reports (i.e., daily IVR) are more accurate than recalled drinking and that recall methods introduce memory decay and memory bias [29]. The extreme variation in the correlation for individual participants between daily IVR and TLFB also indicates that for some individuals, TLFB-derived drinking reports bear little resemblance to drinking reports captured close to their real-time occurrence through daily IVR.

In this study, there was substantial variation in the number of daily IVR reports provided, ranging from 12 to 84. The data missing from some individuals are potentially problematic, but no more so than in studies in which the TLFB is used retrospectively to measure drinking behavior. A common problem in alcohol treatment studies is the participants' failure to return to the treatment site and to comply with efforts to collect TLFB information via the telephone. An advantage of daily data collection via IVR is that multi-level modeling, the use of which is enhanced by daily reports, is robust with regard to missing data and that it weights each participant's data based on the amount of data supplied by that individual.

A potential limitation of this study is that some individuals may have been intoxicated at the time they provided daily reports. Using blood alcohol concentration and collateral reports, Perrine et al. [30] demonstrated the reliability of daily IVR reports. Furthermore, we found no evidence of outlying reporting days during which it took more time than usual to complete the daily protocol, which would suggest intoxication in the respondent.

Another potential problem with intensive self-monitoring using daily data collection is that it could influence the behaviors being measured [31] by increasing awareness of the temporal contingencies between behavior and internal or environmental triggers, by initiating self-focused attention, or by interrupting distraction and related coping efforts [29]. Efforts to minimize measurement reactivity include recording more than one behavior [32] and limiting recording to once a day [33]. In a series of daily process studies, including studies of heavy drinkers, we [18, 23, 34] found no significant evidence of trending or temporal changes in within-person associations that might signal measurement reactivity. Similarly, Hufford et al. [35] reported that electronic monitoring via palmtop computers appeared to show no noticeable reactivity effects.

Daily reports also provide an opportunity to study the moderating effect of treatment on relations between drinking and subjective states, such as mood [17, 36, 37]. Daily measurement of mood or desire to drink may help to elucidate the mechanism of treatment effects on drinking behavior, including the effects of pharmacotherapies [33, 37].

Biological measures and collateral informant reports of drinking

Concerns that alcoholics may under-report or deny their drinking have sparked efforts to identify sensitive and specific biological indicators of recent alcohol use that do not depend on self-report. Available biochemical tests have the advantage of being specific and reliable, and cannot be biased by motivational or cognitive factors. In addition, biological markers of direct or indirect alcohol effects provide an indication of alcohol's physical toxicity and as such can indicate improvement or deterioration during treatment.

Irwin et al. [38] reported that parallel increases in serum GGT concentrations of 20% or greater, in serum aspartate aminotransferase (ASAT) of 40% or greater, and in serum alanine aminotransferase (ALAT) of 20% or greater differentiate recovering alcoholics who remain abstinent from those who resume drinking with a sensitivity of 100% and a specificity of 82%. In the VA disulfiram study [39], blood was obtained every two months for ASAT measurements to monitor for hepatotoxicity. When ASAT increased from baseline, other evidence indicated that the patient had resumed drinking and toxicity was not the problem.

Conigrave et al. [40] recently reviewed the literature on conventional biological measures (including ASAT, ALAT, and GGT), which despite their limited sensitivity and specificity are clinically widely available. A number of pharmacotherapy studies (e.g., [41, 42]) have used these measures to complement or validate self-report information obtained from the patient. Unfortunately, GGT and other liver function tests are subject to large individual differences (e.g., according to age and gender), and are affected by many factors other than alcohol (e.g., liver infections and medication use). In addition, many heavy drinkers do not have abnormal GGT levels, and the variability among individuals with abnormal levels is so great that the use of parametric statistical tests may not be appropriate.

CDT may be marginally better than GGT [43], though it has very poor sensitivity and specificity, when used to screen for heavy drinking in community samples or general medical practice, and it may perform less well for women than for men [44]. Other limitations of these tests include cost, the feasibility of obtaining blood samples at the time of follow-up evaluations, and a relatively short period of maximum sensitivity (i.e., 2–14 days).

Recent approaches involve combining tests, such as CDT with GGT, to enhance their value in identifying relapse [45, 46]. Javors and Johnson [47] reviewed the findings on newer biological markers (including CDT, total serum sialic acid, sialic acid index of apolipoprotein J, and serum β-hexosaminidase). These authors concluded that of the newer tests, CDT is the only one with demonstrated value as a marker for alcohol consumption. They call for additional research on the newer markers and on combining these with more established markers to optimize the detection of heavy drinking.

The limitations of both self-report measures of drinking and biological indicators have led investigators to corroborate self-report data with information collected from collateral informants, who are knowledgeable about the patient's daily activities. Typically, parents, spouses, or friends are interviewed about the patient's drinking and general functioning. Collateral reports, when available, generally corroborate client self-reports [11, 12]. However, because of limitations of collateral reports, there has been the recognition that the high degree of effort required to obtain this information may not be justified by its limited value for validating self-reported drinking behavior by individuals participating in the treatment trial.

Collateral informants are difficult to recruit, sometimes lose contact with the client, and often lack detailed information about the quantity or frequency of drinking. And while it is generally assumed that collaterals honestly report what they observe, there is some anecdotal evidence that at times collaterals may collude with the client to deny drinking when it occurs [48] or possibly "punish" clients by over reporting (whether intentionally or not). Connors and Maisto [49], after reviewing the literature on the reliability and accuracy of alcoholics' self-reported alcohol consumption in relation to reports provided by collateral informants, concluded that individuals provide accurate reports about their drinking and alcohol-related consequences. When reports provided by collaterals are discrepant with participants' reports, the latter almost always show higher levels of alcohol consumption. Agreement between participant and collateral reports is most likely to occur when collaterals are in frequent contact with the participant, are spouses or partners, and are confident about the reports they provide [12].

Alcohol pharmacotherapy trials

Lithium

Patients who were enrolled in early trials of medications to treat alcoholism were treatment-seeking patients, who generally were moderate-to-severe in their level of alcohol dependence. Early studies of the efficacy of lithium for the treatment of alcohol dependence predominantly included inpatient alcoholics. The primary outcome measure reported in these studies was complete abstinence from alcohol, though over more than a decade of efforts to examine the efficacy of this medication, a number of methodological advances appeared. Kline et al. [50] compared lithium with placebo in a sample of inpatient alcoholic veterans. The efficacy outcome employed by these investigators was the occurrence of disabling alcoholic episodes, which occurred when patients' drinking interfered with their daily life, thereby necessitating inpatient detoxification. Merry et al. [51] randomly assigned inpatient alcoholics, stratified on the basis of co-morbid depressive symptoms, to receive either sustained-release lithium or placebo. Following discharge, patients were monitored at 6-week intervals for 1 year. The self-reported outcomes on which the treatment groups were compared were the number of days on which drinking occurred and the number of days on which patients were incapacitated by alcohol, the proportion of patients who were totally abstinent, and the change in depressive symptoms. Pond et al. [52] recruited alcoholics through newspaper advertisements, in an effort to enroll individuals who were not undergoing treatment for their alcoholism. They randomized participants to receive lithium or placebo using a crossover design. Self-reported drinking behavior was used to evaluate the efficacy of treatment, with grams of alcohol consumed per day as the primary outcome measure, which was obtained through weekly interviews. Although information was obtained on the number of tablets consumed, the occurrence of alcohol-related events (e.g., hospitalization) and non-alcohol-related hospitalization or illness, as well as changes in economic, social and psychological status, findings related to these measures were not reported.

In the late 1980s, studies evaluating lithium for alcoholism treatment showed clear methodological improvements, including the use of multiple measures of treatment outcome. Fawcett et al. [53] compared lithium to placebo in a sample of individuals recruited from an inpatient setting. Drinking outcomes were obtained through interviews conducted at least monthly by a nurse with alcoholism treatment experience. For most patients, a significant other provided information at 6-month intervals, which together with the self-report information and medical records was used to make a clinical judgment as to whether the patient had been completely abstinent, whether the patient had received other treatment and whether he or she had been working regularly. In addition to a dichotomous classification as abstinent or not, the time to relapse to any drinking was used in this study as an outcome of treatment. Other outcomes that were examined were abstinence from all abused drugs, severity of depressive symptoms, number of days missed from work, and number of hospitalizations.

Dorus et al. [54] conducted the largest and most methodologically sophisticated study of lithium for alcoholism treatment. The study was a placebo-controlled, multi-center VA Cooperative Study, which enrolled 457 alcoholics recruited as inpatients and stratified on the basis of a lifetime history of depression. The outcomes included total abstinence, number of days of drinking, number of alcohol-related hospitalizations, change in the rating of alcoholism severity and change in the severity of depression. These measures were obtained at 4-week intervals by interviewing the patient and a collateral informant. Based on the study's failure to demonstrate an advantage for lithium over placebo in either the depressed or non-depressed subgroups, it is now generally conceded that except for individuals with co-morbid bipolar disorder, lithium has no role in the treatment of alcoholism [55].

Alcohol-sensitizing medications

Early studies of disulfiram were, in many respects, similar to the later studies of lithium. Fuller and Roth [56] used total abstinence and the number of drinking days as the primary outcomes in the first, large-scale, controlled trial of disulfiram for alcoholism treatment. However, these investigators also compared treatment groups on the number of days of work attended by study participants, their family stability, and adherence to the schedule of study visits as secondary measures of efficacy. A similar approach was taken with the multicenter VA Cooperative Study of disulfiram [57], which, in addition to total abstinence and the number of drinking days as primary treatment outcome measures, included a measure of the time to first drinking episode. In both of these studies, self-reported drinking behavior was supplemented by collateral informant report. The disulfiram studies were, however, methodologically superior to the studies of lithium in that they used a riboflavin marker, which made it possible to measure medication compliance.

In a placebo-controlled study of the efficacy of the alcohol-sensitizing drug calcium carbimide, Peachey et al. [58] employed the following self-report measures of drinking behavior: drinking days, total quantity, and typical daily quantity. This study was unique, insofar as it required participants to return a urine sample daily, which made it possible to measure the concentrations of both alcohol and riboflavin (with which both the active and placebo medications were formulated), thereby providing objective measures of drinking behavior and medication compliance. The validity of daily reports of drinking behavior and medication compliance were thus evaluated against these criterion measures.

Acamprosate

A recent comprehensive meta-analysis of clinical trials of acamprosate [59] identified a total of 20 randomized, placebo-controlled studies, all but one of which are published. The studies were conducted in 12 European countries, the US, and Korea. Seventeen studies met the criteria for inclusion in the meta-analysis, which represent a total of 4087 patients (2160 who were treated with acamprosate and 1927 who received placebo). These studies enrolled patients with a diagnosis of alcohol abuse or dependence, so that there was variation in the severity of alcoholism across the studies, though there was an abstinence-oriented treatment orientation in all. The most common primary outcome measure was cumulative abstinent days (n = 9), followed by "drinking behavior" (n = 3), continuous abstinence (n = 2), and time to first relapse (n = 2, both of which had cumulative abstinent days as a co-primary outcome measure). Relapse rate at each visit, serum GGT activity, and time to treatment failure were each used as a primary outcome in only one of the studies reviewed. Although continuous abstinence rates were not the primary outcome measure in the studies reviewed, Mann et al. [59] were provided access by the pharmaceutical sponsor to data for these studies, and they were able to examine the overall effect of acamprosate on this outcome measure. These authors found a significant advantage for acamprosate over placebo on continuous abstinence rates. The effect sizes, though modest, increased progressively as treatment duration increased from three to six and then to 12 months.

Chick et al. [60] conducted a meta-analysis that included data from 15 of the same studies as those included in the Mann et al. [59] meta-analysis. Although the method of collecting drinking data at the assessment points varied among the studies, data available at each assessment made it possible to examine the overall effect of acamprosate on average daily alcohol consumption, number of days drinking, and the proportion of patients drinking an average of five or more drinks per day. Chick and colleagues acknowledged, however, lack of precision in the consumption estimates for the studies that used categories for the mean reported number of drinks per drinking day and the mean number of drinking days per week.

Overall, in contrast to studies of either lithium or disulfiram, there is some consistency in the outcomes assessed in the acamprosate studies, presumably because a single pharmaceutical company sponsored all of these studies. Nonetheless, there were a number of primary outcomes examined, particularly in view of the wide diversity of the countries in which the studies were conducted and the diversity of requirements for registration in those countries.

Opioid antagonists

A recent meta-analysis of the effects of opioid antagonists on treatment of alcohol dependence [61] identified a total of 14 randomized, placebo-controlled studies of naltrexone and two studies of nalmefene. These studies varied in duration from 4 weeks to one year, with a total of 1688 subjects, 868 who received active medication (including 84 treated with nalmefene) and 820 who received placebo. As shown in Table 2, these studies employed a total of 10 different measures of treatment outcome, the most common of which (94%) was the rate of study discontinuation.

Table 2. Outcome measures reported in randomized, placebo-controlled studies of opioid antagonists (n = 16)

Outcome measure	Number of studies
Discontinuation rate	15
Number of standard drinks	11
Number relapsed to drinking	9
Craving scale score	8
Number of abstinent days	7
Percentage or number of drinking days	6
Functioning score	1
Number relapsed to alcohol dependence	1
Time until relapse	1
Treatment adherence duration	1

Both naltrexone [13, 62] and nalmefene [14, 63] have been shown to reduce the risk of heavy drinking. As a consequence of these findings, subsequent studies of the oral formulation of naltrexone in alcoholics and in problem drinkers [18] and of long-acting naltrexone formulations in alcohol-dependent individuals [64, 65] have used the frequency of heavy drinking as a primary outcome measure.

Outcome measures for alcohol treatment trials: future directions

A decade ago, in anticipation of analyzing data from Project MATCH, investigators from that study [9] addressed the measurement of outcomes in alcohol treatment research. Based on empirical findings from four data sets, the group recommended the use of measures of both frequency (days drinking) and intensity (drinks per drinking day). They found these measures to be relatively uncorrelated and hence argued that they represent independent dimensions of drinking behavior. Furthermore, they pointed out that these measures are easily and inexpensively assessed using self-report techniques and are differentially sensitive to different kinds of treatment.

Recently, in an effort to develop an optimal measure for use in alcohol treatment trials, NIAAA convened a panel of scientists with expertise in the assessment of drinking behavior, particularly as it relates to clinical studies [5]. The panel recommended that percent of days heavy drinking (defined as ≥ 4 drinks in a day for women and ≥ 6 drinks in a day for men) be used as the single outcome measure in alcohol treatment trials. The rationale for choosing this measure was that it is a good predictor of acute alcohol-related problems. Furthermore, use of abstinence as a categorical measure of treatment outcome is limited by the small number of patients who are able to achieve it over an extended period of time and by its inability to show improvement over time.

Abstinence (though traditionally viewed as the most appropriate goal of alcohol treatment [59]), is not a suitable treatment goal for harm reduction efforts or for individuals with levels of severity that do not require abstinence as a treatment goal. Harm reduction approaches, in which reduced drinking frequency and intensity, rather than abstinence from alcohol, are the goals of treatment, have assumed greater importance in both clinical practice and as the focus of alcohol treatment trials, particularly pharmacotherapy research.

The NIAAA panel also recommended use of the TLFB for collecting data on percent of days heavy drinking. The TLFB is the most psychometrically sound and most widely used method of data collection for alcohol treatment trials. Although it is acknowledged that CR methods may provide more accurate measures of drinking, they argue that under most circumstances CR methods cannot be used to estimate pretreatment drinking and thus are not the preferred methods for comparing pre- and post-treatment outcomes. However, this begs the question of whether a hybrid approach, in which subjects report pretreatment drinking using the TLFB or another DDE approach and prospectively report drinking using a CR approach, may incorporate the best aspects of both approaches. As discussed earlier, use of the TLFB or a similar DDE method also ignores the utility of the CR approach for gathering data on mood, desire to drink and other factors that vary day to day. An approach that incorporates these fine-grained data and employs modern analytic tools makes it possible to detect very specific effects of treatment through increased statistical power [4].

The search for a single, "optimal" treatment outcome measure may also prove to be futile. In underscoring the importance of statistical power in treatment outcome studies, Stout [4] argues that a more efficient alternative to increasing sample size is the use of improved measures of treatment outcome and analytic methods tailored to the predicted effects of treatment. Because these effects differ among studies, no single outcome measure is likely to be useful for all studies, a point that contrasts with the efforts of the NIAAA panel.

The choice of outcome measures in the Combine Study, a large multi-center study comparing naltrexone, acamprosate and their combination with placebo, with all medication conditions delivered with either medical management or a more intensive psychotherapy [66], may influence the design of pharmacotherapy research over the next decade. In this study, the primary outcome measures are percent days abstinent and the number of days to first heavy drinking episode, as measured using the Form 90. Secondary outcome measures include number of heavy drinking days, a composite outcome measure that integrates both alcohol consumption and alcohol-related problem variables, biological markers of heavy drinking (e.g., CDT), level of alcohol craving, and presence of DSM-IV alcohol dependence.

As the pharmaceutical industry has come increasingly to recognize the potentially lucrative market that medications to treat alcoholism could represent, there has been growing interest in the conduct of multi-center clinical trials. These allow rapid recruitment of the large study samples required to provide adequate statistical power. However, the measurement of drinking outcomes in multi-center clinical trials presents a number of challenges in terms of feasibility, sensitivity, and conceptualization of the dependent variable. Traditionally, practical considerations have favored the use of self-report measures because of their lower cost and ease of use. Although alternative measures of drinking, such as biochemical tests and collateral informant reports, have been proposed, they have not gained wide acceptance as independent outcome measures. If collected at all in outcome studies, they are more likely to be used to provide circumstantial evidence for the accuracy of self-report information, rather than as a specific validity check on a given individual. Even under the optimal conditions of a large-scale clinical trial, it is more difficult to obtain blood samples and collateral informants' reports than client self-reports. All measures decline in completeness as the time from the baseline assessment increases.

Prospective daily monitoring of drinking behavior using a method such as IVR has the potential to serve as a centralized method for use in multi-center trials. We would speculate that the intersite variability in outcome measurement involving methods such as the TLFB or Form 90 represents a major source of error variance that can threaten the internal validity of multi-center trials. IVR-based administration of a common script for eliciting data on drinking behavior could reduce some of the site effects, thereby representing an important methodological advance in the conduct of multi-center trials. This approach may also provide insights into the mechanisms by which pharmacological treatment effects occur [37].

References

1 Breslin FC, Sobell SL, Sobell LC, Sobell MB (1997) Alcohol treatment outcome methodology: state of the art 1989–1993. *Addict Behav* 22: 145–155
2 Del Boca F, Darkes J (2003) The validity of self-reports of alcohol consumption: state of the science and challenges for research. *Addiction* 98 Suppl 2: 1–12
3 Finney J, Moyer A, Swearingen C (2003) Outcome variables and their assessment in alcohol treatment studies: 1968–1998. *Alcohol Clin Exp Res* 27: 1671–1679
4 Stout RL (2003) Methodological and statistical considerations in measuring alcohol treatment effects. *Alcohol Clin Exp Res* 27: 1686–1691
5 Sobell LC, Sobell MB, Connors GJ, Agrawal S (2003) Assessing drinking outcomes in alcohol treatment efficacy studies: selecting a yardstick of success. *Alcohol Clin Exp Res* 27: 1661–1666
6 Kranzler H, Mason B, Pettinati H, Anton R (1997) Methodological issues in pharmacotherapy trials with alcoholics. In: M Hertzman, D Feltner (eds): *The Handbook of Psychopharmacology Trials*. New York University Press, New York, 213–245
7 Kranzler H, McLellan A, Bohn M (1995) Pharmacotherapies for alcoholism: Theoretical and methodologic perspectives. In: H Kranzler (ed.): *The Pharmacology of Alcohol Abuse*. Springer-Verlag, New York, 1–10
8 Linnoila M, Fawcett J, Fuller R, Harford T, Martin P, Meyer R, Romach M, Sellers E (1994) Evaluation of pharmacologic treatments for alcohol abuse, dependence, and their complications. In: R Prien, D Robinson (eds): *Clinical Evaluation of Psychotropic Drugs: Principles and Guidelines*. Raven Press, New York, 625–649

9 Babor TF, Longabaugh R, Zweben A, Fuller RK, Stout RL, Anton RF, Randall CL (1994) Issues in the definition and measurement of drinking outcomes in alcoholism treatment research. *J Stud Alcohol Suppl* 12: 101–111
10 Moyer A, Finney JW, Swearingen CE (2002) Methodological characteristics and quality of alcohol treatment outcome studies, 1970–98: an expanded evaluation. *Addiction* 97: 253–263
11 Babor TF, Stephens RS, Marlatt GA (1987a) Verbal report methods in clinical research on alcoholism: response bias and its minimization. *J Stud Alcohol* 48: 410–424
12 Babor TF, Steinberg K, Anton R, Del Boca F (2000) Talk is cheap: measuring drinking outcomes in clinical trials. *J Stud Alcohol* 61: 55–63
13 O'Malley SS, Jaffe AJ, Chang G, Schottenfeld RS, Meyer RE, Rounsaville B (1992) Naltrexone and coping skills therapy for alcohol dependence. A controlled study. *Arch Gen Psychiatry* 49: 881–887
14 Mason BJ, Salvato FR, Williams LD, Ritvo EC, Cutler RB (1999) A double-blind, placebo-controlled study of oral nalmefene for alcohol dependence. *Arch Gen Psychiatry* 56: 719–724
15 Sobell L, Sobell M (1992) Timeline follow-back: a technique for assessing self-reported alcohol consumption. In: R Litten, J Allen (eds): *Measuring alcohol consumption: Psychosocial and Biochemical Methods*. Human Press, Clifton NJ, 41–42
16 Armeli S, Tennen H, Affleck G, Kranzler HR (2000) Does affect mediate the association between daily events and alcohol use? *J Stud Alcohol* 61: 862–871
17 Kranzler H, Abu-Hasaballah K, Tennen H, Feinn R, Young K (2004) Using daily interactive voice recording to measure drinking and related behaviors in a pharmacotherapy study of problem drinkers. *Alcohol Clin Exp Res* 28: 1060–1064
18 Kranzler H, Armeli S, Tennen H, Blomqvist O, Oncken C, Petry N, Feinn R (2003) Targeted naltrexone for early problem drinkers. *J Clin Psychopharmacol* 23: 294–304
19 Maisto SA, Sobell LC, Cooper AM, Sobell MB (1982) Comparison of two techniques to obtain retrospective reports of drinking behavior from alcohol abusers. *Addict Behav* 7: 33–38
20 Sobell MB, Sobell LC, Klajner F, Pavan D, Basian E (1986) The reliability of a timeline method for assessing normal drinker college students' recent drinking history: utility for alcohol research. *Addict Behav* 11: 149–161
21 Sobell LC, Brown J, Leo GI, Sobell MB (1996) The reliability of the Alcohol Timeline Followback when administered by telephone and by computer. *Drug Alcohol Depend* 42: 49–54
22 Sobell LC, Agrawal S, Annis H, Ayala-Velazquez H, Echeverria L, Leo GI, Rybakowski JK, Sandahl C, Saunders B, Thomas S et al. (2001) Cross-cultural evaluation of two drinking assessment instruments: alcohol timeline followback and inventory of drinking situations. *Subst Use Misuse* 36: 313–331
23 Carney MA, Tennen H, Affleck G, Del Boca FK, Kranzler HR (1998) Levels and patterns of alcohol consumption using timeline follow-back, daily diaries and real-time "electronic interviews". *J Stud Alcohol* 59: 447–454
24 Alemi F, Stephens R, Parran T, Llorens S, Bhatt P, Ghadiri A, Eisenstein E (1994) Automated monitoring of outcomes: application to treatment of drug abuse. *Med Decis Making* 14: 180–187
25 Searles JS, Helzer JE, Rose GL, Badger GJ (2002) Concurrent and retrospective reports of alcohol consumption across 30, 90 and 366 days: interactive voice response compared with the timeline follow back. *J Stud Alcohol* 63: 352–362
26 Bardone AM, Krahn DD, Goodman BM, Searles JS (2000) Using interactive voice response technology and timeline follow-back methodology in studying binge eating and drinking behavior: different answers to different forms of the same question? *Addict Behav* 25: 1–11
27 Searles JS, Helzer JE, Walter DE (2000) Comparison of drinking patterns measured by daily reports and timeline follow back. *Psychol Addict Behav* 14: 277–286
28 Searles JS, Perrine MW, Mundt JC, Helzer JE (1995) Self-report of drinking using touch-tone telephone: extending the limits of reliable daily contact. *J Stud Alcohol* 56: 375–382
29 Tennen H, Affleck G (1996) Daily processes in coping with chronic pain: Methods and analytic strategies. In: M Zeidner, N Endler (eds): *Handbook of Coping*. Wiley, New York, 151–180
30 Perrine MW, Mundt JC, Searles JS, Lester LS (1995) Validation of daily self-reported alcohol consumption using interactive voice response (IVR) technology. *J Stud Alcohol* 56: 487–490
31 Helzer JE, Badger GJ, Rose GL, Mongeon JA, Searles JS (2002) Decline in alcohol consumption during two years of daily reporting. *J Stud Alcohol* 63: 551–558
32 Hayes S, Cavior N (1980) Multiple tracking and the reactivity of self-monitoring: II. Positive behaviors. *Behav Assess* 2: 283–296

33 Tennen H, Affleck G, Armeli S, Carney MA (2000) A daily process approach to coping. Linking theory, research, and practice. *Am Psychol* 55: 626–636
34 Armeli S, Carney MA, Tennen H, Affleck G, O'Neil TP (2000) Stress and alcohol use: a daily process examination of the stressor-vulnerability model. *J Pers Soc Psychol* 78: 979–994
35 Hufford MR, Shields AL, Shiffman S, Paty J, Balabanis M (2002) Reactivity to ecological momentary assessment: an example using undergraduate problem drinkers. *Psychol Addict Behav* 16: 205–211
36 Collins RL, Morsheimer ET, Shiffman S, Paty JA, Gnys M, Papandonatos GD (1998) Ecological momentary assessment in a behavioral drinking moderation training program. *Exp Clin Psychopharmacology* 6: 306–315
37 Kranzler HR, Armeli S, Feinn R, Tennen H (2004) Targeted naltrexone treatment moderates the relations between mood and drinking behavior among problem drinkers. *J Consult Clin Psychol* 72: 317–327
38 Irwin M, Baird S, Smith TL, Schuckit M (1988) Use of laboratory tests to monitor heavy drinking by alcoholic men discharged from a treatment program. *Am J Psychiatry* 145: 595–599
39 Iber FL, Lee K, Lacoursiere R, Fuller R (1987) Liver toxicity encountered in the Veterans Administration trial of disulfiram in alcoholics. *Alcohol Clin Exp Res* 11: 301–304
40 Conigrave KM, Davies P, Haber P, Whitfield JB (2003) Traditional markers of excessive alcohol use. *Addiction* 98 Suppl 2: 31–43
41 Kranzler HR, Burleson JA, Del Boca FK, Babor TF, Korner P, Brown J, Bohn MJ (1994) Buspirone treatment of anxious alcoholics. A placebo-controlled trial. *Arch Gen Psychiatry* 51: 720–731
42 Volpicelli JR, Rhines KC, Rhines JS, Volpicelli LA, Alterman AI, O'Brien CP (1997) Naltrexone and alcohol dependence. Role of subject compliance. *Arch Gen Psychiatry* 54: 737–742
43 Anton RF, Moak DH (1994) Carbohydrate-deficient transferrin and gamma-glutamyltransferase as markers of heavy alcohol consumption: gender differences. *Alcohol Clin Exp Res* 18: 747–754
44 Allen JP, Litten RZ, Anton RF, Cross GM (1994) Carbohydrate-deficient transferrin as a measure of immoderate drinking: remaining issues. *Alcohol Clin Exp Res* 18: 799–812
45 Sillanaukee P, Olsson U (2001) Improved diagnostic classification of alcohol abusers by combining carbohydrate-deficient transferrin and gamma-glutamyltransferase. *Clin Chem* 47: 681–685
46 Anton RF, Lieber C, Tabakoff B (2002) Carbohydrate-deficient transferrin and gamma-glutamyltransferase for the detection and monitoring of alcohol use: results from a multisite study. *Alcohol Clin Exp Res* 26: 1215–1222
47 Javors MA, Johnson BA (2003) Current status of carbohydrate deficient transferrin, total serum sialic acid, sialic acid index of apolipoprotein J and serum beta-hexosaminidase as markers for alcohol consumption. *Addiction* 98 Suppl 2: 45–50
48 Edwards G (1985) A later follow-up of a classic case series: D. L. Davies's 1962 report and its significance for the present. *J Stud Alcohol* 46: 181–190
49 Connors GJ, Maisto SA (2003) Drinking reports from collateral individuals. *Addiction* 98 Suppl 2: 21–29
50 Kline NS, Wren JC, Cooper TB, Varga E, Canal O (1974) Evaluation of lithium therapy in chronic and periodic alcoholism. *Am J Med Sci* 268: 15–22
51 Merry J, Reynolds CM, Bailey J, Coppen A (1976) Prophylactic treatment of alcoholism by lithium carbonate. A controlled study. *Lancet* 1: 481–482
52 Pond SM, Becker CE, Vandervoort R, Phillips M, Bowler RM, Peck CC (1981) An evaluation of the effects of lithium in the treatment of chronic alcoholism. I. Clinical results. *Alcohol Clin Exp Res* 5: 247–251
53 Fawcett J, Clark DC, Aagesen CA, Pisani VD, Tilkin JM, Sellers D, McGuire M, Gibbons RD (1987) A double-blind, placebo-controlled trial of lithium carbonate therapy for alcoholism. *Arch Gen Psychiatry* 44: 248–256
54 Dorus W, Ostrow DG, Anton R, Cushman P, Collins JF, Schaefer M, Charles HL, Desai P, Hayashida M, Malkerneker U et al. (1989) Lithium treatment of depressed and nondepressed alcoholics. *JAMA* 262: 1646–1652
55 Calabrese JR, Shelton MD, Bowden CL, Rapport DJ, Suppes T, Shirley ER, Kimmel SE, Caban SJ (2001) Bipolar rapid cycling: focus on depression as its hallmark. *J Clin Psychiatry* 62 Suppl 14: 34–41
56 Fuller RK, Roth HP (1979) Disulfiram for the treatment of alcoholism. An evaluation in 128 men. *Ann Intern Med* 90: 901–904

57 Fuller RK, Branchey L, Brightwell DR, Derman RM, Emrick CD, Iber FL, James KE, Lacoursiere RB, Lee KK, Lowenstam I et al. (1986) Disulfiram treatment of alcoholism. A Veterans Administration cooperative study. *JAMA* 256: 1449–1455
58 Peachey JE, Annis HM, Bornstein ER, Sykora K, Maglana SM, Shamai S (1989) Calcium carbimide in alcoholism treatment. Part 1: A placebo-controlled, double-blind clinical trial of short-term efficacy. *Br J Addict* 84: 877–887
59 Mann K, Lehert P, Morgan MY (2004) The efficacy of acamprosate in the maintenance of abstinence in alcohol-dependent individuals: results of a meta-analysis. *Alcohol Clin Exp Res* 28: 51–63
60 Chick J, Lehert P, Landron F (2003) Does acamprosate improve reduction of drinking as well as aiding abstinence? *J Psychopharmacol* 17: 397–402
61 Srisurapanont M, Jarusuraisin N (2002) Opioid antagonists for alcohol dependence. *Cochrane Database Syst Rev* 2: CD001867
62 Volpicelli JR, Alterman AI, Hayashida M, O'Brien CP (1992) Naltrexone in the treatment of alcohol dependence. *Arch Gen Psychiatry* 49: 876–880
63 Mason BJ, Ritvo EC, Morgan RO, Salvato FR, Goldberg G, Welch B, Mantero-Atienza E (1994) A double-blind, placebo-controlled pilot study to evaluate the efficacy and safety of oral nalmefene HCl for alcohol dependence. *Alcohol Clin Exp Res* 18: 1162–1167
64 Kranzler HR, Modesto-Lowe V, Nuwayser ES (1998) Sustained-release naltrexone for alcoholism treatment: a preliminary study. *Alcohol Clin Exp Res* 22: 1074–1079
65 Kranzler H, Wesson D, Billot L (2004) For the DAS Naltrexone Depot Study Group. Naltrexone depot for treatment of alcohol dependence: A multi-center, randomized, placebo-controlled clinical trial. *Alcohol Clin Exp Res* 28: 1051–1059
66 CombineStudyResearchGroup (2003) Testing combined pharmacotherapies and behavioral interventions in alcohol dependence: rationale and methods. *Alcohol Clin Exp Res* 27: 1107–1122

Drugs for Relapse Prevention of Alcoholism
Edited by R. Spanagel and K. Mann

Disulfiram (Antabuse®): the first medication to stop drinking

Dieter Ladewig[1] and Ulrich von Bardeleben[2]

[1] *Department of Substance Use Disorders, Psychiatric University Hospital, Wilhelm Klein-Strasse 27, CH-4025 Basel, Switzerland*
[2] *Psychiatrische Klinik Meiringen, CH-3860 Meiringen/BE, Switzerland*

Introduction

Disulfiram (Antabuse®) is historically one of the first medications found to stop drinking. This compound inhibits acetaldehyde dehydrogenase, so that alcohol consumption leads to an accumulation of acetaldehyde. This in turn results in distressing symptoms, including flushing, syncope, a decrease in blood pressure, and vomiting as well as diarrhea. Disulfiram is available in different dosages with 200 mg/day being a standard dose. A supervised exposition to a minimal amount of the preferred alcohol beverage may help the patient to experience the possible disulfiram-alcohol reaction under a controlled treatment management.

Treatment with disulfiram arouses controversy among the alcohol treatment community. Specialists in some countries neglect this kind of approach, whereas in other countries Antabuse® is an essential part of the treatment of alcoholism. Treatment with disulfiram may be effective, if it is part of an individualized therapy program. This *ex-cathedra* statement is based on a long clinical experience in Switzerland and the Scandinavian countries. Denmark, for instance, has a long tradition for its use. The often quoted finding that placebo-controlled studies are rarely in favor of this medication [1, 2] is in part in contrast to clinical experiences. Table 1 gives an overview of the methodology in 24 studies of oral disulfiram administration. In recent years, several treatment modifications with disulfiram have been described that may improve outcome variables such as duration of abstinence or the amount of alcohol consumed per day in non-abstinent phases. This implies that treatment does not consist only of a medication. One simple modification of treatment with disulfiram is supervised intake [3]. Supervised disulfiram means controlled delivery of the soluble substance in an adequate setting. This means that some form of psychosocial intervention is part of the treatment. Another approach focuses on the importance of a relationship or an accompanying behavior or marital therapy [4]. Both modifications increase the effectiveness of disulfiram.

Table 1. Methodology in 24 studies of oral disulfiram

Total number of subjects	2836	
Mean number of subjects per study (range)	118	(16–605)
Number of studies using objective criteria* (%)	20	(83)
Number of studies of supervised disulfiram (%)	15	(62.5)
Number of single blind studies (%)	4	(17)
Number of double blind studies (%)	1	(4)
Number of studies with matched controls (%)	13	(54)
Number of randomized studies (%)	13	(54)
Number of studies with statistical analysis† (%)	16	(67)
Mean follow-up in months		
(a) on treatment in 23 studies (range)	12.9	(2–24)
(b) post-treatment in six studies (range)	13.7	(0.75–48)
Mean percentage lost to follow-up per study (range)	19.6%	(0–92%)

* but not always explicit nor comprehensive; †not always extensive.

Although, for some clinicians, disulfiram is still the first medication to stop drinking, the spectrum of pharmacological and psychotherapeutic interventions nowadays available is such that in daily practice and in research, combinations have much more a chance of success than a mono-intervention.

Randomized open efficacy study

In addition to disulfiram (Antabuse®), acamprosate (Campral®) and naltrexone (Revia/Nemexin®) have been approved as pharmacological support for maintaining alcohol abstinence. With disulfiram, the consumption of alcohol produces the typical symptoms of flushing, nausea, tachycardia, hypotension and dyspnea- aversive effects, which may prevent the patient from further drinking. Acamprosate has been found to reduce the neural hyperexcitability of the glutamatergic system, which becomes up-regulated under the chronic action of alcohol [5] (see also Chapter 19). The clinically observed and significantly higher abstinence rate under acamprosate [6, 7] is probably due to an inhibition of conditioned withdrawal phenomena, which may appear as "craving" and function as a trigger for relapse [8]. The opioid receptor antagonist naltrexone was found to reduce the activating influence of alcohol on endogenous opioid systems [8, 9] (see also Chapter by Cowen). The release of endogenous opioids, in particular of β-endorphin, in the mesolimbic reward system plays a key role in substance dependence. Several clinical trials have identified a reduction in alcohol consumption under naltrexone [10–12]. Whereas the effect of acamprosate and naltrexone is based on biological findings, the effect of disulfiram is primarily a psychological one. In most studies on the maintenance of alcohol abstinence using the aforementioned medications, pharmacotherapy was combined with various types of psychotherapy. To evaluate whether there are any differences in the course of treatment by using acamprosate, naltrexone or disulfiram in combination with a standardized cognitive behavioral group therapy, we started a study with a sample of alcoholics.

The total sample included 100 patients diagnosed with alcohol dependence according to DSM-IV. 71% of the patients who came for detoxification to the University of Basel, Department of Psychiatry, were men and 29% were women. The mean age of the sample was 44.8 (±9.2) years. Patients were recruited after at least 7 days (max. 14 days) of alcohol abstinence following in-patient detoxification. They were randomly assigned to one of the three medication groups (N = 25 each) with acamprosate, naltrexone, disulfiram, or the "non-medication" group. The patients received trial medication in fixed dosages (1800 mg/d acamprosate; 50 mg/d naltrexone; 200 mg/d disulfiram) when entering the study (12 weeks) in combination with a manualized cognitive behavioral group therapy program for relapse prevention in outpatients. This modified version of Monti's [13] group therapy program was conducted by a clinical psychologist and a psychiatrist. The 12 modules of the program were highly structured in their content and form and included the following elements: 1) An introduction to coping strategies in alcohol dependence, 2) coping with alcohol craving, 3) dealing with thoughts about alcohol, 4) general problem-solving strategies, 5) the ability to refuse the offer of alcohol consumption (saying no), 6) dealing with seemingly insignificant decisions, 7) dealing with high risk and relapse situations, 8) supporting one's own needs, 9) dealing with criticism, 10) dealing with depressed mood, 11) dealing with anger, 12) relaxation techniques.

The four groups studied showed no differences in relation to the sociodemographic data determined at the beginning of the study, such as gender, age, education, income, living and family conditions. Neither were there any differences between the groups with regard to family stress, the frequency of earlier withdrawals or patients' drinking behavior in the three months prior to entering the study. The craving for alcohol was measured on a visual analogue scale (VAS) with higher values indicating greater craving. These measurements were taken weekly. Though there was a slight difference at baseline with lower values in the acamprosate group, no significant differences between groups treated with medication were found during and at the end of the treatment. But there was a significant difference in comparison with the no-drug group. Generally, all the drugs were well tolerated with the acamprosate group reporting initial diarrhea more frequently and the naltrexone group reporting sleep disturbances more frequently. These side-effects did not cause any patient to end treatment. Concerning standard laboratory parameters, which were determined at intervals of 4 weeks, no differences between groups were found for gamma-glutamyl transferase and mean erythrocyte volume. 60% of the patients completed the 12-week study. This corresponds to a dropout rate of 40%, which is fairly low for treatment studies in substance dependence. There were no significant differences between groups in this respect (see Tab. 2).

Table 2. Participation rates in group comparison (N = 100)

	No medication	Disulfiram	Acamprosate	Naltrexone	$\%^2$ or ANOVA
Regular termination of study (%)	48	52	64	68	ns
Duration of standardized observation (weeks; x ± SD)	9.2 ± 3.8	9.2 ± 3.9	9.3 ± 4.2	9.5 ± 3.3	ns
Group meetings attended (x ± SD)	7.5 ± 3.4	7.2 ± 3.6	7.3 ± 3.6	6.4 ± 3.3	ns

ns: not significant; SD: standard deviation

Even in the patients who had dropped out, it was possible and on a weekly basis to assess the amount of drinking per day. The main reason for dropping out was relapse. All missing data from patients that had dropped out of the study were statistically accounted for as full relapses. In the disulfiram group, patients stopped medication before relapse. Although 52% of the patients reported having consumed alcohol at least once during the 12-week treatment, the proportion of abstinent days was 90% overall (see Tab. 3). The four groups showed no significant differences in regard to the percentage of abstinent days. There was a significant difference between the groups in relation to the time period until the first drink; the patients in the disulfiram group had a longer period of abstinence compared to patients in the acamprosate group ($p < 0.05$) or patients in the naltrexone group ($p < 0.05$). With regard to the time period until the first relapse (>40/60 g of alcohol for women/men per drinking day), there were no significant differences between the groups. The same applied for the time period until the first serious relapse (>40/60 g of alcohol for women/men per drinking day on at least 3 successive days) for the three groups on medication, whereas the group with no medication had a significantly shorter period of abstinence. When a drinking event did occur, a comparison between groups revealed the following differences in the quantity of alcohol consumed on a drinking day: the acamprosate group drank significantly lower quantities of alcohol per drinking day compared to the disulfiram group (76 ± 45 g *versus* 152 ± 69 g: x ± SD; $p < 0.05$), and the naltrexone group drank even less (57 ± 41 g; $p < 0.05$). Overall, patients attended, on the average, 61% of all group meetings, which corresponds to a good acceptance of the group therapy program. A semi-structured interview of the participants in this regard revealed that, according to their own estimation, 60% of them had acquired new knowledge with regard to their alcohol problem and had experienced modifications in their behavior. The relatively rigid formal structure of the group therapy was named as the main reason for the acceptance of the program by more than 55% of the participants.

Table 3. Results of relapse prevention program in group comparison (N = 100)

	No medication	Disulfiram	Acamprosate	Naltrexone	No med. *vs.* D	No med. *vs.* A	No med. *vs.* N	D *vs.* A	D *vs.* N
Abstinent days (%: x ± SD)	89.4 ± 18.3	90.1 ± 18.9	91.6 ± 17.2	82.3 ± 19.6	ns	ns	ns	ns	ns
Weeks until first consumption of alcohol (median)	5	12	7	8	0.04*	ns	ns	0.04*	0.04*
Weeks until first serious relapse (>60/40 g alcohol for men/women on 3 consecutive days, median)	8	12	11	11	0.04*	0.04*	0.04*	ns	ns
Alcohol quantity/drinking day (g; x ± SD)	98 ± 77	152 ± 69	76 ± 45	57 ± 41	ns	ns	0.04*	0.04*	0.02*

In summary, the relatively high proportion of abstinent days during treatment may be due to the effects of the different medications as well as the effects of the cognitive-behavioral therapy. Concerning the results of the disulfiram group, patients were aware of the adverse effects when consuming alcohol while taking this medication. This aspect has also been discussed in other studies on disulfiram [1]. Regarding the time period to the first relapse, the three drugs produced comparable results. However, the time period to the first drink was longest in the disulfiram group. On the other hand, the amount of alcohol consumed per day during relapse was highest in the group treated with disulfiram. This is possibly due to the known abstinence-violation effect, which may be of more importance in disulfiram treatment than in treatment with the other two substances. This might also be an effect of the anti-craving properties ascribed to these substances. Yet, all these minor differences between the three medications must not be overvalued, since the group sizes were very small. Nevertheless, outpatient treatment combining psychotherapy and pharmacotherapy seems to be a promising tool in clinical practice. This is also reflected in the remarkably high acceptance rate of this program.

Finally, expectations concerning improvement are important in any treatment, especially in alcoholism treatment. Self-confidence in achieving abstinence was measured with the KAZ [14]. We found that already before the beginning of any treatment, patients with serious relapse and those without relapse (>40 g/60 g of alcohol for women/men per drinking day on at least 3 successive days) could be differentiated. This means that alcoholics with low self-confidence should be identified and treatment with any drug should then be organized with caution. This again demonstrates the importance of psychological factors.

Conclusions

Studies with disulfiram have produced mixed results. However, there are advocates of disulfiram treatment programs. Heather [15] recommended disulfiram as a way of providing some patients with a respite from ravages of heavy drinking. Chick and Brewer [2, 16, 17] warned of the under-use of disulfiram despite the fact that the studies with a sufficient methodology are rare [18]. In particular, the study of Fuller et al. [1] implies that, at least in the short run, disulfiram can reduce the number of drinking days and the amounts of consumed alcohol – a similar finding was reproduced in our study. Some aspects of treatment raise particular methodological concerns. These include patient-treatment matching, the impossibility of double blindness (will be broken by value of the disulfiram-alcohol reaction), the use of disulfiram as part of a treatment package and the problem of compliance. A great variety of patient characteristics has been suggested, which offers potential for matching with disulfiram treatment. The strongest predictor of willingness to take medication was a belief that the medication would be helpful [19], comparable to our findings with the KAZ-35. The list of positive and negative predictive factors is long and in parts contradictory. This means that it is impossible to put all of them in an outcome study which fulfills adequate methodological criteria. Knowing the differences between the right and the wrong patient for disulfiram would clinically be crucial but it remains a long-term objective.

We think that disulfiram may be used to promote self-help – this means that disulfiram medication might be interpreted as a help to increase self-control – and may be used as part of a treatment package knowing that this is methodologically difficult to be proven. The combination with psychotherapeutic interventions like marital therapy, behavioral therapy or social skills training [20] is most promising for establishing compliance and abstinence. The combination of several pharmacotherapies like disulfiram and acamprosate [21] or acamprosate and naltrexone [22] and comparisons with other pharmacotherapies are other fruitful approaches [23]. This established a process of learnt abstinence [24], which was very intensively taken up by Ehrenreich et al. [25].

Years back, disulfiram (Antabuse®) was historically the first medication used to stop drinking. Today, treatment is dependent on the overall treatment setting, deciding on whether treatment starts with disulfiram, acamprosate, naltrexone, or a combination of these compounds. The strategy of combining different "anticraving" substances deserves further clinical attention and there are already several ongoing studies tackling this issue.

References

1 Fuller RK, Branchey L, Brightwell DR (1986) Disulfiram treatment of alcoholism: a veterans administration co-operative study. *JAMA* 256: 1449–1455

2 Chick J, Gough K, Falkowski N (1992) Disulfiram treatment of alcoholism. *Br J Psych* 161: 84–89

3 Hughes JC, Cook CH (1997) The efficacy of disulfiram: a review of outcome studies. *Addiction* 92: 381–395

4 Azrin NH, Sisson RW, Meyers R, Godley M (1982) Alcoholism treatment by disulfiram and community reinforcement therapy. *J Behav Ther and Exp Psychiatry* 13: 105–112

5 Zieglgänsberger W, Hauser C, Wetzel C, Putzke J, Siggins GR, Spanagel R (1996) Actions of Acamprosate on neurons of the central nervous system. In: M Soyka (ed.): *Acamprosate in relapse prevention of alcoholism*. Springer, Heidelberg, 65

6 Sass H, Soyka M, Mann K (1996) Relapse prevention by Acamprosate: results from a placebo-controlled study on alcohol dependence. *Arch Gen Psych* 53: 673–680

7 Mann K, Lehert P, Morgan MY (2004) The efficacy of acamprosate in the maintenance of abstinence in alcohol-dependent individuals: results of a meta-analysis. *Alcohol Clin Exp Res* 28: 51–63

8 Spanagel R, Zieglgänsberger W (1997) Anti-craving compounds for ethanol: new pharmacological tools to study addictive processes. *Trends Pharmadcol Sci* 18: 54–59

9 Herz A (1997) Endogenous opioid systems and alcohol addiction. *Psychopharmacology* 129: 99–111

10 Volpicelli JR, Alterman AI, Hayashida M, O'Brien CP (1992) Naltrexone in the treatment of alcohol dependence. *Arch Gen Psych* 49: 876–880

11 O'Malley S, Jaffe AJ, Chang G, Schottenfeld RS, Meyer RE, Rounsaville B (1992) Naltrexone and coping skills therapy for alcohol dependence. *Arch Gen Psych* 49: 881–887

12 Carmen B, Angeles M, Ana M, Maria AJ (2004) Efficacy and safety of naltrexone and acamprosate in the treatment of alcohol dependence: a systematic review. *Addiction* 99: 811–828

13 Monti PM, Abrams DB, Kadden RM, Cooney NL (eds) (1989) *Treating alcohol dependence. A coping skills training guide.* Guilford Press, New York
14 Körkel J, Lauer G (1995) Rückfälle Alkoholabhängiger: Ein Überblick über die neuen Forschungsergebnisse und -trends. In: J Körkel, G Lauer, R Scheller (eds): *Sucht und Rückfall. Brennpunkte deutscher Rückfallforschung.* Enke, Stuttgart, 154–185
15 Heather N (1989) Disulfiram treatment for alcoholism: deserves re-examination. *Br Med J* 299: 471–472
16 Brewer C (1993) Invited review: recent developments in disulfiram treatment. *Alcohol Alcoholism* 28 (4): 383–395
17 Brewer C, Meyers RJ, Johnsen J (2000) Does disulfiram help to prevent relapse in alcohol abuse? *CNS Drugs* 14 (5): 329–341
18 Garbutt JC, West SL, Carey TS, Lohr KN, Crews FT (1999) Pharmacological treatment of alcohol dependence. A review of the evidence. *JAMA* 281 (14): 1318–1325
19 Swift RM, Duncan D, Nirenberg T, Femino J (1998) Alcoholic patients' experience and attitudes on pharmacotherapy for alcoholism. *J Addict Diseases* 17 (3): 35–47
20 O'Farrell TJ, Hooley J, Fals-Stewart W, Cutter HS (1998) Expressed emotion and relapse in alcoholic patients. *J Consult Clin Psychology* 66 (5): 744–752
21 Besson J, Aeby F, Kasas A, Lehert P, Potgieter A (1998) Combined efficacy of acamprosate and disulfiram in the treatment of alcoholism: a controlled study. *Alcohol Clin Exp Res 22* (3): 573–579
22 Kiefer F, Jahn H, Tarnaske T, Helwig H, Briken P, Holzbach R, Kämpf PH, Stracke R, Bahr M, Naber D et al. (2003) Comparing and combining naltrexone and acamprosate in relapse prevention of alcoholism. *Arch Gen Psychiatry* 60: 92–99
23 Rubio G, Jimenez-Arriero MA, Ponce G, Palomo T (2001) Naltrexone *versus* acamprosate: one year follow-up of alcohol dependence treatment. *Alcohol Alcoholism* 36 (5): 419–425
24 Brewer C, Streel E (2003) Learning the language of abstinence in addiction treatment: some similarities between relapse-prevention with disulfiram, naltrexone, and other pharmacological antagonists and intensive "immersion" methods of foreign language teaching. *Subst Abuse* 24 (3): 157–173
25 Ehrenreich H, Mangholz A, Schmitt M, Lieder P, Völkel ER, Poser W (1997) OLITA: an alternative in the treatment of therapy-resistant chronic alcoholics. First evaluation of a new approach. *Eur Arch Psychiatry Clin Neurosci* 247: 51–54

Naltrexone: preclinical data

Michael S. Cowen

Howard Florey Institute, University of Melbourne, VIC 3010, Australia

Introduction

Naltrexone (N-cylopropylmethylnoroxymorphone) is a potent opiate antagonist with only very weak agonist properties [1]. The drug is well tolerated [2] and orally available [1]. Naltrexone has been shown to precipitate withdrawal from morphine in rats [3] and monkeys [4], and to reverse morphine-induced antinociception against noxious heat and pressure stimuli in rats [5]. Naltrexone was also shown to reverse the antinociception induced by the selective κ-opioid receptor agonist U-69, 593 with slightly higher doses than for morphine-induced antinociception [5], reflecting the somewhat higher affinity of naltrexone for μ-opioid receptors compared with κ- and δ-opioid receptors [6, 7]. Prior to the publication of the first clinical studies examining the effect of naltrexone in alcohol dependence [8, 9], most pre-clinical studies examining the interaction of ethanol with opioids used the shorter-acting, 'pure' opioid antagonist naloxone [10, 11] rather than naltrexone [12–15]. However, since that time a large body of data has been gathered examining the effect of naltrexone on ethanol consumption in various pre-clinical models.

Dosage

In general, naltrexone has been shown to cause a dose-dependent decrease in ethanol consumption and preference in a range of animal models and species, including mice, rats and monkeys. The range of effective doses appears broadly similar across species (although see below), but the effectiveness of naltrexone in decreasing ethanol consumption varies depending on the experimental model used. Thus, naltrexone has been shown to be effective at doses below 1 mg per kg body weight (0.1–0.6 mg/kg) when the period of access to ethanol is short (e.g., 1 h) and naltrexone is injected shortly prior to the access period, such as in limited access schedules [16–20] or in operant self-administration protocols [21–31]. When there is a delay between naltrexone injection and access to ethanol, the doses of naltrexone required to cause a significant reduction in ethanol consumption are much higher [22].

As the period of access to ethanol increases, the dose of naltrexone required to cause a significant suppression of ethanol consumption also increases. When ethanol is available under a 24-h continual access regime [13, 15, 18, 32] or a 23-h operant self-administration session [33], a dose of 10 mg/kg has caused a significant reduction in ethanol consumption in some [13, 15, 18, 32, 33] but not all [33, 34] studies. Although in most of these studies dose-dependency was not established (i.e., the minimum effective dose was not determined), several other studies using doses up to 5 mg/kg [35–38] have shown no significant effect of naltrexone alone under continual access conditions. There are, however, two studies of continual access or near-continual access to ethanol consumption, in which a relatively low dose of naltrexone produced a significant reduction in ethanol consumption. Thus, in the study by Hölter and Spanagel [33], whereas naltrexone up to a dose of 10 mg/kg had no significant effect on operant ethanol self-administration by rats under basal conditions (23-h sessions), following a period of enforced abstinence from ethanol (1 week), a dose of 0.1 mg/kg naltrexone caused a significant reduction in operant ethanol self-administration over the following 23 h period. The data may be explained by the differential motivation to consume ethanol in the two paradigms, deprivation-induced ethanol consumption presumably having a significant opioidergic component. Although this suggestion would appear to be borne out by the studies of Kornet et al. [39, 40], in which a low dose of 0.17 mg/kg naltrexone caused a significant reduction in ethanol consumption by rhesus monkeys in the 17 h following a period of enforced abstinence (48 h), basal ethanol consumption by these monkeys was also suppressed by a low dose of 0.5 mg/kg naltrexone. The discrepancy between the monkey studies [39, 40] and the studies in rats and mice outlined above may therefore also indicate a species difference in the effect of naltrexone on ethanol consumption; unfortunately, the effect of naltrexone on ethanol self-administration by monkeys over such long time periods does not appear to have been examined by any other group.

Williams et al. [31] have pointed out that the doses of naltrexone needed to decrease operant ethanol self-administration by rhesus monkeys are higher than those needed for reversal of morphine-induced antinociception [5], presumably a μ-opioid receptor mediated effect [41]. This may in part reflect the different time frames of these experiments (minutes *versus* hours) and the nature of the drug examined (morphine interacting directly with opioid receptors, ethanol apparently modulating [42, 43] or acting indirectly [44, 45] on opioid receptors). In the only study of selective opioid antagonists in primates [46], whereas μ- and δ-opioid receptor selective antagonists had no effect on ethanol self-administration, the κ-opioid receptor antagonist nor-binaltorphimine was shown to decrease ethanol self-administration. Since nor-binaltorphimine was effective only on the day of injection (whereas its antagonism of κ-opioid receptors was expected to continue over days), the authors [46] were led to suggest that the effect of naltrexone was not mediated by opioid receptors. However, a combination of selective antagonists was not tested, i.e., the involvement of multiple opioid receptor subtypes cannot be ruled out.

Continuous and repeated dosing

In the studies of longer ethanol access periods, the effects of naltrexone have been observed to be most pronounced in the initial 1–4 h post-injection [33, 36, 39]. Since the effect of naltrexone appears to diminish relatively rapidly post-injection, continuous delivery systems have been tested, using either a subcutaneously planted naltrexone pellet [47–49] or osmotic minipump [33, 37, 50]. However, particularly poor outcomes are obtained using such systems: frequently no effect [33, 37, 47] or an increase in ethanol consumption [49] is observed. Any significant decrease normally occurs only immediately (1–2 days) post-surgery [48, 50]. Cowen et al. [50] have argued that the lack of efficacy of such delivery systems may be due to the marked and rapid upregulation of opioid receptors induced by naltrexone over the experimental time period; however, the exact mechanisms remain unclear.

In a few cases using repeated daily injections of naltrexone and discrete ethanol access periods, the initial dose of naltrexone is ineffective and the effect of naltrexone on ethanol consumption appears only after repeated dosing [12, 51], or the effect of naltrexone on ethanol consumption increases over time [52], suggesting an extinction by naltrexone of ethanol-reinforced behavior [53]. In contrast, there are several reports of tolerance, or decreasing efficacy over time, with repeated daily injections of naltrexone [37, 49, 54–56], particularly when the initial dose was effective (1–6 mg/kg in a range of experimental paradigms). Since other studies using similar [52, 57] or higher (10 mg/kg [58, 59]) doses of naltrexone report no development of tolerance over equivalent time periods, the interaction between the dose of naltrexone and the development of tolerance is complex and involves other factors. Undoubtedly, however, as for continuous delivery systems, tolerance can develop to repeated daily injections of naltrexone and this has clear clinical implications.

Acquisition and reinstatement of ethanol seeking behavior

Naltrexone has been shown to significantly retard the acquisition of ethanol consumption [49, 57, 60]; however, in these studies ethanol consumption did increase over time, probably reflecting the difficulty in maintaining opioid receptor blockade with repeated naltrexone dosing and the development of tolerance. Ethanol consumption following a period of enforced deprivation from ethanol – the so-called alcohol deprivation effect (ADE) – is also sensitive to naltrexone, as noted previously [33, 39, 54], indicating a significant opioidergic component in this motivational state. Further, the reinstatement of ethanol-seeking behavior by cues associated with the availability of ethanol has been shown to be sensitive to naltrexone [51, 61–64]. The doses used in inhibiting cue-mediated ethanol-seeking behavior are similar to those used to inhibit operant ethanol self-administration (the protocol under which cue-mediated ethanol-seeking behavior is learnt); however, stress-induced reinstatement of ethanol-seeking behavior is not affected by naltrexone at the same doses [62, 64], suggesting an important role for opioidergic signalling in cue- but not stress-induced reinstatement. One study has reported that the post-shock increase in ethanol consumption can be overcome by naltrexone [15]; however, in this study, unlike the reinstatement studies, the animals were able to sample ethanol; therefore, the effect of naltrexone may have been a post-ingestive effect.

Specificity of action

The original toxicology study for naltrexone indicated that chronic naltrexone, within the doses used to decrease ethanol consumption, was well tolerated in rats and monkeys with no difference in body weight relative to control animals in a timescale of weeks to months [2]. However, over a shorter timescale naltrexone has been shown to cause a reduction in food and fluid intake at doses that are germane to the effects on ethanol consumption [65–71], although interestingly, the reductions tended to be in situations of high motivation, such as in fluid intake subsequent to deprivation [65] or injection of hypertonic saline [66], or food consumption subsequent to deprivation [69] or in response to a highly palatable diet (but not a control diet; [67]). Whereas some selectivity of naltrexone for ethanol consumption compared with water consumption has been demonstrated on occasion [25, 72]; several reports have indicated that when water or a sweetened solution are available concurrently with ethanol, naltrexone has caused a significant decrease in both ethanol and water/sweetened solution consumption [22, 30, 57, 58], although in most of these studies [30, 57, 58] water/sweetened solution consumption recovered over days. In an interesting study by Williams and Woods [27], the concentration of ethanol was varied to alter the relative preference by monkeys for ethanol *versus* water. When ethanol was the preferred solution, ethanol consumption was significantly reduced by naltrexone; when water was the preferred solution, water consumption was significantly reduced by naltrexone. The data clearly suggest that under these conditions, naltrexone is most effective against the most reinforcing or salient solution.

Naltrexone has been shown to decrease ethanol self-administration under operant conditions with no effect on cocaine self-administration [73]. However, in a recent set of studies, whereas naltrexone was ineffective in decreasing food and phencyclidine self-administration when these were available only during operant sessions (while having a significant impact on ethanol and saccharin self-administration) [26], naltrexone was able to decrease food and phencyclidine responding when these were available after the self-administration session [28], an “open economy”. In other words, the post-session availability weakened the apparent reinforcing value of these substances, thus “allowing naltrexone to have greater effectiveness as an intervention” [28].

This result has some resonance with other, previous studies; thus naltrexone was shown to facilitate extinguishment of ethanol-seeking behavior (no ethanol availability following responding, therefore a 'weakened' stimulus; [51]). In contrast, the data would appear to be difficult to reconcile with the relatively greater effect of naltrexone on food, water and ethanol consumption under high motivational states [33, 67, 69]. Clearly, several levels of 'need' or 'want' are involved in these consummatory behaviors, not all of which can be clearly delineated at this point in time but some of which appear to be sensitive to naltrexone. In total, the data indicate that naltrexone may diminish several motivated consummatory behaviors, not solely ethanol consumption, and the relative impairment may depend on the motivational state and the extent to which opioidergic signalling plays a role in mediating that specific behavior within the brain.

Combination pharmacotherapy

In order to enhance the effect of naltrexone on ethanol consumption, several attempts have been made to use naltrexone in combination with some other drug to facilitate a decrease in ethanol consumption. Thus naltrexone has been used in combination with the 5-HT_3 receptor antagonists ondansetron [16] and ICS 205-930 [74], the 5-HT_{2A} receptor antagonist amperozide [72], the calcium channel inhibitor isradipine [54, 57], the selective serotonin reuptake inhibitor fluoxetine [55], with fluoxetine and the thyrotropin-releasing hormone analogue TA-0910 [38] and acamprosate [19]. In some of these studies there was no additional benefit conferred by using naltrexone in combination with another drug, possibly because ethanol consumption was already significantly reduced by naltrexone – a 'floor' effect [19, 55, 57]. In contrast, in the studies by Le and Sellers [16] and Rezvani et al. [38], the combination pharmacotherapy produced a significant decrease in ethanol consumption whereas individually the drugs had no significant impact. Such combinations may have an advantage in overcoming the non-specific effects of naltrexone on food and water consumption, and minimizing the development of tolerance, although the timeframes under which the combinations were examined were usually quite short.

Other interactions of naltrexone and ethanol

Naltrexone has been shown to cause a decrease in ethanol-induced place preference [37], indicating an interaction with the reinforcing properties of ethanol. Interestingly, naltrexone has also been shown to counteract ethanol induced locomotor activity [75] and the hypnotic effects of ethanol in one strain of mouse (the BALB/c mouse strain) but not in two others, the C57BL/6 and DBA/2 mouse strains [75]. However, the interactions between the behavioral effects of ethanol and naltrexone on the one hand, and naltrexone's effect on ethanol consumption on the other, remain uncertain. Although naltrexone has been shown to cause an increase in aversive responses to the taste of ethanol [20, 76, 77], whether this plays a role in naltrexone's effect on ethanol consumption is still unclear, as naltrexone has been shown to significantly decrease intravenous self-administration of ethanol [12, 46]. Further, Linseman [78] demonstrated that the peripherally-acting opioid antagonist methylnaltrexone had no effect on ethanol consumption by rats in a limited access paradigm, whereas naltrexone at the same doses caused a significant decrease in ethanol consumption (although not precluding the possibility that naltrexone alters the palatability of ethanol via a central nervous system mechanism). An obvious test would be to examine the effect of methylnaltrexone on taste responsiveness to ethanol. Naltrexone has also been shown to cause a small decrease in the absorption of ethanol [79].

Conclusions

Naltrexone has been shown to decrease ethanol consumption in a range of animal models. Naltrexone appears to be particularly effective in animal models of relapse such as the alcohol deprivation effect or cue-induced reinstatement of ethanol-seeking behavior. The effects of naltrexone are centrally mediated and involve decreasing the rewarding value of ethanol (subsequent to consumption), disrupting the neural signalling involved in mediating the connection between cues and ethanol-seeking behavior and increasing the perceived aversiveness of ethanol's flavor. The pre-clinical data indicating the development of tolerance and lack of specificity for ethanol consumption (i.e., effects on food and water intake) may indicate that naltrexone is best used in a lower dose range in combination with other drugs; however, further work is needed in this regard.

Acknowledgements
I would like to acknowledge the support of the National Health & Medical Research Council (NHMRC), Australia and E. Krstew for her critical comments.

References

1 Blumberg H, Dayton HB (1973) Naloxone, naltrexone, and related noroxymorphones. *Adv Biochem Psychopharmacol* 8: 33–43
2 Braude MC, Morrison JM (1976) Preclinical toxicity studies of naltrexone. *NIDA Res Monogr* 9: 16–26
3 Blasig J, Hollt V, Herz A, Paschelke G (1976) Comparison of withdrawal precipitating properties of various morphine antagonists and partial agonists in relation to their stereospecific binding to brain homogenates. *Psychopharmacologia* 46: 41–51
4 Valentino RJ, Katz JL, Medzihradsky F, Woods JH (1983) Receptor binding, antagonist, and withdrawal precipitating properties of opiate antagonists. *Life Sci* 32: 2887–2896

5 Millan MJ (1989) Kappa-opioid receptor-mediated antinociception in the rat. I. Comparative actions of mu- and kappa-opioids against noxious thermal, pressure and electrical stimuli. *J Pharmacol Exp Ther* 251: 334–341
6 Gillan MG, Kosterlitz HW, Paterson SJ (1980) Comparison of the binding characteristics of tritiated opiates and opioid peptides. *Br J Pharmacol* 70**:** 481–490
7 Magnan J, Paterson SJ, Tavani A, Kosterlitz HW (1982) The binding spectrum of narcotic analgesic drugs with different agonist and antagonist properties. *Naunyn Schmiedebergs Arch Pharmacol* 319: 197–205
8 O'Malley SS, Jaffe AJ, Chang G, Schottenfeld RS, Meyer RE, Rounsaville B (1992) Naltrexone and coping skills therapy for alcohol dependence. A controlled study. *Arch Gen Psychiatry* 49: 881–887
9 Volpicelli JR, Alterman AI, Hayashida M, O'Brien CP (1992) Naltrexone in the treatment of alcohol dependence. *Arch Gen Psychiatry* 49: 876–880
10 Samson HH, Doyle TF (1985) Oral ethanol self-administration in the rat: effect of naloxone. *Pharmacol Biochem Behav* 22: 91–99
11 Froehlich JC, Harts J, Lumeng L, Li TK (1990) Naloxone attenuates voluntary ethanol intake in rats selectively bred for high ethanol preference. *Pharmacol Biochem Behav* 35: 385–390
12 Altshuler HL, Phillips PE, Feinhandler DA (1980) Alteration of ethanol self-administration by naltrexone. *Life Sci* 26: 679–688
13 Critcher EC, Lin CI, Patel J, Myers RD (1983) Attenuation of alcohol drinking in tetrahydroisoquinoline-treated rats by morphine and naltrexone. *Pharmacol Biochem Behav* 18: 225–229
14 Myers RD, Borg S, Mossberg R (1986) Antagonism by naltrexone of voluntary alcohol selection in the chronically drinking macaque monkey. *Alcohol* 3: 383–388
15 Volpicelli JR, Davis MA, Olgin JE (1986) Naltrexone blocks the post-shock increase of ethanol consumption. *Life Sci* 38: 841–847
16 Le AD, Sellers EM (1994) Interaction between opiate and 5-HT3 receptor antagonists in the regulation of alcohol intake. *Alcohol Alcohol Suppl* 2: 545–549
17 Franck J, Lindholm S, Raaschou P (1998) Modulation of volitional ethanol intake in the rat by central delta-opioid receptors. *Alcohol Clin Exp Res* 22: 1185–1189
18 Koistinen M, Tuomainen P, Hyytia P, Kiianmaa K (2001) Naltrexone suppresses ethanol intake in 6-hydroxydopamine-treated rats. *Alcohol Clin Exp Res* 25: 1605–1612
19 Stromberg MF, Mackler SA, Volpicelli JR, O'Brien CP (2001) Effect of acamprosate and naltrexone, alone or in combination, on ethanol consumption. *Alcohol* 23: 109–116
20 Coonfield DL, Hill KG, Kaczmarek HJ, Ferraro FM, 3rd, Kiefer SW (2002) Low doses of naltrexone reduce palatability and consumption of ethanol in outbred rats. *Alcohol* 26: 43–47
21 Gonzales RA, Weiss F (1998) Suppression of ethanol-reinforced behavior by naltrexone is associated with attenuation of the ethanol-induced increase in dialysate dopamine levels in the nucleus accumbens. *J Neurosci* 18: 10663–10671
22 June HL, Grey C, Warren-Reese C, Durr LF, Ricks-Cord A, Johnson A, McCane S, Williams LS, Mason D, Cummings R et al. (1998) The opioid receptor antagonist nalmefene reduces responding maintained by ethanol presentation: preclinical studies in ethanol-preferring and outbred Wistar rats. *Alcohol Clin Exp Res* 22: 2174–2185
23 Stromberg MF, Casale M, Volpicelli L, Volpicelli JR, O'Brien CP (1998) A comparison of the effects of the opioid antagonists naltrexone, naltrindole, and beta-funaltrexamine on ethanol consumption in the rat. *Alcohol* 15: 281–289
24 Williams KL, Winger G, Pakarinen ED, Woods JH (1998) Naltrexone reduces ethanol- and sucrose-reinforced responding in rhesus monkeys. *Psychopharmacology (Berl)* 139: 53–61
25 Middaugh LD, Kelley BM, Cuison ER, Jr, Groseclose CH (1999) Naltrexone effects on ethanol reward and discrimination in C57BL/6 mice. *Alcohol Clin Exp Res* 23: 456–464
26 Rodefer JS, Campbell UC, Cosgrove KP, Carroll ME (1999) Naltrexone pretreatment decreases the reinforcing effectiveness of ethanol and saccharin but not PCP or food under concurrent progressive-ratio schedules in rhesus monkeys. *Psychopharmacology* 141: 436–446
27 Williams KL, Woods JH (1999) Naltrexone reduces ethanol- and/or water-reinforced responding in rhesus monkeys: effect depends upon ethanol concentration. *Alcohol Clin Exp Res* 23: 1462–1467
28 Carroll ME, Cosgrove KP, Campbell UC, Morgan AD, Mickelberg JL (2000) Reductions in ethanol, phencyclidine, and food-maintained behavior by naltrexone pretreatment in monkeys is enhanced by open economic conditions. *Psychopharmacology* 148: 412–422

29 Middaugh LD, Lee AM, Bandy AL (2000) Ethanol reinforcement in nondeprived mice: effects of abstinence and naltrexone. *Alcohol Clin Exp Res* 24: 1172–1179
30 Shelton KL, Grant KA (2001) Effects of naltrexone and Ro 15-4513 on a multiple schedule of ethanol and Tang self-administration. *Alcohol Clin Exp Res* 25: 1576–1585
31 Williams KL, Kane EC, Woods JH (2001) Interaction of morphine and naltrexone on oral ethanol self-administration in rhesus monkeys. *Behav Pharmacol* 12: 325–333
32 Myers RD, Lankford MF (1996) Suppression of alcohol preference in high alcohol drinking rats: efficacy of amperozide *versus* naltrexone. *Neuropsychopharmacology* 14: 139–149
33 Holter SM, Spanagel R (1999) Effects of opiate antagonist treatment on the alcohol deprivation effect in long-term ethanol-experienced rats. *Psychopharmacology* 145: 360–369
34 Goodwin FL, Campisi M, Babinska I, Amit Z (2001) Effects of naltrexone on the intake of ethanol and flavored solutions in rats. *Alcohol* 25: 9–19
35 McMillen BA, Williams HL (1995) Volitional consumption of ethanol by fawn-hooded rats: effects of alternative solutions and drug treatments. *Alcohol* 12: 345–350
36 Davidson D, Amit Z (1997) Effect of ethanol drinking and naltrexone on subsequent drinking in rats. *Alcohol* 14: 581–584
37 Middaugh LD, Bandy AL (2000) Naltrexone effects on ethanol consumption and response to ethanol conditioned cues in C57BL/6 mice. *Psychopharmacology* 151: 321–327
38 Rezvani AH, Overstreet DH, Mason GA, Janowsky DS, Hamedi M, Clark E, Jr, Yang Y (2000) Combination pharmacotherapy: a mixture of small doses of naltrexone, fluoxetine, and a thyrotropin-releasing hormone analogue reduces alcohol intake in three strains of alcohol-preferring rats. *Alcohol Alcoholism* 35: 76–83
39 Kornet M, Goosen C, Van Ree JM (1991) Effect of naltrexone on alcohol consumption during chronic alcohol drinking and after a period of imposed abstinence in free-choice drinking rhesus monkeys. *Psychopharmacology* 104: 367–376
40 Kornet M, Goosen C, Van Ree JM (1992) Opioid modulation of alcohol intake in monkeys by low doses of naltrexone and morphine. *Ann N Y Acad Sci* 654: 469–471
41 Romberg R, Sarton E, Teppema L, Matthes HW, Kieffer BL, Dahan A (2003) Comparison of morphine-6-glucuronide and morphine on respiratory depressant and antinociceptive responses in wild type and micro-opioid receptor deficient mice. *Br J Anaesth* 91: 862–870
42 Charness ME, Gordon AS, Diamond I (1983) Ethanol modulation of opiate receptors in cultured neural cells. *Science* 222: 1246–1248
43 Tabakoff B, Hoffman PL (1983) Alcohol interactions with brain opiate receptors. *Life Sci* 32: 197–204
44 Olive MF, Koenig HN, Nannini MA, Hodge CW (2001) Stimulation of endorphin neurotransmission in the nucleus accumbens by ethanol, cocaine, and amphetamine. *J Neurosci* 21: RC184
45 Marinelli PW, Quirion R, Gianoulakis C (2003) A microdialysis profile of beta-endorphin and catecholamines in the rat nucleus accumbens following alcohol administration. *Psychopharmacology (Berl)* 169: 60–67
46 Williams KL, Woods JH (1998) Oral ethanol-reinforced responding in rhesus monkeys: effects of opioid antagonists selective for the mu-, kappa-, or delta-receptor. *Alcohol Clin Exp Res* 22: 1634–1639
47 Nation JR, Horger BA, Pugh CK, Bratton GR, Rowe LD (1990) The effects of naltrexone on cadmium-induced increases in oral ethanol self-administration. *Alcohol* 7: 17–20
48 Iso H, Brush FR (1991) Opposite effects of naltrexone on ETOH intake by Syracuse high and low avoidance rats. *Alcohol* 8: 443–448
49 Phillips TJ, Wenger CD, Dorow JD (1997) Naltrexone effects on ethanol drinking acquisition and on established ethanol consumption in C57BL/6J mice. *Alcohol Clin Exp Res* 21: 691–702
50 Cowen MS, Rezvani AH, Jarrott B, Lawrence AJ (1999) Ethanol consumption by Fawn-Hooded rats following abstinence: effect of naltrexone and changes in mu-opioid receptor density. *Alcohol Clin Exp Res* 23: 1008–1014
51 Bienkowski P, Kostowski W, Koros E (1999) Ethanol-reinforced behaviour in the rat: effects of naltrexone. *Eur J Pharmacol* 374: 321–327
52 Stromberg MF, Volpicelli JR, O'Brien CP (1998) Effects of naltrexone administered repeatedly across 30 or 60 days on ethanol consumption using a limited access procedure in the rat. *Alcohol Clin Exp Res* 22: 2186–2191
53 Sinclair JD (1990) Drugs to decrease alcohol drinking. *Ann Med* 22: 357–362
54 Gardell LR, Reid LD, Boedeker KL, Liakos TM, Hubbell CL (1997) Isradipine and naltrexone in

combination with isradipine interact with a period of abstinence to reduce rats' intakes of an alcoholic beverage. *Alcohol Clin Exp Res* 21: 1592–1598
55 Gardell LR, Whalen CA, Chattophadyay S, Cavallaro CA, Hubbell CL, Reid LD (1997) Combination of naltrexone and fluoxetine on rats' propensity to take alcoholic beverage. *Alcohol Clin Exp Res* 21: 1435–1439
56 Boyle AE, Stewart RB, Macenski MJ, Spiga R, Johnson BA, Meisch RA (1998) Effects of acute and chronic doses of naltrexone on ethanol self-administration in rhesus monkeys. *Alcohol Clin Exp Res* 22: 359–366
57 Cramer CM, Gardell LR, Boedeker KL, Harris JR, Hubbell CL, Reid LD (1998) Isradipine combined with naltrexone persistently reduces the reward-relevant effects of cocaine and alcohol. *Pharmacol Biochem Behav* 60: 345–356
58 Gardell LR, Hubbell CL, Reid LD (1996) Naltrexone persistently reduces rats' intake of a palatable alcoholic beverage. *Alcohol Clin Exp Res* 20: 584–588
59 Reid LD, Gardell LR, Chattopadhyay S, Hubbell CL (1996) Periodic naltrexone and propensity to take alcoholic beverage. *Alcohol Clin Exp Res* 20: 1329–1334
60 Davidson D, Amit Z (1997) Naltrexone blocks acquisition of voluntary ethanol intake in rats. *Alcohol Clin Exp Res* 21: 677–683
61 Katner SN, Magalong JG, Weiss F (1999) Reinstatement of alcohol-seeking behavior by drug-associated discriminative stimuli after prolonged extinction in the rat. *Neuropsychopharmacology* 20: 471–479
62 Le AD, Poulos CX, Harding S, Watchus J, Juzytsch W, Shaham Y (1999) Effects of naltrexone and fluoxetine on alcohol self-administration and reinstatement of alcohol seeking induced by priming injections of alcohol and exposure to stress. *Neuropsychopharmacology* 21: 435–444
63 Ciccocioppo R, Martin-Fardon R, Weiss F (2002) Effect of selective blockade of mu(1) or delta opioid receptors on reinstatement of alcohol-seeking behavior by drug-associated stimuli in rats. *Neuropsychopharmacology* 27: 391–399
64 Liu X, Weiss F (2002) Additive effect of stress and drug cues on reinstatement of ethanol seeking: exacerbation by history of dependence and role of concurrent activation of corticotropin-releasing factor and opioid mechanisms. *J Neurosci* 22: 7856–7861
65 Holtzman SG (1975) Effects of narcotic antagonists on fluid intake in the rat. *Life Sci* 16: 1465–1470
66 Brown DR, Holtzman SG (1980) Evidence that opiate receptors mediate suppression of hypertonic saline-induced drinking in the mouse by narcotic antagonists. *Life Sci* 26: 1543–1550
67 Apfelbaum M, Mandenoff A (1981) Naltrexone suppresses hyperphagia induced in the rat by a highly palatable diet. *Pharmacol Biochem Behav* 15: 89–91
68 Goldberg SR, Morse WH, Goldberg DM (1981) Acute and chronic effects of naltrexone and naloxone on schedule-controlled behavior of squirrel monkeys and pigeons. *J Pharmacol Exp Ther* 216: 500–509
69 Lowy MT, Yim GK (1981) The anorexic effect of naltrexone is independent of its suppressant effect on water intake. *Neuropharmacology* 20: 883–886
70 Lang IM, Strahlendorf JC, Strahlendorf HK, Lutherer LO, Barnes CD (1982) The effects of chronic administration of naltrexone on appetite and water exchange in rats. *Pharmacol Biochem Behav* 16: 909–913
71 Sanger DJ, McCarthy PS (1982) A comparison of the effects of opiate antagonists on operant and ingestive behavior. *Pharmacol Biochem Behav* 16**:** 1013–1015
72 Lankford MF, Myers RD (1996) Opiate and 5-HT2A receptors in alcohol drinking: preference in HAD rats is inhibited by combination treatment with naltrexone and amperozide. *Alcohol* 13: 53–57
73 Stromberg MF, Sengpiel T, Mackler SA, Volpicelli JR, O'Brien CP, Vogel WH (2002) Effect of naltrexone on oral consumption of concurrently available ethanol and cocaine in the rat. *Alcohol* 28: 169–179
74 Mhatre M, Pruthi R, Hensley K, Holloway F (2004) 5-HT3 antagonist ICS 205-930 enhances naltrexone´s effects on ethanol intake. *Eur J Pharmacol* 491: 149–156
75 Kiianmaa K, Hoffman PL, Tabakoff B (1983) Antagonism of the behavioral effects of ethanol by naltrexone in BALB/c, C57BL/6, and DBA/2 mice. *Psychopharmacology (Berl)* 79: 291–294
76 Hill KG, Kiefer SW (1997) Naltrexone treatment increases the aversiveness of alcohol for outbred rats. *Alcohol Clin Exp Res* 21: 637–641
77 Ferraro FM, 3rd, Hill KG, Kaczmarek HJ, Coonfield DL, Kiefer SW (2002) Naltrexone modifies

the palatability of basic tastes and alcohol in outbred male rats. *Alcohol* 27: 107–114
78 Linseman MA (1989) Central *versus* peripheral mediation of opioid effects on alcohol consumption in free-feeding rats. *Pharmacol Biochem Behav* 33**:** 407–413
79 Linseman MA, Le AD (1997) Effects of opioids on the absorption of alcohol. *Pharmacol Biochem Behav* 58: 79–84

Drugs for Relapse Prevention of Alcoholism
Edited by R. Spanagel and K. Mann

Naltrexone: clinical data

Charles P. O'Brien, Helen M. Pettinati and David W. Oslin

University of Pennsylvania, Department of Psychiatry/Philadelphia Veterans Affairs Medical Center, Treatment Research Center, 3900 Chestnut Street, Philadelphia, PA 19104-6178, USA

History

Most medications used in the treatment of psychiatric disorders were discovered by clinical serendipity rather than as laboratory models translated to clinical use. Naltrexone took a different path. It was the subject of intense study during the 1970s as scientists scrambled to understand the functions of the newly discovered endogenous opioid system by specifically blocking opiate receptors with naloxone or naltrexone. Both drugs had been developed to reverse or block the effects of externally administered opiates such as heroin, but they could also block internally produced opioids that act at the same receptors. Altshuler presented an interesting study at the 1979 meeting of the College on Problems of Drug Dependence [1] showing that naltrexone blocked the self-administration of alcohol in rhesus monkeys. Interestingly, only 8 of 22 monkeys spontaneously self-administered intravenous alcohol, but these alcohol-preferring monkeys consistently decreased their alcohol taking in a dose-dependent manner when treated with naltrexone. This study was consistent with the hypothesis that alcohol may release endogenous opioid peptides that activate opiate receptors and contribute significantly to the reinforcement produced by alcohol. The fact that alcohol was spontaneously self-administered by only a minority of monkeys suggested genetic variation in this phenomenon.

There were other studies beginning in the early 1980s that further contributed to the development of a hypothesis concerning a linkage between endogenous opioids and alcohol reinforcement and perhaps alcoholism and these are reviewed in more detail elsewhere in this volume (Chapter by Cowen). The view that condensation products of dopamine and alcohol-derived aldehyde (tetrahydroisoquinolines) might be causing opiate effects was never supported by data and did not play a role in motivating the first clinical study. By 1983, there were sufficient clues in the animal literature for one of us (CO'B) to request and receive permission to add alcoholism studies to his Investigational New Drug Licence (IND) for naltrexone. The initial open label experience treating alcoholics with naltrexone was encouraging; thus a formal protocol was submitted to the Philadelphia Veteran's Administration (VA)

Medical Center human subjects review board. Initial efforts to obtain grant funding from the manufacturer or from the National Institutes of Health (NIH) were unsuccessful. Thus the formal study began with some support from a VA Medical Research Center Grant (CO'B, PI) and a postdoctoral fellowship program. Conducting the study proved difficult because clinical staff at the time resisted the concept of a double-blind medication trial among alcoholics who were also engaged in an intensive 12-step recovery process. Few patients were enrolled until Joseph Volpicelli completed psychiatric training in 1985 and joined the center as a new, enthusiastic post-doctoral fellow. Recruitment of patients immediately increased and the first formal study was completed.

The first naltrexone study in alcoholics

Several unusual aspects of our first study should be mentioned. The patients were all male veterans engaged in a day hospital rehabilitation program at the Philadelphia VA Medical Center. Thus they came to the hospital five days per week for the first month totaling about 25 h per week of time in the clinic. Although they averaged about 20 years of heavy drinking, they were considered moderately well motivated because they were accepted for this intensive program. The dose of 50 mg naltrexone per day was arbitrarily chosen because this was the dose that we were using since 1973 for the prevention of relapse to heroin addiction. We had no idea whether the same dose of naltrexone that blocked the average dose of heroin could work on endogenous opioids theoretically released by alcohol. The major outcome measure, relapse to clinically significant drinking, was carefully chosen and defined. This was important because many prior studies of alcoholism counted a single drink as a relapse. If we had defined relapse as a single drink, there would have been no significant difference between naltrexone and placebo. In this study relapse was defined as 5 or more drinking days within one week, five or more drinks on a single drinking occasion or coming to the clinic with a blood alcohol concentration above 100 mg/dl. Over the 3-month course of the study, 54% of those patients in the intensive day hospital program who received placebo met criteria for relapse. Only 23% of those assigned to the naltrexone group relapsed (Fig. 1). Craving for alcohol was significantly reduced in the patients randomized to naltrexone and liver enzyme levels were lower but not significantly different (P = . 08) than levels in control patients. These results were published in a preliminary version [2] and in a complete version [3].

These results were initially met with great skepticism and publication of the paper was delayed by negative reviews. Fortunately, Roger Meyer, a member of the University of Pennsylvania Scientific Advisory Board at the time was impressed by the original data and he joined with the team at Yale led by Stephanie O'Malley to attempt a replication. O'Malley et al. [4] studied a sample of DSM III-R defined alcoholics who were recruited by newspaper advertisements for once-weekly therapy using one of two manual guided therapies:

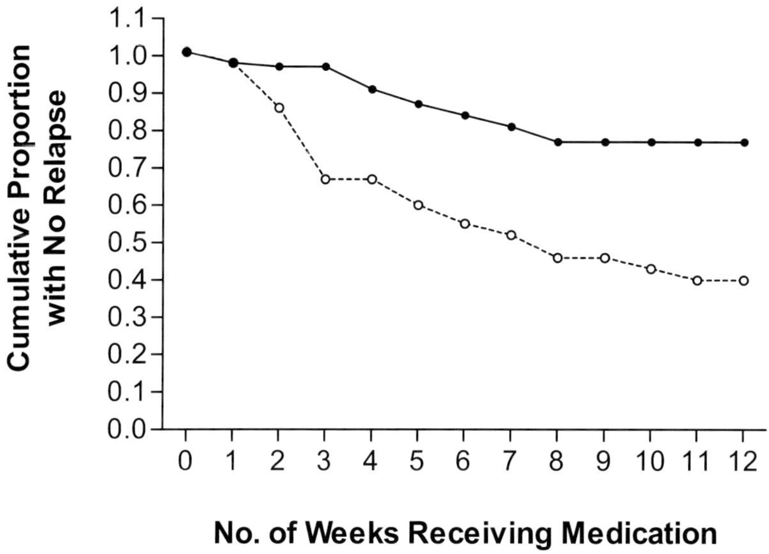

Figure 1 Relapse rates (as defined in the text) for the naltrexone hydrochloride- (closed circles) and placebo-treated (open circles) groups across the 12 weeks of study [3].

coping skills/relapse prevention or supportive therapy. They also received either naltrexone 50 mg daily or placebo in a two by two design. The sample consisted of 97 patients who received medication for at least one week. They were 74% male with an average age of 40 years and moderately severe alcoholism. During the three-month trial, naltrexone was found to be clearly superior to placebo on measures of alcohol consumption. There was an interaction between medication and type of therapy with the combination of naltrexone and supportive therapy having the highest abstinence rate. If the patient did any drinking, the coping skills group with naltrexone had a lower relapse rate. As with the prior study at the University of Pennsylvania, the overall naltrexone group had a significantly lower relapse rate and reported less craving for alcohol. At end point, naltrexone-treated patients had significantly lower aspartate aminotransferase (AST) levels (indicative of alcohol-related liver dysfunction), $P < .05$ with a trend toward lower alanine aminotransferase (ALT) levels ($P < .10$), a finding consistent with their lower alcohol consumption.

Food and Drug Administration approval

The history of naltrexone's approval by the Food and Drug Administration (FDA) is unusual in that neither of the two above efficacy studies were initiated or funded by the drug's manufacturer. There was no program to obtain FDA approval for alcoholism, but a senior scientist at Dupont, Leonard Cook, was scheduled to retire shortly after the studies' publication. One of his retirement awards was a golfing afternoon with the company president. Dr. Cook took this opportunity to point out to the president that one of his medications was effective against alcoholism. This prompted the president to direct his staff to review the studies and initiate the appropriate requests to the FDA. Based on these two studies, the FDA added alcoholism to the indications for naltrexone and since it was then a generic drug, it granted three years of exclusivity to the company. The medication was briefly promoted for the treatment of alcoholism in the United States, but even now few American physicians are aware of this treatment option and fewer still use it regularly. Indeed most prescriptions written in clinical practice are only filled for 30 days with no refills [5]. Clinical researchers, however, have continued to study naltrexone for the treatment of alcoholism and much additional knowledge has been developed.

Clinical trials

Since those first two studies, at least 25 additional controlled clinical trials have been reported, in seven different countries using many different protocols. While 22 of these trials showed a significant benefit for alcoholics randomized to naltrexone, clearly not all patients showed a response. Two of the largest trials showed no benefit for naltrexone over placebo [6, 7]. A third trial was reported as negative based on self reports of drinking, but the group randomized to naltrexone showed significantly lower gamma glutamyl transferase levels, an objective sign of decreased alcohol drinking [8]. The 27 published controlled trials are the subject of an intense review, analyzing their strengths and weaknesses [9]. Most were relatively small sample size (<100 patients per cell) and of relatively short duration (12 weeks). A review of the data from these studies gives some direction to clinicians seeking guidelines for pharmacotherapy of alcoholism.

Some authors have characterized the literature on naltrexone as inconsistent and the effect size as "modest." That is a fair characterization if one is comparing naltrexone for alcoholism to penicillin for pneumococcal pneumonia (before the appearance of resistant strains). Perhaps a more appropriate comparison would be to major depression. Both depression and alcoholism are behavioral disorders that are strongly influenced by environmental factors and respond to psychotherapy to a significant degree. One significant difference in the development of these two classes of agents is the funding support for the research. A substantial number of the clinical trials for anti-depressant research are funded by private industry, which has a significant financial reward for favorable research findings. There is concern that this financial or business incentive can lead to suppression of clinical trials that are not favorable to the company. Even with the potential incentive to suppress negative findings, "failed trials" where the active medication fails to beat placebo are common [10, 11]. No one knows the proportion of trials that "fail," but one estimate cited by a well-known depression researcher is 50% [12]. Thus the trials of anti-depressants that are published are heavily weighted toward positive trials and any attempt to calculate an effect size based on published trials would give a highly inflated estimate. Naltrexone, in contrast, has a history of development outside of the pharmaceutical industry as described above and investigators tend to publish everything. A second significant difference in the development of medication for addiction is the use of standard behavioral intervention for both the placebo group and active medication group. In essence these trials test the medication as adjunctive treatment and this has a consequence of producing seemingly high "placebo" response rates. In contrast, depression trials are seldom done in combination with a behavioral intervention. Thus, the naltrexone trials have not been carefully designed to show the drug in the best light.

Calculation of a standard effect size in the case of a medication with this much variability appears to be of dubious value. In a given patient sample, some alcoholics find naltrexone to be a life-changing medication that enables them for the first time to stop compulsive drinking. Others go right on drinking with no apparent effect from the medication. In most patient samples, there is a sufficient number of responders such that there is a difference in favor of naltrexone that meets the 5% significance level or better. Why is there such variability in response to naltrexone among alcoholics? We will begin by discussing heterogeneity in response to alcohol.

Heterogeneity in alcoholism

Alcoholism is well known as a heterogeneous disorder with variability in age of onset, drinking pattern, family history, associated diagnoses and clinical course. There is also a large variation in response to alcohol itself. Gamma-aminobutyric acid (GABA), glutamate, dopamine, serotonin and opioid peptides are involved in alcohol reinforcement [13]. It appears that the endogenous opioid system plays an important role in some but not all individuals. This was first noted in animal studies and was the reason that naltrexone was tested in the first clinical trial. The existing data suggest that there is wide variability in the degree to which alcohol activates endogenous opioids. Naltrexone is a very specific medication that has little affinity for any receptor system except opiate receptors. It will have no effect in alcoholics whose illnesses do not significantly involve the endogenous opioid system.

Both the animal and the human data are consistent with a straightforward hypothesis concerning the mechanism of naltrexone's action in alcoholism: *some but not all individuals react to alcohol by an activation of endogenous opioids that have both peripheral and central effects.* One central effect is the production of reward or euphoria. Alcohol produces a specific activation of reward systems when self-administered by animals. This is manifested by increased extra-cellular dopamine measured by microdialysis in the nucleus accumbens [14].

The primary transducer that translates the alcohol signal into release of opioids is still unknown. We have learned from animal models that opioid peptides inhibit GABA inhibitory neurons in the ventral tegmental area (VTA). This inhibition of the inhibitors of dopamine neurons in the VTA results in an activation of these neurons and a release of dopamine in limbic reward pathways where these neurons project. The increase in dopamine in the nucleus accumbens that occurs after alcohol self-administration in rats is blocked by naltrexone pre treatment. The animal also stops taking alcohol, presumably because the reward has been blocked.

In human subjects, endogenous opioids in the brain cannot be measured directly. Gianoulakis [15] has studied plasma beta endorphin (BE) and has shown a low baseline level in both abstinent alcoholics and in non-alcoholics with a high risk of developing alcoholism because of a strong family history of the disorder. This finding is supported by the report of Genazzani [16] showing BE levels in the cerebrospinal fluid three-fold lower in alcoholics compared to controls. The high-risk subjects showed a dose-related BE response to alcohol in the laboratory (Fig. 2), an effect not seen in the low-risk subjects [17]. In animals, both the central and pituitary endogenous opioid response to alcohol has been observed, but only the pituitary response has been verified in humans. Another study [18] showed that alcohol produces more simulation (high) in non-alcoholics with a family history of alcoholism than in subjects with no history of alcohol in the family. Pre-treatment with naltrexone blocks this high during laboratory administration of alcohol under double-blind conditions. This finding is consistent with patient reports during double-

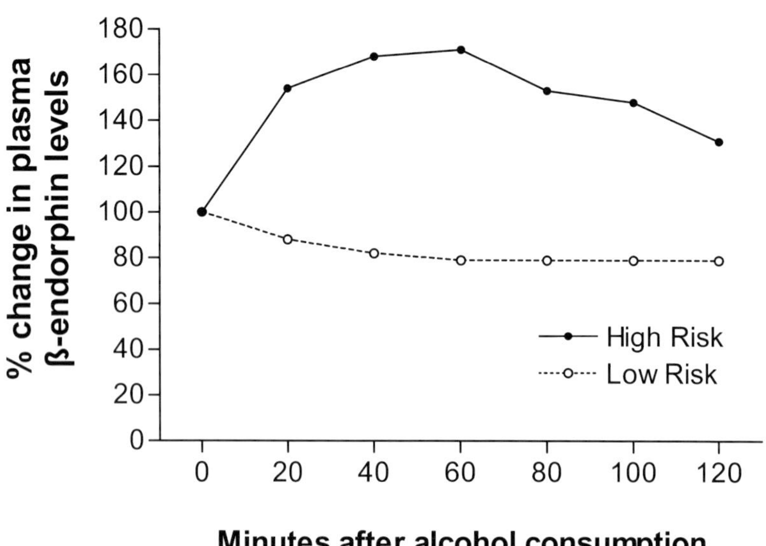

Figure 2. Change in β-endorphin levels after alcohol consumption [17].

blind clinical trials stating that the expected high from drinking alcohol is diminished when the patient is receiving naltrexone but not placebo [19]. There are additional data from human laboratory studies showing that pre-treatment with naltrexone lessens desire for alcohol and quantity consumed [20, 21].

Adherence to the medication regimen

A major variable in any clinical trial involving medication is the degree to which the patients actually take the medication. Obviously the medication cannot be expected to help if patients do not take it regularly. Most patients beginning treatment for alcoholism are somewhat ambivalent about their motivation to abstain from all alcohol. Thus missed doses are common. Several of the studies reported as showing a significant benefit to the patients randomized to naltrexone would not be positive on a "strict intent to treat" analysis [22, 23]. Of course, any system to measure adherence must be applicable to both drug and placebo groups.

Another factor is medication side-effects. In one of the negative clinical trials, adverse subjective effects seemed to play a major role [6]. Indeed, up to 10% of alcoholics may report nausea and this may be related to a kind of "endogenous opioid precipitated withdrawal." It is well known that naltrexone precipitates withdrawal in opiate addicts even days after the last dose of an opiate. Alcohol causes the release of endogenous opioids that may be displaced by naltrexone due to its high affinity for opiate receptors. There is some evidence that the more recent the last alcohol intake and the higher the dose of alcohol, the more likely that nausea will occur with naltrexone ingestion [24]. There is also evidence that there may be an inherited sensitivity to opiate antagonists, perhaps due to differences in baseline tone of the endogenous opiate system. Non-alcoholic family members of alcoholics show increased sensitivity to naloxone as measured by cortisol response [25].

A major recent advance in the use of naltrexone has been the development of a depot preparation that provides effective blood levels for 30–40 days. Three such products are currently in clinical trials and only preliminary results have been published [26, 27]. The largest trial was just completed involving 624 subjects treated for 12 weeks [28]. It showed a highly significant effect in male alcoholics (48% reduction, $p < .0001$) but no benefit for females. Noteworthy was the minimal rate of side-effects. The slow release formulation gives long lasting blood levels but no peak effect as is seen within the first 2 h of an oral dose. Thus few patients complained of nausea and less than 2% had an injection site reaction. It is hoped that a depot formulation will soon be approved for general clinical use. With this advance, the problem of adherence should be greatly improved.

Pharmacology of naltrexone

Naltrexone was developed in the early 1970s as a treatment for heroin addiction. The early clinical studies showed that it was a competitive antagonist of heroin and other opioids. As information about the opiate receptor system was developed, it was found that naltrexone had highest affinity for μ receptors, but also affinity for κ and ∂ receptors (Tab. 1) [29]. This broad-spectrum effect on opiate receptors may be important to its effect on alcohol drinking. Using a rat model of drinking, Stromberg [30] showed that specific antagonists at μ and ∂ receptors were not as effective individually as naltrexone, which acts on all three types of opiate receptors.

Naltrexone also has a long duration of activity at brain receptor sites. Although the plasma half-life of naltrexone and its active metabolite Beta naltrexol is only 10–12 h, two studies using C^{11} carfentanil showed that one 50 mg dose blocks brain μ receptors for 48–72 h [31, 32]. This duration of action is an advantage in that one or two missed doses would not necessarily leave the patient unprotected. Of course the availability of a depot preparation will add much more consistency to the long-term use of naltrexone.

Table 1. Naltrexone affinity at opioid receptor subtypes

	Receptor binding Ki (nM)		
	Mu	Delta	Kappa
Antagonist:			
Naltrexone	0.37	9.4	4.8
Agonists:			
Morphine (μ)	38	510	1,900
DADL-enkephalin (δ)	150	1.8	>10,000
(–)-EKC (κ)	2.3	5.2	2.2

Schmidt, WK et al., *Drug Alcohol Depend*, 1985:14:339–362 [29]

Anti-craving effect

The first two clinical trials of naltrexone in the treatment of alcoholism measured craving for alcohol. Patients randomly assigned to naltrexone reported significantly less craving and this predicted less alcohol use and a lower likelihood of relapse. Other studies that measured craving as a secondary outcome measure also reported a similar effect. It is not clear how naltrexone could have this effect. Craving is a subjective phenomenon which is also controversial [33] Some have argued that craving has no meaning, but it is at least very interesting that so often in these double-blind trials, craving reduction predicted decreased alcohol consumption.

Animal models have shown that placing a rat in a chamber where the animal has previously self-administered alcohol results in a conditioned increase in dopamine (DA) before any alcohol has been received. Could this be a model of craving? While naltrexone in this model has been found to block the DA increase seen after alcohol ingestion, it did not block the conditioned DA increase seen prior to alcohol availability.

Naltrexone has been reported to reduce craving elicited by cues that have been previously associated with alcohol during a clinical trial [34]. In a laboratory study, O'Malley [35] elicited craving by a priming dose of alcohol and then offered alcoholics additional drinks. Those alcoholics pre treated with naltrexone showed less craving and consumed fewer drinks than those pre treated with placebo. The naltrexone group also showed more activation of the hypothalamo-pituitary-adrenal axis. Naltrexone elevated cortisol in the baseline period and higher cortical levels were significantly correlated with lower craving throughout the experiment.

Psychotherapy

One of the major variables across studies was the type of psychosocial intervention used in conjunction with naltrexone. Only one study referred patients to primary care physicians with no psychotherapy or counseling [36]. The first study [2, 3] consisted of patients in an intensive outpatient program or day hospital. Abstinence was the goal and 12-step groups were a part of treatment, but the program took pains to avoid having patients feel guilty over slips. In the second study [4], patients received either coping skills or supportive therapy and there was an interaction. Supportive therapy with naltrexone had the highest abstinence rate, but of those who drank, coping skills with naltrexone had the lowest relapse rate.

In reviewing all of the studies, it appears that tolerance of some drinking is an important aspect of counseling. Those patients randomized to naltrexone may remain completely abstinent, but commonly some drinking occurs. Since the patients perceive less reward, they are less likely to continue drinking heavily and thus they "slip" without a relapse. If this behavior is regarded in a strongly negative manner by the therapist as might occur in a strict 12-step program, the patient is more likely to become discouraged or even drop out of treatment. Overall, the best results seem to occur in programs where therapy is supportive and oriented toward medication adherence [37].

Genetics

Alcoholism has long been known to have a strong heredity component. In addition to the elegant adoption studies over the past 30 years, there are more recent studies dealing with non-alcoholic relatives of alcoholics that provide relevant information concerning the effects of naltrexone. The studies of Gianoulakis and colleagues reviewed above demonstrate differences in sensitivity of the plasma BE response to graded doses of alcohol. Blood alcohol levels were similar between high-risk and low-risk subjects, but the BE response to the same alcohol blood level was much greater in the high-risk subjects [15]. In a similar vein, King [18] found that high-risk subjects were more likely to report stimulation (euphoria) than low-risk subjects with similar alcohol blood levels in the laboratory and this stimulation was attenuated by naltrexone pretreatment, but not placebo. Family history has also been found to be a predictor of clinical response to naltrexone in alcoholics [38, 39]

More recently, the $A_{+118}G$ polymorphism of the μ opiate receptor gene was examined in alcoholics who had participated in naltrexone clinical trials [40]. In subjects of European descent, individuals with one or two copies of the Asp40 allele treated with naltrexone had significantly lower rates of relapse ($p = 0.045$) and a longer time to return to heavy drinking ($p = 0.040$) than those homozygous for the Asn40 allele. There were no differences in overall abstinence rates ($p = 0.668$), nor were there differences in relapse rates between patients of the two genotypes among those assigned to placebo. The $A_{+118}G$ polymorphism is of interest because functional differences have been demonstrated both *in vitro* and *in vivo*. Bond and colleagues showed that, in cell culture, mu-opioid receptors encoded by the Asp40 variant bind beta-endorphin and activate G- protein coupled protein potassium ion channels with three times greater potency than receptors encoded by the Asn40 variant. Both Wand et al. and Hernandez-Avila et al. [41, 42] found that individuals with one or two copies of the Asp40 allele had altered hypothalamic-pituitary-adrenal (HPA) axis activation induced by naloxone, while Smolka et al. [43] showed that individuals with the Asp40 variant display greater dopaminergic sensitivity during acute alcohol withdrawal. The Asp40 allele has also been shown to have a dose response association with frequency of drinking among alcohol-dependent patients such that those homozygous for the Asp40 allele drink more often than those heterozygous or homozygous for the Asn40 allele [44].

Medication side effects

The most common side-effect of naltrexone is nausea and sometimes vomiting. There is some evidence that this is related to quantity and recency of alcohol consumption as mentioned above. Beyond nausea, there is great concern over the possibility of hepatic toxicity. When alcoholism was added to the indications for naltrexone in 1995, the FDA was concerned about a report of increased liver enzymes reported in a trial of naltrexone in the treatment of obesity. The patients were receiving 350 mg daily or seven times the normal daily dose. The increase in enzymes was reversible and did not result in liver damage, but it caused the inclusion of a "black box" warning about potential

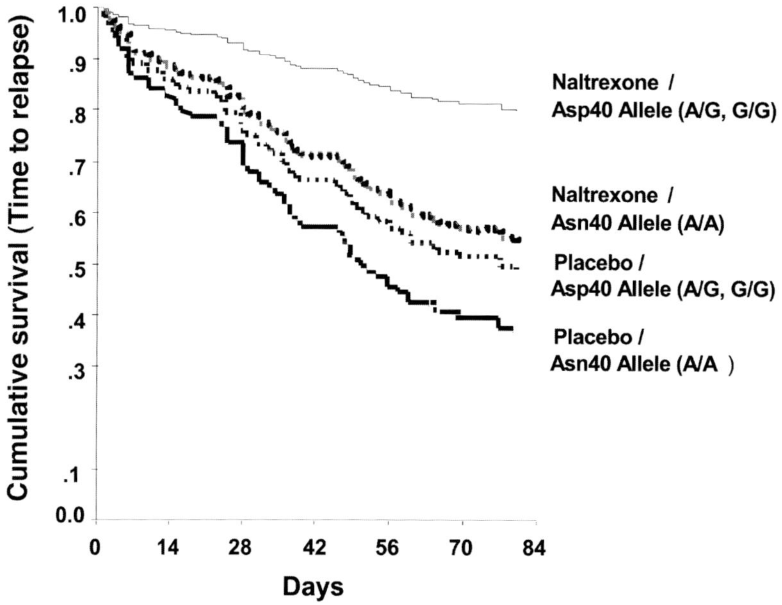

Figure 3. Survival analyses for time to relapse in subjects with one or two copies of the Asp40 allele vs those homozygous for the Asn40 allele by medication group.

liver toxicity. Subsequently, naltrexone has not been found to produce hepatic toxicity in clinical trials of alcoholism or heroin addiction. Actually the converse is often true. Naltrexone produces a decrease in alcohol consumption and with decreased alcohol, a true hepatic toxin, the liver profile improves.

Selection of patients

Faced with the treatment of a patient with alcoholism, a clinician must consider whether a medication should be included in the treatment plan. Of course detoxification is the first goal, but a long-term plan for prevention of relapse is essential. Ideally there should be a psychiatric evaluation that includes assessment of need for treatment in all the domains of the Addiction Severity Index [45]. If a co-existing disorder is present such as depression, appropriate medication should be prescribed. For specific relapse prevention, medication in conjunction with psychotherapy should always be considered. Alcoholism is a potentially fatal disease and it makes no sense to withhold effective therapy. The choices in the United States at present are naltrexone, disulfiram and the recently approved acamprosate. Depending on the circumstances and willingness of the patient, disulfiram may be ideal. Many patients, however, are unwilling to accept disulfiram. Naltrexone is generally more acceptable and has the advantage of possibly reducing craving for alcohol. Furthermore, if the patient does drink, the pleasant effects of the alcohol may be less, but there will be no adverse consequences of drinking such as those seen in patients receiving disulfiram. Beginning in July 2004, acamprosate became another possibility and more studies comparing the two medications are needed [46]. There is preliminary evidence [47] that specific types of alcoholics are likely to respond better to naltrexone or to acamprosate and thus prospective studies of the two medications involving well-characterized alcoholics are in order.

Based on clinical experience and evidence from clinical trials, naltrexone seems to have greater efficacy in those with a family history of alcoholism. Additional studies involving genotyping of alcoholics in clinical trials are in order to determine whether the Oslin study described above can be replicated and extended so that more precise patient selection can be based on genotype. A large trial of depot naltrexone [28] found a strong gender effect with naltrexone being highly effective in males but not in females. Other selection factors seem to relate to medication adherence. Older alcoholics were more likely to take medication regularly, come to appointments and show a strong drug/placebo difference [48].

References

1 Altshuler HL Phillips PA, Feinhandler DA (1980) Alternation of enthanol self-administration by naltrexone. *Life Sciences* 26: 679–688
2 Volpicelli JR, Alterman AI, Hayashida M, O'Brien CP (1992) Naltrexone in the treatment of alcohol dependence. *Archives of General Psychiatry* 49: 876–880
3 Volpicelli JR, O'Brien CP, Alerman AI, Hayashida M (1990) Naltrexone and the treatment of alcohol dependence: initial observations. In: LB Reid (ed.): *Opioids, Bulimia, Alcohol Abuse and Alcoholism,* Springer-Verlag, New York, 195–214
4 O'Malley SS, Jaffe, AJ, Chang G, Schottenfeld RS, Meyer RE, Rounsaville B (1992) Naltrexone and coping skills therapy for alcohol dependence: a control study. *Arch Gen Psychiat* 49: 881–887
5 Harris KM, DeVries A, Dimidjian K (2004) Datapoints: Trends in naltrexone use among members of a large private health plan. *Psychiatric Services* 55(3): 221
6 Kranzler HR, Modesto-Lowe V, Van Kirk J (2000) Naltrexone *versus* nefazodone for treatment of alcohol dependence. A placebo-controlled trial. *Neuropsychopharmacology* 22(5): 493–503
7 Krystal JH, Cramer JA, Krol WF, Kirk GF, Rosenheck RA, Veterans Affairs Naltrexone Cooperative Study 425 Group (2001) Naltrexone in the treatment of alcohol dependence. *N Engl J Med* 345(24): 1734–1739
8 Gastpar M, Bonnet U, Boning J. Mann K. Schmidt LG. Soyka M. Wetterling T. Kielstein V. Labriola D. Croop R (2002) Lack of efficacy of naltrexone in the prevention of alcohol relapse: results from a German multicenter study. *J Clin Psychopharmacol* 22(6): 592–598
9 Pettinanti HM, Oslin D, O'Brien CP (in preparation) A review of all naltrexone clinical trials for alcoholism
10 Robinson DS, Rickels K (2000) Concerns about clinical drug trials. *Journal of Clinical Psychopharmacology* 20(6): 593–596
11 Chan A-W, Hrobjartsson A, Haahr M, Gotzsche PC, Altman DG (2004) Empirical evidence for selective reporting of outcomes in randomized trials. *JAMA* 291(20): 2457–2465
12 Thase M (2003) Issues and Ethics. Presented at the *Annual American College of Neuropsychopharmacology Annual Meeting*, San Juan, Puerto Rico
13 Koob GF, Rassnick S, Heinrichs S, Weiss F (1994) Alcohol, the reward system and dependence. *Exs* 71: 103–114
14 Gonzales RA, Weiss F (1998) Suppression of ethanol-reinforced behavior by naltrexone is associated with attenuation of the ethanol-induced increase in dialysate dopamine levels in the nucle-

us accumbens. *The Journal of Neuroscience* 18(24): 10663–10671
15 Gianoulakis C (1996) Implications of endogenous opioids and dopamine in alcoholism: human and basic science studies. *Alcohol and Alcoholism* 31(1): 33–42
16 Genazzani AR, Nappi G, Facchinetti F, Mazzella GL, Parrini D, Sinforiani E, Petraglia F, Savoldi F (1982) Central deficiency of beta-endorphin in alcohol addicts. *J Clin Endocrinol Metab* 55(3): 583–586
17 Gianoulakis C (1990) Characterization of the effects of acute ethanol administration on the release of beta-endorphin peptides by the rat hypothalamus. *Eur J Pharmacol* 180: 21–29
18 King A, Volpicelli J, Frazer A, O'Brien CP (1997) Effect of naltrexone on subjective alcohol response in subjects at high and low risk for future alcohol dependence. *Psychopharmacology* 129: 15–22
19 Volpicelli JR, Clay KL, Watson NT, O'Brien CP (1995) Naltrexone in the treatment of alcoholism: predicting response to naltrexone. *Journal of Clinical Psychiatry* 56: 39–44
20 Swift RM, Whelihan W, Kuznetsov O, Buongiorno G, Hsuing H (1994) Naltrexone-induced alterations in human ethanol intoxication. *American Journal of Psychiatry* 151(10): 1463–1467
21 Davidson D, Palfai T, Bird C, Swift R (1999) Effects of naltrexone on alcohol self-administration in heavy drinkers. *Alcohol Clin Exp Res* 23(2): 195–203
22 Volpicelli JR, Rhines KC, Rhines, JS, Volpicelli LA, Alterman, AI, O'Brien CP (1997) Naltrexone and alcohol dependence. Role of subject compliance. *Arch Gen Psychiatry* 54(8): 737–742
23 Chick J., Anton R, Checinski K. Croop R. Drummond DC. Farmer R. Labriola D. Marshall J. Moncrieff J. Morgan MY. Peters T. Ritson B. (2000) A multicentre, randomized, double-blind, placebo-controlled trial of naltrexone in the treatment of alcohol dependence or abuse. *Alcohol and Alcoholism* 35(6): 587–593
24 O'Malley SS, Krishnan-Sarin S, Farren C, O'Connor PG (2000) Naltrexone-induced nausea in patients treated for alcohol dependence: clinical predictors and evidence for opioid-mediated effects. *J Clin Psychopharmacol* 20(1): 69–76
25 Wand GS, Mangold, D, El Deiry S, McCaul ME, Hoover D (1998) Family history of alcoholism and hypothalamic opioidergic activity. *Arch Gen Psychiatry* 55(12): 1114–1119
26 Kranzler HR, Modesto V, Nuwayser ES (1998) A sustained-release naltrexone preparation for treatment of alcohol dependence. *Alcoholism: Clinical and Experimental Research* 22(5): 1074–1079
27 Kranzler HR, Wesson DR, Billot L for the Drug Abuse Sciences Naltrexone Depot Study Group (2004) Naltrexone depot for treatment of alcohol dependence: a multicenter, randomized, placebo-controlled clinical trial. *Alcoholism: Clinical and Experiment Research* 28(7): 1051–1059
28 Silverman B (2004) Clinical Trial of Depot Naltrexone. Presented at the *American Psychiatry Association Annual Meeting*, New York City, New York
29 Schmidt WK, Tam SW, Shotzberger GS, Smith DH Jr, Clark R, Vernier VG (1985) Nalbuphine. *Drug Alcohol Depend* 14(3–4): 339–362
30 Stromberg MF, Casale M, Volpicelli L, Volpicelli JR, O'Brien CP (1998) A comparison of the effects of the opioid antagonists naltrexone, naltrindole, and beta-funaltrexamine on ethanol consumption in the rat. *Alcohol* 15(4): 281–289
31 Lee MC, Wagner HN, Tanada S, Frost JJ, Bice AN, Dannals RF (1988) Duration of occupancy of opiate receptors by naltrexone. *J Nuclear Med* 29(7): 1207–1211
32 McCaul ME, Kim WK Wand GS, Bencherif B, Dannals RF, Frost JJ (2004) PET Measurement of μ- and d-opioid receptor availability before and during naltrexone treatment in alcohol dependent subjects (poster). Presented at the *Winter Conference on Brain Research,* Copper Mountain, Colorado
33 Tiffany ST (1999) Cognitive concepts of craving. *Alcohol Res Health* 23(3): 215–224
34 Monti PM, Rohsenow DJ, Hutchison KE, Swift RM, Mueller TI, Colby SM, Brown RA, Gulliver SB, Gordon A, Abrams DB (1999) Naltrexone's effect on cue-elicited craving among alcoholics in treatment. *Alcohol Clin Exp Res* 23(8): 1386–1394
35 O'Malley SS, Krishnan-Sarin S, Farren C, Sinha R, Kreek MJ (2002) Naltrexone decreases craving and alcohol self-administration in alcohol dependent subjects and activates the hypothalamo-pituitary-adrenocortical axis. *Psychopharmacology* 160(1): 19–29
36 Latt NC, Jurd S, Houseman J, Wutzke SE (2002) Naltrexone in alcohol dependence: a randomized controlled trial of effectiveness in a standard clinical setting. *The Medical Journal of Australia* 176(11): 530–534
37 Pettinati HM, Volpicelli JR, Pierce JD, O'Brien CP, Pierce JD Jr, (2000) Improving naltrexone

response: an intervention for medical practitioners to enhance medication compliance in alcohol dependent patients. *J Addict Dis* 19(1): 71–83
38 Jaffe AJ, Rounsaville B, Chang G, Schottenfeld RS, Meyer RE, O'Malley SS (1996) Naltrexone, relapse prevention, and supportive therapy with alcoholics: an analysis of patient treatment matching. *J Consult Clin Psychol* 64(5): 1044–1053
39 Monterosso JR, Flannery BA, Pettinati HM, Oslin DW, Rukstalis M, O'Brien CP, Volpicelli JR (2001) Predicting treatment response to naltrexone: the influence of craving and family history. *Am J Addict* 10(3): 258–268
40 Oslin DW, Berrettini W, Kranzler HR, Pettinati H, Gelernter J, Volpicelli JR, O'Brien CP (2003) A functional polymorphism of the μ-opioid receptor gene is associated with naltrexone response in alcohol-dependent patients. *Neuropsychopharmacology* 28: 1546–1552
41 Wand GS, Mangold E, El Deiry S, McCaul ME, Hoover D (1998) Family history of alcoholism and hypothalamic opioidergic activity. *Arch Gen Psychiatry* 55: 1114–1119
42 Hernandez-Avila C, Wand G, Luo X, Gelernter J, Kranzler H (2003) Association between the cortisol response to opioid blockade and the Asn40Asp polymorphism at the mu-opioid receptor locus (OPRM1). *Am J Med Genet* 118B: 60–65
43 Smolka M. Sander T, Schmidt L, Samochowiec J, Rommelspacher H, Gschiedel N, Wendel B. Hoehe MR (1999) Mu-opioid receptor variants and dopaminergic sensitivity in alcohol withdrawal. *Psychoneuroendocrinology* 24: 629–638
44 Kim SG, Kim C-M, Kang D-H, Kim J-K, Byun W-T, Kim S-Y, Park J-M, Kim M-J, Oslin DW (2004) Association of functional opioid receptor genotypes with alcohol dependence in Koreans. *Alcohol Clin Exp Res* 28(7): 986–990.
45 McLellan AT, Luborsky L, Woody GE, O'Brien CP (1980) An improved diagnostic evaluation instrument for substance abuse patients: The Addiction severity Index. *J Nervous and Mental Disease* 168: 26–33
46 Kiefer F, Jahn H, Tarnaske T, Helwig H, Briken P, Holzbach R, Kampf P, Stracke R, Baehr M, Naber D, Widemann K (2003) Comparing and combining naltrexone and acamprosate in relapse prevention of alcoholism: a double-blind, placebo-controlled study. *Arch Gen Psychiatry* 60(1): 92–99
47 Lesch, OM (2004) How to translate the biological basic glutamate mechanisms in preclinical research to human trials and therapy approaches. Presented at the *European Winter Conference on Brain Research*, Les Arcs, France
48 Oslin DW, Pettinati H, Volpicelli JR (2002) Alcoholism treatment adherence: older age predicts better adherence and drinking outcomes. *Am J Geriatr Psychiatry* 19(6); 740–747

Drugs for Relapse Prevention of Alcoholism
Edited by R. Spanagel and K. Mann

Acamprosate: preclinical data

Philippe De Witte[1], Daniel Bachteler[2] and Rainer Spanagel[2]

[1] *UCL – Biologie du Comportement, Place Croix du Sud,1, 1348 Louvain-la-Neuve, Belgium*
[2] *Department of Psychopharmacology, Central Institute of Mental Health, University of Heidelberg (CIMH), J5, 68159 Mannheim, Germany*

Introduction

Acamprosate (calcium-bis-acetyl-homotaurinate, Campral®) was introduced to the European market in the time between 1990 and 2000 and since then it has been successfully applied in clinical relapse prevention as an anti-craving/anti-relapse drug in detoxified alcoholics. Numerous preclinical studies on alcohol drinking and relapse behavior have been performed with acamprosate [1, 2]. Besides the demonstration of its anti-relapse and anti-craving properties it has been demonstrated that acamprosate neither produces discriminative stimulus properties [3] nor does it produce place conditioning [4]. This implies that acamprosate is lacking in any rewarding properties and cannot be used as a substitution drug (see Chapter by Bachteler and Spanagel). The primary site of action is the glutamatergic system [1, 2, 5] and it has been shown that N-methyl-D-aspartate (NMDA) receptor subunits are modulated by acamprosate [6]. Until the present, the detailed mode of action is still not clear; however, a functional interplay with other glutamate receptors has recently been described. Thus, Harris et al. [7, 8] described a novel site of acamprosate action at metabotropic glutamate receptors, in particular at mGluR5. Furthermore, a series of *in vivo* microdialysis studies on the interaction of ethanol and acamprosate with brain taurine levels suggest that acamprosate reduces the preference for alcohol in dependent animals through the release of taurine [9–13]. The following paragraphs review animal studies on the behavioral effects of acamprosate and provide insights into the latest research on acamprosate, especially its relation to brain taurine and its action on the glutamate system.

Animal studies – general characterization

As a therapeutic drug, acamprosate combines important characteristics: it neither exerts its actions by imitating alcohol's reinforcing effects, nor does it alter ethanol-induced hypothermia, taste-aversion or motor impairment [14].

Rewarding effects of the compound itself could not be demonstrated in conditioned place preference experiments in rats [4, 15], nor could it substitute for or antagonize ethanol effects in drug discrimination experiments [3]. Furthermore, its effects are not psychotropic, sedative, anti-depressant or anxiolytic [16]. However, it is suggested that acamprosate reduces craving and relapse that are associated with a hyper-glutamatergic state in the brain and the phenomenon of conditioned withdrawal [17] and thus could be used as an anti-craving/anti-relapse drug.

Attenuation of ethanol intake and relapse

The initial study on the effects of acamprosate was performed by Boismare et al. [18], reporting a significant decrease in voluntary ethanol intake in rats, daily treated with 0.26 and 0.52 mMol/kg Ca-acetylhomotaurine, respectively. This effect could be inhibited by simultaneous administration of the γ-aminobutyric acid (GABA) antagonist bicuculline, thus leading to the assumption of an involvement of the GABAergic system. Acamprosate injections had no consequences on total fluid intake, but selectively affected alcohol preference.

These first results were confirmed by other studies [19, 20]. Although different experimental protocols were used, all studies consistently reported a significant, dose-dependent reduction of ethanol consumption after chronic treatment with doses of 100 and 200 mg/kg or acute treatment with 200 and 400 mg/kg acamprosate, respectively. A dose of only 50 mg/kg produced no significant effect [21]. Water intake during chronic treatment remained unaffected. However, Heyser et al. [22] reported an increasing water consumption after an acute high dose of 400 mg/kg. The aforementioned effects were confirmed in more recent studies under operant and non-operant conditions [21, 23–26].

A generally accepted method for assessing relapse-like drinking behavior in rats is the alcohol-deprivation model. Ethanol consumption in long-term ethanol-experienced animals is measured after a period of imposed abstinence, leading to a strong transient increase in alcohol intake, termed the alcohol deprivation effect (ADE) [27, 28] (see also Chapter by Spanagel). Using this model, administration of acamprosate to long-term alcohol-drinking rats dose-dependently reduced the alcohol deprivation effect, and at the highest dose (200 mg/kg), alcohol intake even dropped significantly below baseline levels [29]. It should be emphasized that alcohol consumption after a deprivation effect is more effectively reduced than during baseline drinking [25]. Furthermore, chronic injections of acamprosate were more effective in reducing the alcohol deprivation effect than acute single administrations [22]. In conclusion, acamprosate reduces not only voluntary ethanol self-administration under free-choice and operant paradigms, but is also effective in reducing relapse behavior, as measured by the alcohol deprivation effect.

Only recently the anti-craving effect of acamprosate was demonstrated by Bachteler et al. [30] in the reinstatement model, which in contrast to the ADE model (relapse), measures the animals' motivation to get the drug and thus reflects craving for alcohol [31]. A stimulus is said to reinstate drug seeking, if the animal restarts to lever press for the drug after an extinction phase, but without the primary reinforcer "ethanol" being available. Administration of acamprosate (100, 200 mg/kg) significantly and selectively attenuated responding for ethanol at both doses after presentation with the ethanol-related stimulus. It should however be noted that in the aforementioned reinstatement paradigm, only the behavioral outcome of alcohol craving can be assessed. The subjective state, associated with an incentive motivation to administer the drug, remains unknown. In addition, it should be kept in mind that not all patients are responding to acamprosate treatment, leading to relapse despite medication. The potential role of acamprosate as a neuroprotective agent in humans could furthermore support its use, as already proposed in different *in vitro* models [32–34], but also needs further investigation.

Involvement of the glutamatergic system in the action of acamprosate

The exact mechanism as to how acamprosate diminishes alcohol consumption and reduces the likelihood of relapse is still not clear. Different neurobiological pathways have been implicated in the etiology of alcohol dependence and one pathway seems to involve the glutamatergic system [1, 35–37], where chronic alcohol intake leads to compensatory changes. It is suggested that acamprosate acts mainly on a hyper-glutamatergic state, yet having only little effect on a "normal" glutamatergic state [17, 33, 38, 39].

Depending on the brain region and the rat strain, low doses of ethanol can increase glutamate levels in the brain, whereas high intoxicating doses can decrease glutamate levels [40]. The mechanism behind the inhibitory effect of high intoxicating doses of ethanol on glutamate release is not clear. Although multiple mechanisms have been implicated in this action, considering the general inhibitory influence of GABA on glutamatergic neurotransmission, it may be suggested that the inhibitory effect of ethanol on glutamate is due to an initial increase of GABA release, which in turn inhibits the release of glutamate. Whatever the mechanism might be, the inhibitory effect of high intoxicating doses of ethanol leads to several adaptive responses within the glutamatergic system following its chronic administration. Indeed, extracellular glutamate levels are enhanced during withdrawal [39, 41] and long-lasting alterations in glutamate release mechanisms following chronic alcohol intake have also been demonstrated. Thus, following a period of abstinence, ethanol-conditioned stimuli can induce an increase in extracellular glutamate levels in the amygdala [42], suggesting that conditioned responses to extracellular glutamate may participate in environmental cue-induced craving and relapse behavior.

In several *in vivo* microdialysis studies, it has been shown that acamprosate can reduce enhanced glutamate levels. Thus, rats which were alcoholized by ethanol inhalation, exhibited enhanced extracellular glutamate levels in the nucleus accumbens, whereas rats which were simultaneously alcoholized and treated orally by acamprosate (400 mg/kg/day), failed to present the increase in glutamate during ethanol withdrawal [41]. Acamprosate was also able to decrease augmented glutamate release in the hippocampus following repeated withdrawal episodes [39]. On the behavioral level, acamprosate reduced context-dependent ethanol effects [43]. In conclusion, enhanced glutamate levels which occur during withdrawal or conditioned withdrawal and which reflect a hyper-glutamatergic state of the brain can be effectively blocked by acamprosate. This hypothesis has recently received further support by studying transgenic mice, which exhibit a hyper-glutamatergic state in their brain and show enhanced alcohol consumption compared to control wild-type mice [44]. A threshold dose of acamprosate, which had no effect on alcohol drinking in control wild-type mice, produced a strong reduction of alcohol intake in the transgenic mice [44], demonstrating a causal relationship between a hyper-glutamatergic system and enhanced alcohol intake as well as the action of acamprosate.

Up to now it is not clear how acamprosate interferes with enhanced glutamate levels. However, several studies have shown an interaction of acamprosate with glutamate receptors and it has been suggested that acamprosate binds to the polyamine binding site of the NMDA receptor [45, 46] and to the metabotropic mGlu5 receptor [7, 8]. This in turn might induce a cascade of genomic processes [1, 6], which could finally alter glutamate release and re-uptake mechanisms.

Acamprosate and taurine

Acamprosate presents chemical structural similarities with taurine and particularly shares a similar sulfur group (see Appendix: Chemical structures). Interactions between the endogenous amino acid taurine and ethanol in the central nervous system have been identified in recent years [38, 47]. The amino acid taurine is synthesized from cysteine, and is present in cells at relatively high concentrations in the mM range. Taurine is one of the most abundant amino acids in the brain and plays an integral role in physiological processes such as osmoregulation and neuromodulation. Thus, taurine exerts positive allosteric modulatory effects on neuronal ligand-gated chloride channels (i.e., $GABA_A$ and glycine receptors) as well as inhibitory effects on other ligand- and voltage-gated cation channels (i.e., NMDA and Ca^{2+} channels). Neuroprotective and neuromodulatory functions have also been attributed to taurine. Taurine has been shown to modulate cell excitability [48], to prevent neuronal excitotoxicity [49], to protect neuronal membranes against different toxic damages [50] and to regulate calcium homeostasis [51]. Together, all these protective effects result in the membrane stabilization properties of taurine [52]. Behavioral evidence suggests that taurine can alter the locomotor stimulatory, sedating, and motivational effects of ethanol in a dose-dependent manner [38, 47]. Furthermore, microdialysis studies have revealed that in response to ethanol, elevated extracellular levels of taurine can be observed in numerous brain regions, an effect which is shared by acamprosate administration (Figs 1 and 2).

The administration of acute ethanol to rats increases the level of the sulphonated amino acid taurine in many brain regions, including nucleus accumbens [9, 10] (Fig. 1), hippocampus, frontal cortex [41] and amygdala [11, 42]. The increase in taurine microdialysate level seems thus to represent a global answer of the brain to an acute administration of ethanol. Chronic administration of ethanol to rats leads to a progressive increase of the basal level of taurine within the entire brain [10]. Exposure of primary astrocyte culture to iso-osmotic ethanol from 10 to 100 mM leads to cell swelling and the release of taurine. Ethanol-induced cell swelling probably activates volume-sensitive channels, and taurine passively diffuses outside the cells along its concentration gradient [12].

The effect of ethanol on taurine has also been tested in Sardinian ethanol-preferring (sP) and non-preferring (sNP) rats. The sP and sNP rats have been selectively outbred from a population of Wistar rats for their high and low

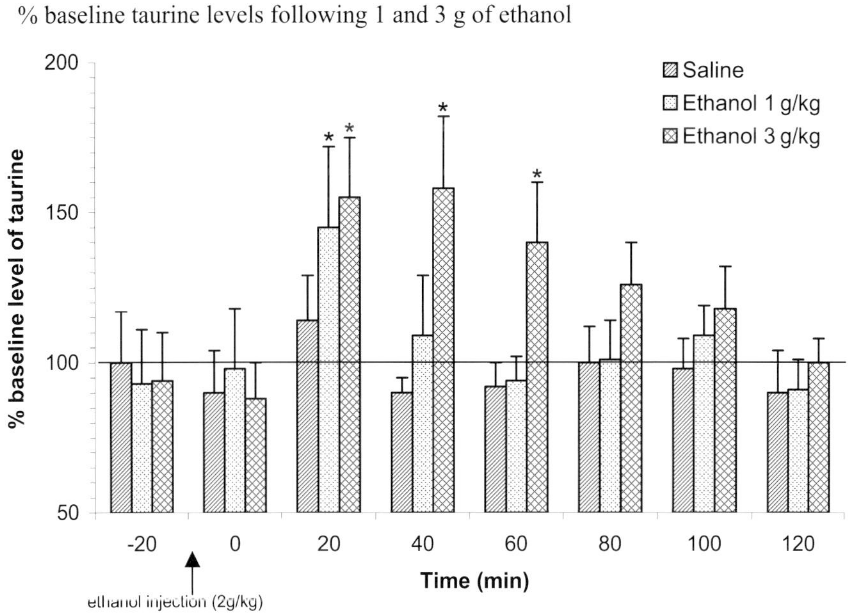

Figure 1. Effects of an acute IP injection of 1 and 3 g/kg ethanol (IP) or saline on extracellular taurine levels in the nucleus accumbens in naïve rats. Data are presented as means + S.E.M. * indicates significant differences to baseline.

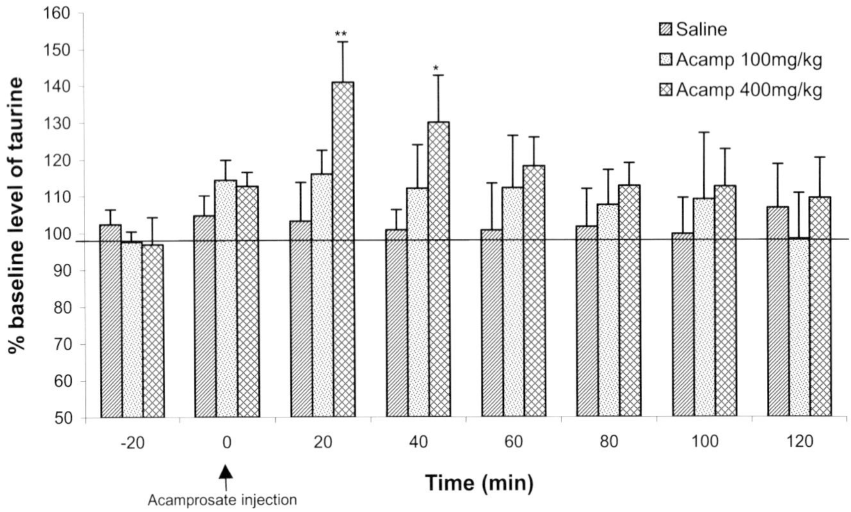

Figure 2. Effects of an acute IP injection of 100 and 400 mg/kg acamprosate (IP) or saline on extracellular taurine levels in the nucleus accumbens in naïve rats. Data are presented as means + S.E.M. * indicates significant differences to baseline.

ethanol consumption respectively [53]. Results indicate that ethanol at high doses (2 g/kg) induced an immediate and significant increase in taurine microdialysate content in the nucleus accumbens of both sP and sNP rats. However, this ethanol-induced taurine release was significantly higher in sNP rats in comparison to sP rats (Fig. 3). Furthermore, the differences in taurine responsiveness to ethanol in sP and sNP rats was not caused by differences in their ethanol absorption, redistribution and elimination, as their blood ethanol elimination followed a similar time course. It is noteworthy that sNP rats displayed an ethanol-induced taurine release similar to that observed in Wistar rats, while the responsiveness of taurine to ethanol administration in sP rats was significantly lower. It is interesting to relate this to the pattern of ethanol consumption in these rats, as Wistar and sNP rats are reluctant to consume ethanol, while sP rats voluntarily drink large quantities of ethanol in a free-choice procedure [53], suggesting that there is an inverse relationship between the intensity of ethanol-induced extracellular taurine release and ethanol preference.

Another *in vivo* microdialysis study was conducted to study the effect of ethanol administration on brain taurine levels in High-Alcohol Sensitive (HAS) and Low-Alcohol Sensitive (LAS) rats. These rat lines were used to test whether the effects of acute ethanol on extracellular concentrations of taurine might be related to genetic differences in ethanol sensitivities. HAS and LAS rats from the University of Colorado were genetically selected according to their differential sensitivities to the hypnotic effects of acute ethanol [54]. Using microdialysis in awake and freely moving animals, HAS rats displayed a reduced ethanol-induced taurine release in comparison to LAS rats (Fig. 4).

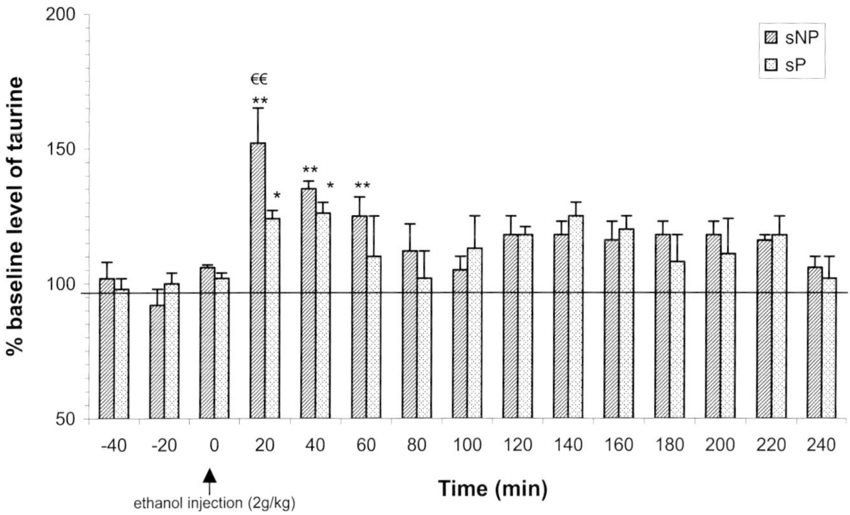

Figure 3. Time course of extracellular taurine content of the microdialysate from the nucleus accumbens before and after an acute intraperitoneal injection of 2.0 g/kg ethanol to either sNP or sP rats. Data are expressed as mean (+ S.E.M.) percentage of baseline level which was calculated for each rat by averaging the concentration of the three sample values before injection. $^{*}P < 0.05$ relative to respective control group injected with saline; $^{**}P < 0.01$ relative to respective control group injected with saline; and €€ $P < 0.01$ relative to sP rats.

Taurine responsiveness to ethanol was therefore inversely related to initial ethanol sensitivity in these genetically selected rat lines. This result is unlikely to be due to pharmacokinetic differences between HAS and LAS rats in ethanol absorption, distribution and elimination, since no differences in blood ethanol concentrations in these two lines of rats were observed [55]. Taurine may contribute to the reduction of several ethanol-adverse effects and this may thus explain why HAS rats show a reduced release of this regulatory amino acid following ethanol administration. On the contrary, in LAS, a higher taurine release may oppose some of the adverse effects of ethanol, thereby contributing to their overall lower sensitivity to ethanol.

With respect to the effects of ethanol on brain taurine levels, similar observations have been made with acamprosate. Thus, the IP injection of 400 mg/kg acamprosate induced a significant increase of extra-cellular taurine concentration in the nucleus accumbens lasting for 40 min (Fig. 2). An IP injection of 1 g/kg acamprosate induced an even larger increase in brain taurine lasting at least 3 h following injection [38]. Pretreatment with acamprosate orally for 30 days in the drinking bottle at 400 mg/kg induced an augmentation of taurine release in both sP and sNP rats. In this experiment, the acamprosate-pretreated sP group obtained brain taurine levels following ethanol injection similar to the brain taurine levels in sNP rats, without a pretreatment of acamprosate suggesting that the release of taurine after acute administration of ethanol in the

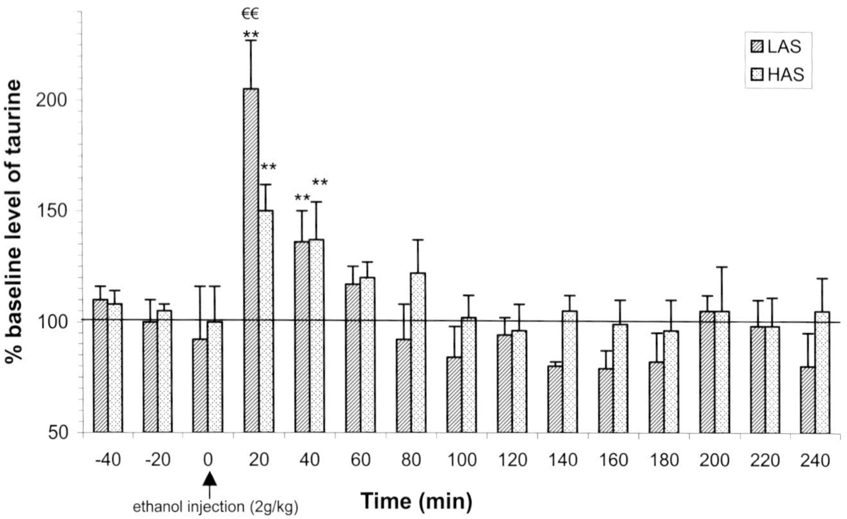

Figure 4. Time course of extracellular taurine content of the microdialysate from the nucleus accumbens before and after an acute intraperitoneal injection of 2.0 g/kg ethanol to either LAS or HAS rats. Data are expressed as mean (+S.E.M.) percentage of baseline level which was calculated for each rat by averaging the concentration of the three sample values before injection. ** $P < 0.01$ relative to respective control group injected with saline; and €€ $P < 0.01$ relative to HAS rats.

sNP group could be involved in their avoidance of alcohol intake. These studies suggest that acamprosate reduces the preference for alcohol in sP rats through the release of brain taurine.

In summary, the endogenous taurine system may be an important modulator of the effects of ethanol on the central nervous system and might counteract some of ethanol´s adverse effects. Externally added taurine, such as in taurine-supplementation studies [13], might therefore modulate ethanol consumption, probably by preventing ethanol-adverse effects. The action of acamprosate on ethanol consumption might also be mediated, at least in part, by taurine. However, it is still unclear how and where taurine acts, as is the relationship between taurine and glutamate.

Conclusion

To achieve its therapeutic effects, acamprosate may act on several neurochemical systems. However, there is a clear interaction with the glutamatergic system and there is convincing evidence that acamprosate effectively reduces a hyper-glutamatergic state, which may trigger, at least in some alcoholic patients, craving and relapse. Whether this dampening effect on a hyper-glutamatergic state is due to binding on the polyamine site of the NMDA receptor or mGluR5 and the thereby induced cascade of genomic processes, remains to be further investigated. Furthermore, the relationship between the effects of acamprosate on the endogenous taurine system and a hyper-glutamatergic state is still not well understood; however, it might be due to taurine producing a protective effect in this respect. A better understanding of the precise mode of action of acamprosate would help in identifying treatment responders (individually adapted pharmacotherapy; see also project PREDICT in the Chapter by Mann) and could lead to compound optimization.

References

1 Spanagel R, Zieglgansberger W (1997) Anti-craving compounds for ethanol: new pharmacological tools to study addictive processes. *Trends Pharmacol Sci* 18(2): 54–59
2 Zornoza T, Cano MJ, Polache A, Granero L (2003) Pharmacology of acamprosate: an overview. *CNS Drug Rev* 9(4): 359–374
3 Spanagel R, Zieglgänsberger W, Hundt W (1996) Acamprosate and alcohol: III. Effects on alcohol discrimination in the rat. *Eur J Pharmacol* 305: 51–56
4 Kratzer U, Schmidt WJ (2003) Acamprosate does not induce a conditioned place preference and reveals no state-dependent effects in this paradigm. *Prog Neuropsychopharmacol Biol Psychiatry* 27: 653–656
5 Littleton J, Zieglgansberger W (2003) Pharmacological mechanisms of naltrexone and acamprosate in the prevention of relapse in alcohol dependence. *Am J Addict* 12 Suppl 1: S3–11
6 Rammes G, Mahal B, Putzke J, Parsons C, Spielmanns P, Pestel E, Spanagel R, Zieglgansberger W, Schadrack J (2001) The anti-craving compound acamprosate acts as a weak NMDA-receptor antagonist, but modulates NMDA-receptor subunit expression similar to memantine and MK-801. *Neuropharmacology* 40(6): 749–760
7 Harris BR, Prendergast MA, Gibson DA, Rogers DT, Blanchard JA, Holley RC, Fu MC, Hart SR, Pedigo NW, Littleton JM (2002) Acamprosate inhibits the binding and neurotoxic effects of trans-ACPD, suggesting a novel site of action at metabotropic glutamate receptors. *Alcohol Clin Exp Res* 26(12): 1779–1793
8 Harris BR, Gibson DA, Prendergast MA, Blanchard JA, Holley RC, Hart SR, Scotland RL, Foster TC, Pedigo NW, Littleton JM (2003) The neurotoxicity induced by ethanol withdrawal in mature oranotypic hippocampal slices might involve cross-talk between metabotropic glutamate type 5 receptors and N-methyl-D-aspartate receptors. *Alcohol Clin Exp Res* 27(11): 1724–1735
9 Dahchour A, Quertemont E, De Witte P (1994) Acute ethanol increases taurine but neither glutamate nor GABA in the nucleus accumbens of male rats: a microdialysis study. *Alcohol Alcoholism* 29: 485–487
10 Dahchour A, De Witte P (1996) Taurine increases in the nucleus accumbens microdialysate after acute ethanol administration to naive and chronically alcoholised rats. *Brain Res* 735: 9–19
11 Quertemont E, Dahchour A, Ward RJ, De Witte P (1999) Ethanol induces taurine release in the amygdala: an *in vivo* microdialysis study. *Addict Biol* 4: 47–54
12 Quertemont E, Devitgh A, De Witte P (2003) Systemic osmotic manipulations modulate ethanol-induced taurine release: a brain microdialysis study. *Alcohol* 29: 11–19
13 Quertemont E, Goffaux V, Vlaminck AM, Wolf C, De Witte P (1998) Oral taurine supplementation modulates ethanol-conditioned stimulus preference. *Alcohol* 16: 201–206
14 Le Magnen J, Tran G, Durlach J (1987) Lack of effects of Ca-acetyl homotaurinate on chronic and acute toxicities of ethanol in rats. *Alcohol* 4(2): 103–108
15 McGeehan AJ, Olive MF (2003) The anti-relapse compound acamprosate inhibits the development of a conditioned preference to ethanol and cocaine but not morphine. *Br J Pharmacol* 138: 9–12
16 Kiefer F, Wiedemann K (2002) Aktueller Stand der pharmakotherapeutischen Rückfallprophylaxe bei Alkoholabhängigkeit mit Acamprosat und Naltrexon. *Nervenheilkunde* 4: 183–192
17 Littleton J (1995) Acamprosate in alcohol dependence: how does it work? *Addiction* 90: 1179–1188

18 Boismare F, Daoust M, Moore ND, Saligaut C, Lhuintre JP, Chretien P, Durlach J (1984) A homotaurine derivate reduces the voluntary intake of ethanol by rats. Are cerebral GABA receptors involved? *Pharmacol Biochem Behav* 21: 787–789
19 Le Magnen J, Tran G, Durlach J, Martin C (1987) Dose-dependent suppression of the high alcohol intake of chronically intoxicated rats by Ca-acetylhomotaurinate. *Alcohol* 4(2): 97–102
20 Gewiss M, Heidbreder C, Opsomer L, Durbin P, De Witte P (1991) Acamprosate and diazepam differentially modulate alcohol-induced behavioural and cortical alterations in rats following chronic inhalation of ethanol vapor. *Alcohol Alcoholism* 26(2): 129–137
21 Stromberg MF, Mackler SA, Volpicelli JR, O´Brien CP (2001) Effect of acamprosate and naltrexone, alone or in combination, on ethanol consumption. *Alcohol* 23(2): 109–116
22 Heyser CJ, Schulteis G, Durbin P, Koob GF (1998) Chronic acamprosate eliminates the alcohol deprivation effect while having limited effects on baseline responding for ethanol in rats. *Neuropsychopharmacology* 18: 125–133
23 Czachowski CL, Legg BH, Samson HH (2001) Effects of acamprosate on ethanol-seeking and self-administration in the rat. *Alcohol Clin Exp Res* 25(3): 344–350
24 Olive MF, Nannini MA, Ou CJ, Koenig HN, Hodge CW (2002) Effects of acute acamprosate and homotaurine on ethanol intake and ethanol-stimulated mesolimbic dopamine release. *Eur J Pharmacol* 437(1–2): 55–61
25 Hölter SM, Landgraf R, Zieglgänsberger W, Spanagel R (1997) Time course of acamprosate action on operant ethanol self-administration after ethanol deprivation. *Alcohol Clin Exp Res* 21(5): 862–868
26 Heyser CJ, Moc K, Koob GF (2003) Effects of naltrexone alone and in combination with acamprosate on the alcohol deprivation effect in rats. *Neuropsychopharmacology* 28(8): 1463–1471
27 Sinclair JD, Senter RJ (1967) Increased preference for ethanol in rats following alcohol deprivation. *Psycho Sci* 8: 11
28 Spanagel R, Hölter SM (1999) Long-term alcohol self-administration with repeated alcohol deprivation phases: an animal model of alcoholism? *Alcohol Alcoholism* 34: 231–243
29 Spanagel R, Hölter SM, Allingham K, Landgraf R (1996) Acamprosate and alcohol: I. Effects on alcohol intake following alcohol deprivation in the rat. *Eur J Pharmacol* 305(1–3): 39–44
30 Bachteler D, Economidou D, Danysz W, Ciccocioppo R, Spanagel R. The effects of acamprosate and neramexane on cue-induced reinstatement of ethanol-seeking behavior in rat. *Neuropsychopharmacology; in press*
31 Shaham Y, Shalev U, Lu L, de Wit H, Stewart J (2003) The reinstatement model of drug relapse: history, methodology and major findings. *Psychopharmacology* 168: 3–20
32 Al Qatari M, Khan S, Harris B, Littleton J (2001) Acamprosate is neuroprotective against glutamate-induced excitotoxicity when enhanced by ethanol withdrawal in neocotrical cultures of fetal rat brain. *Alcohol Clin Exp Res* 25: 1276–1283
33 Koob GF, Mason BJ, De Witte P, Littleton J, Siggins GR (2002) Potential neuroprotective effects of acamprosate. *Alcohol Clin Exp Res* 26(4): 586–592
34 Mayer S, Harris BR, Gibson DA, Blanchard JA, Prendergast MA, Holley RC, Littleton J (2002) Acamprosate, MK-801, and ifenprodil inhibit neurotoxicity and calcium entry induced by ethanol withdrawal in organotypic slice cultures from neonatal rat hippocampus. *Alcohol Clin Exp Res* 26: 1468–1478
35 Tsai G, Coyle JT (1998) The role of glutamatergic neurotransmission in the pathophysiology of alcoholism. *Annu Rev Med* 49: 173–184
36 Le A, Shaham Y (2002) Neurobiology of relapse to alcohol in rats. *Pharmacol Ther* 94(1–2): 137–156
37 Weiss F, Porrino LJ (2002) Behavioral neurobiology of alcohol addiction: recent advances and challenges. *J Neurosci* 22(9): 332–3337
38 Dahchour A, De Witte P (2000) Ethanol and aminoacids in the central nervous system: assessment of the pharmacological actions of acamprosate. *Prog Neurobiol* 60: 343–362
39 Dahchour A, De Witte P (2003) Effects of acamprosate on excitatory amino acids during multiple ethanol withdrawal periods. *Alcohol Clin Exp Res* 27(3): 465–470
40 Spanagel R, Bienkowski P (2002) Glutamatergic mechanisms in alcohol dependence and addiction. In: D Lodge, W Danysz, CG Parsons (eds): *Therapeutic Potential of Ionotropic Glutamate Receptor Antagonists and Modulators*. Graham Publishing Co., Johnson City, TN, USA, 375–403
41 Dahchour A, De Witte P, Bolo N, Nedelec JF, Muzet M, Durbin P, Macher JP (1998) Central effects of acamprosate: Part 1: acamprosate blocks the glutamate increase in the nucleus accum-

bens microdialysate in ethanol withdrawn rats. *Psychiat Res-Neuroimag* 82: 107–114
42 Quertemont E, De Neuville J, De Witte P (1998) Changes in the amygdala amino acid microdialysate after conditioning with a cue associated with ethanol. *Psychopharmacology* 139: 71–78
43 Quertemont E, Brabant C, De Witte P (2002) Acamprosate reduced context-dependent ethanol effects. *Psychopharmacology* 164(1): 10–18
44 Spanagel R, Pendyala G, Abarca C, Zghoul T, Sanchis-Segura C, Magnone MC, Lascorz J, Depner M, Holzberg D, Soyka M et al. (2005) The circadian clock gene Period2 alters the glutamatergic system and thereby modulates alcohol consumption. *Nature Med*; *in press*
45 Naassila M, Hammoumi S, Legrand E, Durbin P, Daoust M (1998) Mechanism of action of acamprosate. Part I. Characterization of spermidine-sensitive acamprosate binding site in rat brain. *Alcohol Clin Exp Res* 22(4): 802–809
46 Popp RL, Lovinger DM (2000) Interaction of acamprosate with ethanol and spermine on NMDA receptors in primary cultured neurons. *Eur J Pharmacol* 394(2–3): 221–231
47 Olive MF (2002) Interactions between taurine and ethanol in the central nervous system. *Amino Acids* 23(4): 345–357
48 Huxtable RJ (1989) Taurine in the central nervous system and the mammalian actions of taurine. *Prog Neurobiol* 32: 471–533
49 Zhao P, Huang YL, Cheng JS (1999) Taurine antagonizes calcium overload induced by glutamate of chemical hypoxia in cultured rat hippocampal neurons. *Neurosci Lett* 268: 25–28
50 Timbrell JA, Seabra V, Waterfield CJ (1995) The *in vivo* and *in vitro* protective properties of taurine. *Gen Pharmac* 26: 453–462
51 El Idrissi A, Trenkea E (1999) Growth factors and taurine protect against excitotoxicity by stabilizing calcium homeostasis and energy metabolism. *J Neurosci* 19: 9459–9468
52 Wright CE, Tallan HH, Lin YY, Gaull GE (1986) Taurine: biological update. *Annu Rev Biochem* 55: 427–453
53 Colombo G (1997) Ethanol drinking behaviour in sardininan alcohol-preferring rats. *Alcohol Alcoholism* 32: 443–453
54 Draski LJ, Spuhler KP, Erwin VG, Baker RC, Deitrich RA (1992) Selective breeding of rats differing in sensitivity to the effects of acute ethanol administration. *Alcohol Clin Exp Res* 16: 48–54
55 Dahchour A, Hoffman A, Deitrich R, De Witte P (2000) Effects of ethanol on extracellular amino acid levels in High and Low-alcohol sensitive rats: a microdialysis study. *Alcohol Alcoholism* 35: 548–553

Drugs for Relapse Prevention of Alcoholism
Edited by R. Spanagel and K. Mann

Acamprosate: clinical data

Karl F. Mann

Department of Addictive Behavior and Addiction Medicine, University of Heidelberg, Central Institute of Mental Health, J 5, 68159 Mannheim, Germany

Development and first testing

Around 1982 French pharmaceutical companies started to synthesize amino acid neuromediators. In a close collaboration with several universities a number of taurine and homotaurine compounds were investigated. The regular screening procedures provided evidence that one molecule in particular, calcium acetylhomotaurinate, provided some interesting pharmacological properties. Acamprosate was consequently tested in an animal model of alcohol relapse at the University of Rouen. Administration of acamprosate to alcohol-preferring rats reduced their alcohol consumption significantly compared to a group of control rats. The results of this study were presented at the second International Society for Biomedical Research on Alcoholism (ISBRA) Conference in Santa Fe [1]. Acamprosate was further tested in another animal model of alcoholism where it again revealed a significant dose-dependent reduction of voluntary alcohol consumption (Le Magnen et al. [2]).

The first clinical studies in patients with alcohol problems were also carried out in the early 1980s in France. The positive results in men were published in the Lancet [3]. A larger multi-center double-blind placebo-controlled study in 569 detoxified alcohol-dependent patients over a three month treatment period also showed a significant benefit of acamprosate over placebo [4]. Upon these results the French health authorities registered acamprosate in 1989. Marketing in France started under the brand name “Aotal”. At the same time the health authorities asked for additional assessments especially on the long-term efficacy of the molecule. At this stage the developing company was joined by Groupe LIPHA (today an affiliate of Merck KgA) for the further development of the compound. A clinical program was launched in different research centers throughout Europe. More than 4000 alcohol-dependent patients have been treated meanwhile in studies of 3, 6 or 12 months duration [5, 6]. Today the compound is registered in about fourty different countries worldwide and more than 50 million patients have been treated. In August 2004 the FDA approved acamprosate for use in the USA, twenty years after the first report in Santa Fe.

Pharmacology of acamprosate

Acamprosate (calcium acetylhomotaurinate; Campral®, Merck-Santé, Lyon, France) has a molecular structure similar to that of several endogenous amino acids. Its mechanism of action as an anti-dipsotropic agent is not entirely clear, although its main interactions appear to be with the glutamatergic system [7–10]. It modulates the NMDA receptor function [11, 12]. There is further evidence of an additional binding site on the metabotropic mGluR5 receptor [13]; this receptor has been implicated in addictive processes, modulates glutamatergic neurotransmission and interacts with the NMDA receptor. It is likely that these effects of acamprosate on the polyamine binding sites and mGluR5 receptors attenuate glutamatergic hyperexcitability [14, 15]. For details of the preclinical studies see the adjacent chapter by Spanagel in this book.

Randomized controlled studies in patients

The effects of a therapy on total abstinence from or a reduction in alcohol consumption are the major determinants of efficacy in alcohol-dependent individuals (Fig. 1). There are a number of ways in which this can be measured (see Kranzler, this book), and several have been used in the published acamprosate trials [6]. *Continuous abstinence,* defined as abstinence from randomization to

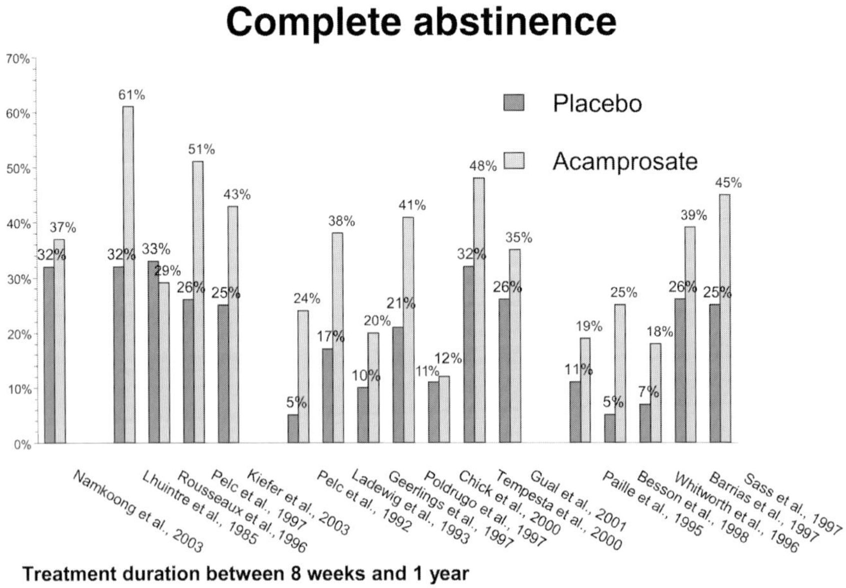

Figure 1. Complete abstinence in 17 randomized controlled trials.

study end, is the most stringent and rigorous of these, and is considered to be the desired treatment goal (Fig. 1) Continuous abstinence also has the advantage of being a simple binary outcome parameter, *viz.*, success or failure. This means that its allocation does not depend on arbitrary definitions of what constitutes a relapse to drinking, or on inter-trial differences in data analysis. However, the required assimilation of patients who exit the trial early, for whatever reason, to treatment failure, reduces the specificity of the measure.

In most of the acamprosate clinical trials, *Cumulative Abstinence Duration* (CAD), which provides an estimate of the proportion of abstinent days over the total length of the trial, has been used as the main efficacy endpoint (Fig. 2). This measure encompasses all patients at all time points during study participation and allows for minor lapses in abstinence without categorical assessment as a treatment failure. However, CAD is a difficult measure to compare between trials, because of lack of standardization in the methods used to collect information on daily alcohol consumption and the methods used to calculate the CAD proportion.

This drug has been evaluated in over 3000 patients in eighteen placebo-controlled clinical trials in Europe and one in South Korea. A further study has been performed in the United States of America, whose results have not yet

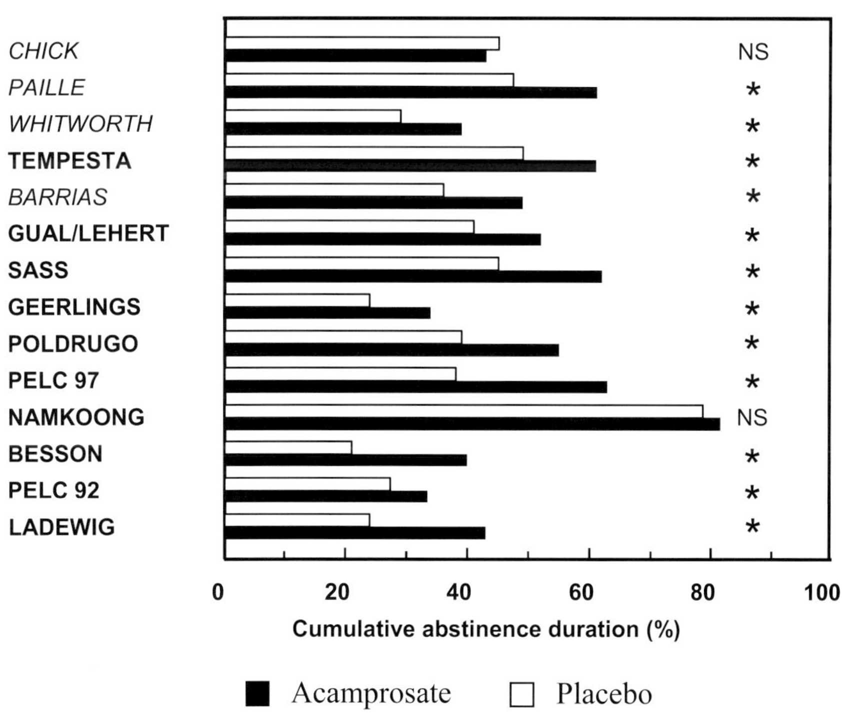

Figure 2. Cumulative abstinence duration (%) in randomized controlled trials (modified after [5]).

been published. Nine of these studies included over one hundred patients per arm. The majority of the studies use the proportion of patients remaining abstinent at the study end (Fig. 1), the cumulative abstinence proportion (i.e., the number of abstinent days expressed as a proportion of total study days (Fig. 2), the time to first drink (Fig. 3), or a combination of these, as the primary outcome variables.

These studies have produced consistent results showing acamprosate treatment to be superior to placebo in maintaining abstinence. In all but three published controlled clinical studies, the proportion of treated patients abstaining at the end of the study was roughly twice as high as for patients receiving placebo. Treatment periods of up to a year were studied. In addition, two studies [16, 17] evaluated long-term abstinence one year after the end of the treatment period and shown the treatment effects to be maintained. Two of the negative studies [18, 19] were small, and could have been underpowered; one also used a two-month treatment period, [19] which was lower than that of all the other studies. The absence of effect in the other negative study by Chick et al. [20], which was the largest of all published trials with acamprosate, may be attributable to the latency in initiating treatment. The study drug was intro-

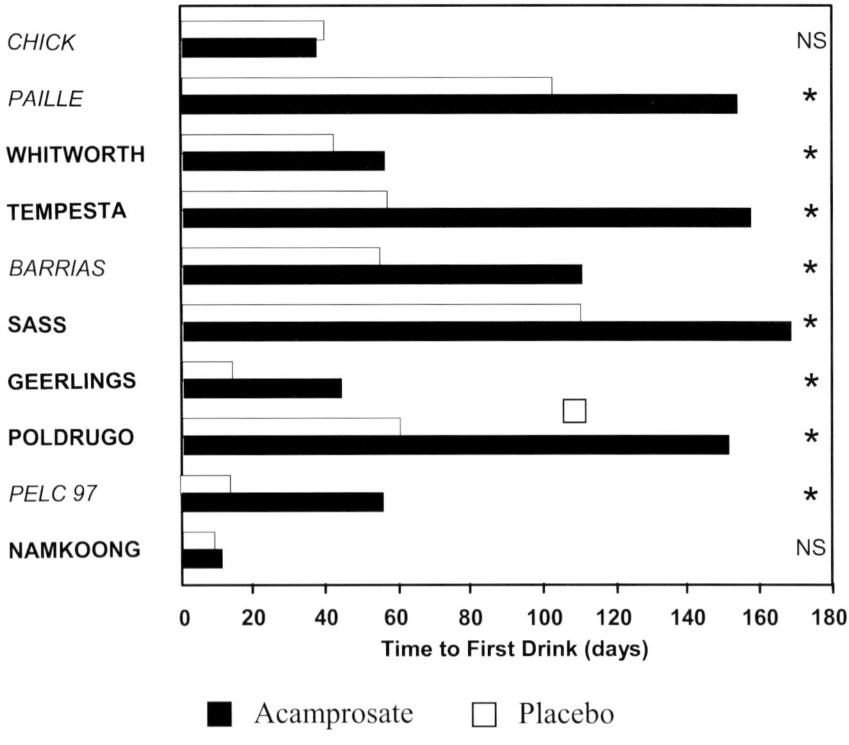

Figure 3. Time to first drink (days) in randomized controlled trials (modified after [6]).

duced after a long stabilisation period (25 days) which followed acute weaning. However, during this period, a substantial proportion of patients had already resumed drinking. This was the only study that used such a design. However, there was some evidence for a reduction of craving in patients treated with acamprosate in this study, as well as in the studies of Pelc et al. [21] and Paille et al. [22].

An evidence-based medicine approach has also been brought to bear on the efficacy data for acamprosate. In the systematic review of Garbutt, which analysed all data available up to 1997 [23], it was concluded that the proof of efficacy for acamprosate was strong, with the most consistent finding being a decrease in drinking frequency, with several studies reporting abstinence rates to be doubled. A literature-based meta-analysis, comparing eleven randomised clinical trials has also concluded that acamprosate is efficacious in reducing alcohol consumption [24]. In a more recent meta-analysis, in which the original clinical trial data from seventeen trials were re-analysed, Mann et al. [6] also concluded that acamprosate was effective, and suggested that the treatment effect could increase with time (see also below).

Five (four earlier and one recent) meta-analyses on the efficacy of acamprosate in the treatment of alcohol dependence have been undertaken, all of which have concluded that acamprosate is effective in maintaining abstinence in detoxified alcohol-dependent individuals [24–28]. However, the first four meta-analyses were limited in their conclusions, as they were based almost exclusively on literature reports.

The aim of the most recent meta-analysis [6] was to undertake a more extensive analysis of the relative benefit of acamprosate in alcohol-dependent individuals using the studies published to date, supplemented, where possible, by data obtained from the manufacturer's in-house reports.

The relative benefit (relative risk) was 1.47% with continuous abstinence rates at 6 months of 36.1% acamprosate and 23.4% placebo. The numbers needed to treat (NNT) for one successful outcome calculated for all 17 studies at 6 months were 7.78 and at 12 months 7.5 (see Fig. 4).

Cost–effectiveness

Data on the cost-effectiveness on different treatment approaches in the alcohol research field are still extremely scarce. First attempts for an economic evaluation of acamprosate treatment compared with placebo showed favourable results for acamprosate [29]. The authors based their analyses primarily on the Austrian double-blind RCT [17]. They came up with net cost saving of 528 €/patient in the 24-month treatment and aftercare period. This would mean a global anticipated net saving of 1.74 million € over two years for the Belgian health insurance system. With reference to the total costs of alcoholism these estimations may represent a significant potential for cost savings [30].

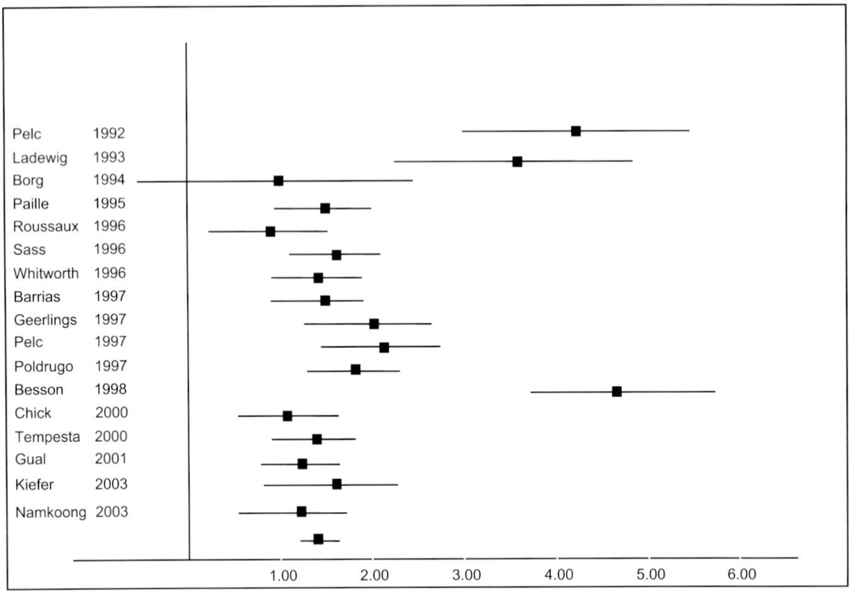

Figure 4. Meta analysis of 17 randomized controlled trials (modified after [6]). Forest diagram of continuous abstinence rates at 6 months and relative benefit ratios and their 95% confidence limits in 17 randomised placebo-controlled trials of acamprosate for the treatment of alcohol dependence determined using the random model of DerSimonian and Laird (1986).

Combining acamprosate and other medications

The effects of acamprosate and naltrexone appear to relate to different aspects of drinking behaviour, with the former stabilising abstinence and the latter decreasing alcohol consumption [5]. There are little direct comparative data on the relative benefits of the two treatments. A recent head-to-head study performed in Spain by Rubio et al. [31] suggested that naltrexone was more efficacious in preventing relapse to heavy drinking. However, the conclusions from this trial should be treated with caution, since the study was not blinded and there was imbalance in the drop-out rate. In another study performed in Germany, which did not have these methodological limitations, a combination of naltrexone and acamprosate was compared to each drug alone and to placebo [32]. Both naltrexone and acamprosate alone increased the time to first drink and the time to first relapse into heavy drinking compared to placebo, and there was no significant difference observed between the effects of the two drugs. The combination of both drugs was about 4 × more effective than placebo. A larger comparative study addressing this issue involving more than 1300 patients, "The COMBINE Research Group" [33], is currently underway in the United States of America. A small (28 patients) short-duration (one month) study in Italy [34] compared acamprosate with fluoxetine and found both agents to reduce alcohol consumption, although the small number of patients involved limit the interpretation of the data. Besson et al. [35] conducted a RCT with acamprosate *versus* placebo. Both groups were stratified for concomitant voluntary use of disulfiram. Acamprosate was significantly better than placebo. Additional disulfiram improved this result further.

Acamprosate and psychotherapy

Although the concomitant use of psychotherapy was not standardised in the early studies of acamprosate, the fact that efficacy was indeed demonstrated suggests that efficacy is manifest irrespective of the psychotherapy regime used. This idea is also supported by a recent open-label study of acamprosate which found that abstinence duration was essentially similar across five different forms of psycho-social support [36], although it should be pointed out that allocation to psychotherapy groups was not randomised. However, a similar result was obtained in another randomized study comparing cognitive behavioural therapy *versus* minimal motivational support [37]. The authors found no additional effect of behavioural therapy over minimal support. So the question remains open of how much and what kind of additional psychotherapy should be offered. The issue is also being addressed in the ongoing COMBINE study [33].

Feeney et al. [38] conducted an interesting study where an established treatment program with Cognitive Behavioural Treatment (CBT) for alcoholics in Australia was compared with the same program and additional administration of acamprosate. Whereas program attendance in both groups was similar, relapse rates occurred significantly sooner and more frequently in the CBT group alone. Alcohol abstinence after 12 weeks was 38% (CBT + acamprosate) compared with 14% (CBT alone)

Conclusions and perspectives

Acamprosate is the most widely validated treatment medication for alcohol-dependent patients. Nevertheless, several questions remain unanswered. (i) The combination issue with naltrexone and other psychotropic medications deserves further attention and study. (ii) Despite the significant benefit, the effect sizes are only in the modest to moderate range. This means that there is certainly room for improvement. One attempt could be to identify potential acamprosate responders *a priori*. Such a study is currently being undertaken in PROJECT PREDICT [39]. Based on the assessment of biological variables, including neuroimaging and neuro-physiological measures as well as questionnaires, the prediction of acamprosate or a naltrexone responder is made. Consequently all 432 patients are randomized to acamprosate, naltrexone or placebo. This study is closely linked to preclinical research, where predictors for treatment response to several drugs are tested [40].

References

1 Boismare F, Daoust M, Moore N, Saligaut C, Lhuintre JP, Chrétien P, Durlach J (1984) an homotaurinate derivative reduces the voluntary intake of ethanol by rats: are cerebral GABA receptors involved? *Pharmacol Biochem Behav* 21: 787–789

2 Le Magnen J, Tran G, Durlach J, Martin C (1987) Dose-dependent suppression of the high alcohol intake of chronically intoxicated rats by Ca-acetyl homotaurinate. *Alcohol Alcoholism* 4: 97–102

3 Lhuintre JP, Daoust M, Moore ND, Chretien P, Saligaut C, Tran G, Boismare F, Hillemand B (1985) Ability of calcium bis acetyl homotaurine, a GABA agonist, to prevent relapse in weaned alcoholics. *Lancet* 1: 1014–1016

4 Hillemand B, Lhuintre JP, Steru L, Boismare F, Daoust M, Moore N et al. (1989) Multicenter trial of calcium bis (acetylhomotaurinate) in treatment of alcohol dependence. *Rev de L'Alcohol* 34: 9–28

5 Mann K (2004) Pharmacotherapy of alcohol dependence: a review of the clinical data. *CNS Drugs* 8: 485–504

6 Mann K, Lehert P, Morgan MY (2004) The efficacy of acamprosate in maintaining abstinence in alcohol-dependent individuals: results of a meta-analysis. *Alcohol Clin Exp Res* 28: 51–63

7 Al Quatari M, Bouchenafa O, Littleton J (1998) Mechanisms of action of acamprosate. Part II. Ethanol dependence modifies effects of acamprosate on NMDA receptor binding in membranes from rat cerebral cortex. *Alcohol Clin Exp Res* 22: 810–814

8 Littleton J, Zieglgansberger W (2003) Pharmacological mechanisms of naltrexone and acamprosate in the prevention of relapse in alcohol dependence. *Am J Addict* 12 (Suppl 1): S3–S11

9 Littleton J (1995) Acamprosate in alcohol dependence: how does it work? *Addiction* 90: 1179–1188

10 Rammes G, Mahal B, Putzke J, Parsons C, Spielmanns P, Pestel E, Spanagel R, Zieglgänsberger W, Schadrack J (2001) The anti-craving compound acamprosate acts as a weak NMDA-receptor antagonist, but modulates NMDA-receptor subunit expression similar to memantine and MK-801. *Neuropharmacology* 40: 749–760

11 Naassila M, Hammoumi S, Legrand E, Durbin P, Daoust M (1998) Mechanisms of action of acamprosate. Part I. Characterization of spermidine-sensitive acamprosate binding site in rat brain. *Alcohol Clin Exp Res* 22: 802–809

12 Popp RL, Lovinger DM (2000) Interaction of acamprosate with ethanol and spermine on NMDA receptors in primary cultured neurons. *Eur J Neurosci* 394: 221–231

13 Harris BR, Prendergast MA, Gibson DA, Rogers DT, Blanchard JA, Holley RC, Fu MC, Hart SR, Pedigo NW, Littleton JM (2002) Acamprosate inhibits the binding and neurotoxic effects of trans-ACPD, suggesting a novel site of action at metabotropic glutamate receptors. *Alcohol Clin Exp Res* 26: 1779–1793

14 Bäckström S, Bachteler D, Koch S, Hyytiä P, Spanagel R (2003) The mGluR5antagonist MPEP prevents alcohol seeking and relapse behavior. *Eur J Neurosci* 29: 921–928

15 Chiamulera C, Epping-Jordan MP, Zocchi A, Marcon C, Cottiny C, Tacconi S, Corsi M, Orzi, F, Conquet F (2001) Reinforcing and locomotor stimulant effects of cocaine are absent in mGluR5 null mutant mice. *Nature Neurosci* 4: 873–874

16 Sass H, Soyka M, Mann K, Zieglgansberger W (1996) Relapse prevention by acamprosate. Results from a placebo-controlled study on alcohol dependence. *Arch Gen Psychiatry* 53: 673–680

17 Whitworth AB, Fischer F, Lesch OM, Nimmerrichter A, Oberbauer H, Platz T, Potgieter A, Walter H, Fleischhacker WW (1996) Comparison of acamprosate and placebo in long-term treatment of alcohol dependence. *Lancet* 347: 1438–1442

18 Rousseaux JP, Hers D, Ferauge M (1996) Does acamprosate influence alcohol consumption of weaned alcoholics? *J Pharm Belg* 51: 65–68

19 Namkoong K, Lee B-O, Lee P-G, Choi M-J, Lee E, and Korean Acamprosate Clinical Trial Investigators (2003) Acamprosate in Korean alcohol-dependent patients: a multi-centre, randomized, double-blind, placebo-controlled study. *Alcohol Alcoholism* 38: 135–141

20 Chick J, Howlett H, Morgan MY, Ritson B for the UKMAS investigators (2000) United Kingdom Multicentre Acamprosate Study (UKMAS): a 6-month prospective study of acamprosate *versus* placebo in preventing relapse after withdrawal from alcohol. *Alcohol Alcoholism* 35: 176–187

21 Pelc I, Verbanck P, Le Bon O, Gavrilovic M, Lion K, Lehert P (1997) Efficacy and safety of acam-

prosate in the treatment of detoxified alcohol-dependent patients. *Br J Psychiat* 170: 73–77
22 Paille FM, Guelfi JD, Perkins AC, Royer RJ, Steru L, Parot P (1995) Double-blind randomized multicentre trial of acamprosate in maintaining abstinence from alcohol. *Alcohol Alcoholism* 30: 239–247
23 Garbutt JC, West S, Carey TS, Lohr KN, Crews FT (1999) Pharmacological treatment of alcohol dependence: a review of the evidence. *JAMA* 281: 1318–1325
24 Kranzler HR, Van Kirk J (2001) Efficacy of naltrexone and acamprosate for alcoholism treatment: a meta-analysis. *Alcohol Clin Exp Res* 25: 1335–1341
25 Hopkins JS, Garbutt JC, Poole CL, West SL, Carey TS (2002) Naltrexone and acamprosate: meta-analysis of two medical treatments for alcoholism. *Alcohol Clin Exp Res* 26 (Suppl): 130A, 751
26 Schoechlin C, Engel RR (2000) Meta-analysis of pharmotherapeutic trials. In: G Zernig, A Saria, M Kurz, S O'Malley (eds): *Handbook of Alcoholism*. Innsbruck University Press, 339–351
27 Slattery J, Chick J, Cochrane M, Craig J, Godfrey C, Macpherson K, Parrott S, Quinn S, Tochel C, Watson H (2003) Prevention of relapse in alcohol dependence. Health Technology Assessment Report 3, Glasgow: Health Technology Board of Scotland
28 Berglund M, Thelander S, Salaspuro M, Franck J, Andreasson S, Ojehagen A (2003) Treatment of alcohol abuse: an evidence-based review. *Alcohol Clin Ex Res* 27: 1645–1656
29 Annemans L, Vanoverbeke N, Tecco J, D'Hooghe D (2000) Economic evaluation of campral (acamprosate) compared to placebo in maintaining abstinence in alcohol-dependent patients. *Eur Addict Res* 6: 71–78
30 Bühringer G, Augustin R, Bergmann E. Bloomfield K, Funk W, Junge B, Kraus L, Merfert-Diete C, Rumpf HJ, Simon R et al (2000) Alkoholkonsum und alkoholbezogene Störungen in Deutschland, Schriftenreihe des Bundesministeriums für Gesundheit Band 128, Nomos, Baden-Baden
31 Rubio G, Jimenez-Arriero MA, Ponce G, Palomo T (2001) Naltrexone *versus* acamprosate: one year follow-up of alcohol dependence of alcohol dependence treatment. *Alcohol Alcoholism* 36: 419–425
32 Kiefer F, Jahn H, Tarnaske T, Helwig H, Briken P, Holzbach R, Kämpf P, Stracke R, Baehr M, Naber D et al. (2003) Comparing and combining naltrexone and acamprosate in relapse prevention of alcoholism: a double-blind, placebo-controlled study. *Arch Gen Psychiatry* 60: 92–99
33 The COMBINE Study Research Group (2003) Testing combined pharmacotherapies and behavioral interventions in alcohol dependence: (The COMBINE Study): a pilot feasibility study. *Alcohol Clin Exp Res* 7: 1107–1122
34 Gerra G, Caccavari R, Delsignore R, Bocchi R, Fertonani G, Passeri M (1992) Effects of fluoxetine and Ca-acetyl-homotaurinate on alcohol intake in familial and non-familial alcoholic patients. *Curr Therap Res Clin Exp* 52: 291–295
35 Besson J, Aeby F, Kasas A, Lehert P, Potgieter A (1998) Combined efficacy of acamprosate and disulfiram in the treatment of alcoholism: a controlled study. *Alcohol Clin Exp Res 22:* 573–579
36 Pelc I, Ansoms C, Lehert P, Fischer F, Fuchs WJ, Landron F, Pires Preto AJ, Morgang MY (2002) The European NEAT program: an integrated approach using acamprosate and psychosocial support for the prevention of relapse in alcohol-dependent patients with a statistical modelling of therapy success prediction. *Alcohol Clin Exp Res* 26: 1529–1538
37 De Wildt WAJM, Schippers GM, Van den Brink W, Potgieter AS, Deckers F, Bets D (2002) Does psychosocial treatment enhance the efficacy of acamprosate in patients with alcohol problems? *Alcohol Alcoholism* 4: 375–382
38 Feeney GF, Young RM, Connor JP, Tucker J, McPherson A (2002) Cognitive behavioural therapy combined with the relapse-prevention medication acamprosate: are short-term outcomes for alcohol dependence improved? *Aust NZ J Psychiatry* 36: 622–628
39 Mann K, Smolka M, Spanagel R, Heinz A. Searching for the acamprosate and naltrexone responders: PROJECT PREDICT. *Alcoholism Clin Exp Res*; *in press*
40 Bachteler D, Economidou D, Danysz W, Ciccocioppo R, Spanagel R (2004) The effects of acamprosate and neramexane on cue-induced reinstatement of ethanol-seeking behavior in rat. *Neuropsychopharmacology*; *in press*

Drugs for Relapse Prevention of Alcoholism
Edited by R. Spanagel and K. Mann

Serotonergic compounds: preclinical data

Anh Dzung Lê[1, 2, 3] and Douglas R. Funk[1]

[1] *Department of Neuroscience, Center for Addiction and Mental Health, Toronto, Ontario, Canada M5S 2S1*
[2] *Departments of Pharmacology and* [3]*Psychiatry, University of Toronto, Toronto, Ontario, Canada M5S 1A8*

Introduction

Evidence for a role of the brain serotonergic system in alcohol dependence was first reported in 1968 by Myers and Veale [1]. Depletion of brain serotonin by treatment with parachlorophenylalanine (pCPA), a serotonin synthesis inhibitor, produced a strong and long-lasting suppression of alcohol consumption in rats. Since that time, the involvement of brain 5-HT in alcohol dependence has been studied extensively. Over the years, a variety of neurobiological approaches have been used to investigate the role of 5-HT in alcohol self-administration. Early work was dominated by studies examining the effects of lesioning the central serotonergic system by the administration of selective neurotoxins such as 5,6- and 5, 7- dihydroxytryptamine (DHT) to destroy 5-HT terminals [2]. These studies were followed by the extensive examination of the effects of raising 5-HT levels with serotonin precursors such as tryptophan or 5-hydroxytryptophan [3] or with selective serotonin reuptake inhibitors (SSRIs) [4]. The vast literature on the effects of 5-HT manipulations on alcohol consumption in experimental animals has been a subject of many excellent reviews [5–7]. In general, it can be said that although there are some inconsistencies in the literature, the preclinical data generated over the last three decades strongly suggest that the brain 5-HT system plays a significant role in alcohol consumption and that manipulations that augment 5-HT function reduce alcohol consumption [5–7].

Over the last few years many advances have been made in understanding the potential mechanisms underlying the role of 5-HT receptors in alcohol consumption. The 5-HT_1, 5-HT_2 and 5-HT_3 receptors and associated subtypes have received the most attention. To a great extent, these advances can be attributed to the development of new agents that have high selectivity for these receptor types and subtypes as well as the availability of transgenic mice with selective deletion of these receptors. The involvement of these 5-HT receptor types and subtypes in alcohol consumption will be the focus of this chapter.

Relapse is the major clinical problem in the treatment of alcohol abuse, as up to one-half of alcoholics relapse shortly after detoxification [7]. Most preclinical studies examining the role of 5-HT in alcohol abuse, however, have focused on the initiation or maintenance of alcohol consumption. The major reason for the absence of studies on relapse was in part due to the absence of appropriate animal models. In addition, there was a common belief that an understanding of the role of 5-HT systems in the initiation and maintenance of alcohol consumption would lead to development of 5-HT-based pharmacotherapies for all aspects of alcohol dependence. Over the last five years, a number of animal models for relapse to alcohol have been developed and some pharmacological studies concerning the effects of 5-HT compounds on relapse to alcohol have been carried out [8].

In this chapter we will first examine the recent preclinical literature on the involvement of 5-HT_1, 5-HT_2, and 5-HT_3 receptor types or subtypes on alcohol consumption. We will then examine existing studies concerning the involvement of the 5-HT system in relapse to alcohol.

5-HT receptor types and subtypes in alcohol consumption

Our understanding of the 5-HT receptor systems has progressed considerably over the last decade. There are seven families of 5-HT receptors, and at least 14 distinct receptor subtypes have been characterized, the majority of which appear to have functional roles in the central nervous system (CNS) [9]. Of these various receptors, the functional involvement of 5-HT_1, 5-HT_2 and 5-HT_3 subtypes in alcohol intake have been examined in detail due to the availability of selective ligands.

5-HT$_{1A}$

The 5-HT_{1A} receptors occur at high densities in 5-HT cell body areas, particularly in the dorsal and the median raphe nuclei where they function as inhibitory autoreceptors. They are also found in various forebrain areas known to be involved in the effects of abused drugs, including the frontal cortex, amygdala, hippocampus and septum where they function as postsynaptic receptors [9, 10]. This distribution is consistent with a role for this receptor in mediating the effects of 5-HT on cortical and limbic structures and also in modulating the activity of 5-HT neurons. Stimulation of 5-HT_{1A} receptors produces complex changes in the activity of 5-HT systems, depending on the doses of drugs employed or where they are administered. For example, peripheral administration of low doses of 8-hydroxy-2-di-n-propylamino tetralin (8-OH-DPAT), a prototype 5-HT_{1A} receptor agonist, or its central injection directly into the median and dorsal raphe, activates 5-HT_{1A} autoreceptors resulting in decreases in 5-HT cell firing and 5-HT release in terminal regions [10].

Evidence for the involvement of 5-HT_{1A} receptors in the regulation of alcohol drinking derives from two experimental strategies: a) evaluation of 5-HT activity and 5-HT_{1A} receptor binding in rats that have been selectively bred for high and low alcohol consumption, and b) pharmacological manipulation of 5-HT_{1A} activity. Lower densities of 5-HT_{1A} receptors were found in the dorsal and the median raphe nuclei of alcohol-preferring P rats compared to alcohol non-preferring NP rats, strains selectively bred for high and low alcohol consumption respectively [11]. This lower density of 5-HT_{1A} autoreceptors in the P rats is consistent with the observation of lower numbers of 5-HT neurons in P compared to NP rats [11].

Treatment with a range of chemically diverse 5-HT_{1A} receptor agonists such as 8-OH-DPAT, ipsapirone or gepirone has been shown to reduce alcohol intake in a variety of experimental procedures used to assess alcohol consumption [12–21]. The most common 5-HT_{1A} receptor agonist, 8-OH-DPAT, in doses ranging from 0.125 to 2.5 mg/kg has been shown to suppress alcohol intake in various animal models of alcohol consumption including two-bottle choice drinking or operant self-administration [12–14, 18, 19] in different strains of rats including those selectively bred for high and low alcohol consumption. Low doses of 8-OH-DPAT (0.03 to 0.06 mg/kg), however, have been shown to stimulate alcohol intake in a limited access drinking procedure when access to alcohol solution was restricted to 60 min per day [13, 14]. Such stimulation of alcohol consumption induced by low doses of 8-OH-DPAT has been attributed to the selective stimulation of 5-HT_{1A} autoreceptors located in the raphe nuclei, as direct infusions of 8-OH-DPAT into the dorsal or median raphe also enhance alcohol intake [12]. The stimulation of alcohol intake induced by 8-OH-DPAT in the dorsal raphe is more selective to ethanol as it did not increase water intake, whereas infusions of 8-OH-DPAT into the median raphe increase both water and alcohol intake. Direct infusion of 8-OH-DPAT into the dorsal raphe, however, has been shown to reduce alcohol intake in the 8 h period following infusion [15]. The discrepancy in these findings might be related to a number of factors ranging from the strain of rat used, the nature of the alcohol drinking procedure and most importantly, the doses of 8-OH-DPAT employed. The effects of 0.1 to 2.5 μg 8-OH-DPAT injected into the dorsal raphe were used to assess alcohol intake in Wistar rats during a 40 min period in the Tomkins et al. study [12], whereas alcohol intake in AA rats selectively bred for high alcohol intake was measured following infusion of 10 μg into the dorsal raphe in the Schreiber et al. study [15].

The highly selective 5-HT_{1A} receptor antagonist WAY 100635 in doses ranging from 0.05 to 0.5 mg/kg has been shown to suppress alcohol intake in female P rats [16]. The suppression of alcohol intake, however, was not dose-dependent, and the effect was not seen at a higher dose of 1 mg/kg. Tomkins and co-workers [17] also found that the 1 mg/kg dose of WAY 100635 did not alter alcohol self-administration in Wistar rats. It is possible that the absence of the effect of WAY 100635 at the high dose might be due to the loss of its specificity at the pre-synaptic 5-HT_{1A} autoreceptor. 8-OH-DPAT or ipsapirone, at doses which reduced operant alcohol self-administration, have also been shown to suppress responding for water or saccharin [18, 19], suggesting that these compounds might reduce alcohol self-administration through a non-specific mechanism. Other studies, however, indicate that 8-OH-DPAT, administered at doses that suppress operant alcohol self-administration [20] or operant responding for a conditioned reinforcer, did not affect locomotor activity or responding on the lever that was not associated with conditioned reinforcement [21], indicating that 8-OH-DPAT might suppress responding for alcohol through a specific mechanism.

The involvement of post-synaptic 5-HT_{1A} receptors in alcohol consumption is unclear. Direct infusions of 5-HT_{1A} agonists to the raphe as well as to other brain structures such as dorsal and ventral hippocampus, amygdala, septum and striatum have been used to dissociate the contribution of pre- and post-synaptic 5-HT_{1A} receptors to their anxiolytic and antidepressant actions [10]. While it is clear that pre-synaptic 5-HT_{1A} receptors can modulate alcohol intake, the nature of the involvement of post-synaptic 5-HT_{1A} receptors in alcohol consumption remains to be determined. To our knowledge, there are no reports in the literature of studies specifically designed to examine the effect of direct infusions of 5-HT_{1A} receptor ligands in limbic brain structures on alcohol intake. Such studies are necessary to clarify the role of post-synaptic 5-HT_{1A} receptors in the regulation of alcohol consumption.

5-HT_{1B}

5-HT_{1B} receptor binding is localized primarily to sites in the basal ganglia, including the ventral pallidum, globus pallidus, substantia nigra and in the limbic system, most notably the dorsal subiculum of the hippocampus [22]. The mRNA for the receptor, on the other hand, is found in the cingulate and entorhinal cortex, nucleus accumbens, striatum, subthalamic nucleus, hippocampus, raphe nuclei and cerebellum [23]. Lesion and electron microscopic studies have localized the 5-HT_{1B} receptors to axons and terminals [23, 24], so it is likely that the mismatch between receptor binding and mRNA reflects the fact that the receptors are transported away from the cell body after synthesis. The localization of the 5-HT_{1B} receptors to terminals is consistent with the results of *in vitro* studies showing that the release of a variety of transmitters is modulated by 5-HT_{1B} receptors, including gamma aminobutyric acid (GABA), substance P, glutamate and acetylcholine [25–28]. 5-HT_{1B} receptors also regulate the release of 5-HT from the terminals of the midbrain raphe projection neurons in the forebrain [29].

A number of approaches provide converging evidence that the 5-HT_{1B} receptor is linked to alcohol dependence. Studies examining the effects of 5-HT_{1B} receptor agonists point to the involvement of 5-HT_{1B} receptors in alcohol preference and intake in laboratory animals. Administration of the mixed 5-$HT_{1B/2C}$ agonist mCPP was reported to reduce both alcohol and water intake in a two-bottle, limited access design [30]. In a more recent study, done using a similar design, mCPP reduced alcohol, water and food intake [31]. These results with mCPP suggest that activation of 5-$HT_{1B/2C}$ receptors may produce global effects on consummatory behaviors. Using a limited access operant design, however, Wilson et al. [20] found that another mixed 5-$HT_{1B/2C}$ agonist, TFMPP, reduced operant responding for alcohol selectively at the lower doses tested; at higher doses, locomotor activity was also decreased. In agreement with these findings, Maurel et al. [31] reported that mCPP and another 5-HT_{1B} agonist, CP-94,253, selectively reduced responding for alcohol. Using a more detailed analysis of the receptors involved, Tomkins and O'Neill [32] found that the mixed agonists at 5-$HT_{1B/1A}$ receptors, CGS12066B and RU24969 reduced operant responding for alcohol. They showed that at lower doses, RU24969 selectively reduced alcohol intake, as these doses did not affect the responding of animals trained to self-administer saline. Importantly, Tomkins and O'Neill [32] demonstrated the dependency of these effects on the 5-HT_{1B} receptor as the selective 5-HT_{1B} antagonist GR127935 blocked the effects of RU24969.

Genetic studies have also provided evidence for the involvement of 5-HT_{1B} receptors in alcoholism, although there are some inconsistent findings. In humans, variations in the gene coding for the 5-HT_{1B} receptor have been linked to alcoholism in a subset of alcoholics with antisocial behavioral traits [33]. In mice, a mapping study showed that a locus associated with alcohol preference was found at the precise location of the 5-HT_{1B} receptor gene [34]. The results of studies on alcohol intake in mice with targeted gene knockouts of the 5-HT_{1B} receptor are controversial. In an initial study, 5-HT_{1B}–/– mice were shown to have greater preference for and spontaneous consumption of alcohol than wild type controls [35]. The same group later failed to replicate this finding [36]. Other studies have also found that 5-HT_{1B}-deficient mice do not differ in these parameters [37]. At present, the reasons for these inconsistencies are not clear.

5-HT_{2A}

5-HT_{2A} receptors are expressed throughout the rat CNS including several regions involved in alcohol- or drug-related behaviors. Studies using receptor autoradiography, immunohistochemistry and *in situ* hybridization have demonstrated that 5-HT_{2A} receptors are expressed at high levels in the frontal cortex, nucleus accumbens, septum, amygdalar nuclei, hippocampus, hypothalamus, ventral tegmental area (VTA) and raphe nuclei [38, 39]. Administration of agonists active at the 5-HT_{2A} receptor such as 2,5-dimethoxy-4-iodoamphetamine (DOI) [40, 41] have been shown to reduce alcohol intake using a 24 h access procedure. Maurel et al. [40] replicated their earlier finding in showing that the 5-HT_{2A} agonist DOI significantly and selectively reduced operant responding for alcohol. Interestingly, antagonists such as amperozide, FG 5974, MDL 100,907, mianserin and ritanserin also reduce alcohol intake [31, 40–42]. A likely reason for this discrepancy is that the drugs have affinity for other subtypes of 5-HT receptors.

Downregulation of 5-HT_{2A} receptors in discrete brain areas with microinjections of antisense have been shown to have significant effects on alcohol intake. Blakley et al. [43] found that chronic infusion of 5-HT_{2A} antisense via osmotic minipumps intracerebroventricularly, or directly into the central amygdala (CeA) significantly reduced alcohol preference and intake in a 24 h access design. Antisense treatment of the prefrontal cortex (PFC), on the other hand, significantly increased alcohol preference and intake. The specificity of these effects is called into question, as changes in saccharin preference in the same direction as the effects on alcohol were observed. Antisense application to the hippocampus or dorsal raphe nucleus did not affect either saccharin or alcohol preference or intake.

5-HT_{2C}

The 5-HT_{2C} receptor is expressed in a number of brain areas related to drug reinforcement including the frontal cortex, septum, hippocampus and amygdala [39, 44]. There are few studies examining the effects of 5-HT_{2C} selective drugs on alcohol preference or consumption. Tomkins et al. [17] showed that the 5-HT_{2C} agonist Ro60-0175 reduced operant responding for alcohol, and that this effect was blocked by the selective 5-HT_{2C} antagonist SB242,084. The antagonist also blocked the reduction of responding for alcohol induced by the 5-HT releaser dexfenfluramine. SB242,084 alone increased responding for alcohol. These compounds also had effects on blood alcohol levels, with the agonist decreasing and the antagonist increasing alcohol concentration.

5-HT_3

Of the 5-HT receptors, the 5-HT_3 receptor is the only one linked to an ion-gated channel. 5-HT_3 receptor binding is found at low levels throughout cortical and subcortical structures as revealed by autoradiographic mapping techniques [45, 46]. *In situ* hybridization and immunohistochemical analysis reveal that 5-HT_3 receptors are expressed in a large proportion of GABA neurons that co-express cholycystokinin immunoreactivity [45]. 5-HT_3 receptors have been shown to regulate the activity of the mesolimbic dopamine system at the level of the VTA as well as the nucleus accumbens [47, 48]. 5-HT_3 receptors have been shown to be involved in many actions of alcohol [46] and currently drugs acting at the 5-HT_3 receptors are targeted as potential pharmacotherapeutic agents for the treatment of alcohol abuse [7].

Preclinical studies examining the effects of various 5-HT_3 receptor antagonists on alcohol consumption have revealed a conflicting pattern. A number of studies have found that administration of 5-HT_3 receptor antagonists such as MDL 72222, ondansetron, zacopride and tropisetron reduce alcohol consumption in rats in a two-bottle choice paradigm [49–54]. The reduction of alcohol consumption induced by 5-HT_3 receptor antagonists has also been observed in other species such as marmosets [54] and mice [53]. While consistent suppression of alcohol consumption induced by a 5-HT_3 receptor antagonist was observed when alcohol consumption was measured in a continuous access, 2-bottle choice design [49–54], conflicting findings were obtained under restricted access conditions, whether or not two-bottle choice [51, 53] or operant procedures were employed [51, 55]. It has been suggested that temporal environmental cues associated with the presentation of alcohol in the limited access drinking conditions might reduce the effectiveness of 5-HT_3 receptor antagonists, as it was shown that the suppression of alcohol consumption by a 5-HT_3 receptor antagonist did not occur when access to alcohol was presented in a random manner [56].

The effects of 5-HT_3 receptor agonists on alcohol consumption have also been investigated. Intraventricular or intra-accumbens injection of the 5-HT_3 receptor agonist 2-methyl-5-HT or m-chlorophenylbiguanide (mCPBG) reduced alcohol consumption without affecting food or water intake in rats [57]. Transgenic mice over-expressing the 5-HT_3 receptors also show reduced alcohol consumption compared to controls [58]. The 5-HT_3 transgenic mice have also been shown to be more sensitive to the effects of ethanol compared to controls and it has been suggested that such differences might account for their low alcohol consumption [59]. Whether the suppression of alcohol intake induced by central administration of 5-HT_3 receptor agonists is due to an enhancement of sensitivity to the effects of ethanol remains to be determined.

5-HT_3 receptor antagonists have been shown to attenuate or block the stimulation of dopamine (DA) release in the nucleus accumbens or somatodendritic DA release in the VTA induced by ethanol [11, 46]. The antagonism of ethanol-stimulated mesolimbic DA activity by 5-HT_3 receptor antagonists has been suggested as the primary mechanism underlying their attenuation of alcohol intake [7, 11, 46]. In support of this, it was observed that in 6-OHDA lesioned rats, the suppression of alcohol intake induced by tropisetron was markedly reduced [60]. A recent study by Rodd-Henricks et al. [61] demonstrated, furthermore, that intracranial self-administration of ethanol in the posterior VTA by Wistar rats can be blocked by co-administration of a 5-HT_3 receptor antagonist.

5-HT and relapse to alcohol

As mentioned earlier, most of the neurobiological investigation into the mechanisms underlying alcohol dependence has focused on the neurochemical mechanisms underlying alcohol reinforcement with little attention being paid to those underlying relapse [8]. Knowledge of the mechanisms underlying drug reinforcement, however, might not be sufficient to explain why individuals are still vulnerable to relapse after a prolonged period of abstinence [62].

The preclinical investigation into the role of serotonergic mechanisms in alcohol dependence is no exception. Although over the last number of years, there have been many advances in the development of animal models to study relapse to alcohol, there has been little attempt to evaluate the effects of serotonergic agents in these models.

Animal models of relapse to alcohol

Currently there are two animal models that have been developed to study relapse to alcohol in experimental animals [8]. The first model utilizes the alcohol deprivation effect (ADE), which is described as the temporary increase in alcohol consumption following a short period of abstinence [63]. The magnitude of the ADE effects increases with repeated cycles of deprivation [64, 65]. The increase in alcohol-taking after abstinence and the increase in the intensity of ADE after repeated cycles of deprivation parallel the effects of repeated deprivation seen on the priming effects of alcohol in humans [8]. The second model that has been used to assess relapse to alcohol is the reinstatement procedure [8]. In this model, animals are trained to lever press for alcohol, the operant response is extinguished and the effects of acute exposure to alcohol [66–68], cues previously associated with alcohol [69] or stressors [67, 68] on the reinstatement of the lever pressing response are examined.

While these two models have been used quite extensively over the last several years to evaluate the effects of various compounds, particularly those acting at opioidergic or corticotrophin-releasing factor receptors [8] on relapse to alcohol, little work has been conducted with serotonergic agents. To our knowledge there has been only one study that examined the effects of 5-HT on relapse to alcohol using the ADE model. Rodd-Henricks et al. [65] found that the 5-HT_3 receptor antagonists MDL 7222 or tropisetron can suppress ADE. The doses of these compounds required to suppress ADE, however, were much higher than those required to reduce baseline alcohol consumption. We have also investigated the effects of serotonergic agents on relapse to alcohol using the reinstatement procedure. In our early studies [68, 69], we have shown that treatment with fluoxetine can attenuate reinstatement of alcohol seeking induced by exposure to foot shock stress, but not by that induced by priming with alcohol. Recently, we have found that paroxetine, another serotonin re-uptake inhibitor, or the 5-HT_3 receptor antagonist tropisetron can also attenuate the reinstatement of alcohol seeking induced by foot shock stress [70].

Conclusions

Our understanding of the role of the serotonergic systems in alcohol dependence has advanced greatly over the last several years. While the availability of highly selective ligands for the various types of 5-HT receptors have led to a clearer picture of the function of the 5-HT system in alcohol intake, one should bear in mind that 5-HT receptor systems interact with one another to regulate behaviors including alcohol intake. A case in point is the elegant study by Haiser and Tescott [71], which employed 5-HT_{2C} knock out mice and a selective 5-HT_{1B} receptor antagonist to evaluate the effect of the 5-$HT_{1B/2C}$ receptor agonist mCPP on locomotor activity. These workers demonstrated that preferential activation of the 5-HT_{2C} receptor can prevent the expression of behaviors mediated by other 5-HT receptor types or subtypes. This work underscores the idea that a combination of experimental strategies is required to elucidate the nature of the 5-HT receptor types modulating alcohol intake.

While little work has been carried out examining the involvement of 5-HT systems in relapse, the existing data suggest that the 5-HT mechanisms mediating relapse and alcohol consumption can be dissociated, as evidenced by the differential dose requirement [65] or the fact that they have differential effects on the reinstatement of ethanol-seeking induced by stressors and priming [68].

Acknowledgements
Supported by grant AA13108 from the NIAAA to Dr. A.D. Lê.

References

1 Myer RD, Veale WL (1968) Alcohol preference in the rat: reduction following depletion of brain serotonin. *Science* 160: 1469–1471
2 Melchior CL, Myers RD (1976) Genetic differences in ethanol drinking of the rat following injection of 6-OHDA, 5,6-DHT or 5,7-DHT into the cerebral ventricles. *Pharmacol Biochem Behav* 5: 63–72
3 Zabik JE, Sprague JE, Binkerd J (1994) Central and peripheral components of the inhibitory actions of 5-HTP on ethanol consumption in the rat. *Pharmacol Biochem Behav* 47: 547–551
4 Gill C, Amit Z (1989) Serotonin re-uptake inhibitors and alcohol consumption. A review of recent studies. *Rec Dev Alcohol* 7: 225–248
5 Sellers EM, Higgins GA, Sobell MB (1992) 5-HT and alcohol abuse. *Trends Pharmacol Sci* 13: 69–75
6 LeMarquand D, Pihl RO, Benkelfat C (1994) Serotonin and alcohol intake, abuse, and dependence: findings of animal studies. *Biol Psychiatry* 36: 395–421
7 Johnson BA, Ait-Daoud N (2000) Neuropharmacological treatments for alcoholism: scientific basis and clinical findings. *Psychopharmacology* 149: 327–344
8 Lê AD, Shaham Y (2000) Neurobiology of relapse to alcohol in rats. *Pharmacol Ther* 94: 137–156
9 Hoyer D, Hannon JP, Martin GR (2002) Molecular, pharmacological and functional diversity of 5-HT receptors. *Pharmacol Biochem Behav* 71: 533–554
10 De Vry J (1995) 5-HT_{1A} receptor agonists: recent developments and controversial issues, *Psychopharmacology* 121: 1–26
11 McBride WJ, Li T-K (1998) Animal models of alcoholism: neurobiology of high alcohol drinking behavior in rodents. *Critical Rev Neurobiol* 12: 339–369
12 Tomkins DM, Sellers EM, Fletcher PJ (1994) Median and dorsal raphe injections of the 5-HT_{1A} receptor agonist, 8-OH-DPAT, and the GABAA agonist Muscimol, increase voluntary ethanol intake in Wistar rats. *Neuropharmacology* 33: 349–358
13 Ciccocioppo R, Panocka I, Polidori C, Dourish CT, Massi M (1997) Blockade of pre- and post-synaptic 5-HT_{1A} receptors does not modify the effect of fluoxetine or 5-hydroxytryptophan on ethanol and food intake in rats. *Psychopharmacology* 134: 55–63
14 Tomkins DM, Higgins GA, Sellers EM (1994) Low doses of the 5-HT1A agonist 8-hydroxy-2-

(di-n-propylamino)-tetralin (8-OH DPAT) increase ethanol intake. *Psychopharmacology* 115: 173–179
15 Schreiber R, Opitz K, Glaser T, De Wry J (1993) Isapirone and 8-OH-DAPT reduce ethanol preference in rats: involvement of presynatic 5-HT$_{1a}$ receptors. *Psychopharmacology* 112: 100–110
16 Zhou FC, McKinzie DL, Patel TD, Lumeng L, Li TK (1998) Additive reduction of alcohol drinking by 5-HT1A antagonist WAY 100635 and serotonin uptake blocker fluoxetine in alcohol-preferring P rats. *Alcohol Clin Exp Res* 22: 266–269
17 Tomkins DM, Joharchi N, Tampakeras M, Martin JR, Wichmann J, Higgins GA (2002) An investigation of the role of 5-HT(2C) receptors in modifying ethanol self-administration behaviour. *Pharmacol Biochem Behav* 71: 735–744
18 Roberts AJ, McArthur RA, Hull EE, Post C, Koob GF (1998) Effects of amperozide, 8-OH-DPAT, and FG 5974 on operant responding for ethanol. *Psychopharmacology* 137: 25–32
19 Schreiber R, Manze B, Haussels A, De Vry J (1999) Effects of the 5-HT$_{1A}$ receptor agonist ipsapirone on operant self-administration of ethanol in the rat. European. *Neuropsychopharmacology* 10: 37–42
20 Wilson AW, Neill JC, Costall B (1997) An investigation into the effects of 5-HT agonists and receptor antagonists on ethanol self-administration in the rat. *Alcohol* 16: 249–270
21 Wilson AW, Costall B, Neill JC (2000) Manipulation of operant responding for an ethanol-paired conditioned stimulus in the rat by pharmacological alteration of the serotonergic system. *J Psychopharmacol* 14(4): 340–346
22 Sari Y, Miquel MC, Brisorgueil MJ, Ruiz G, Doucet E, Hamon M, Verge D (1999) Cellular and subcellular localization of 5-hydroxytryptamine1B receptors in the rat central nervous system: immunocytochemical, autoradiographic and lesion studies. *Neuroscience* 88: 899–915
23 Bruinvels AT, Landwehrmeyer B, Gustafson EL, Durkin MM, Mengod G, BranchekTA, Hoyer D, Palacios JM (1994) Localization of 5-HT1B, 5-HT1D alpha, 5-HT1E and 5-HT1F receptor messenger RNAin rodent and primate brain. *Neuropharmacology* 33: 367–386
24 Boschert U, Amara DA, Segu L, Hen R (1994) The mouse 5-hydroxytryptamine1B receptor is localized predominantly on axon terminals. *Neuroscience* 58: 167–182
25 Arvieu L, Mauborgne A, Bourgoin S, Oliver C, Feltz P, Hamon M, Cesselin F (1996) Sumatriptan inhibits the release of CGRP and substance P from the rat spinal cord. *Neuroreport* 12: 1973–1976
26 Boeijinga PH, Boddeke HW (1996) Activation of 5-HT1B receptors suppresses low but not high frequency synaptic transmission in the rat subicular cortex *in vitro*. *Brain Res* 721: 59–65
27 Johnson SW, Mercuri NB, North RA (1992) 5-hydroxytryptamine1B receptors block the GABAB synaptic potential in rat dopamine neurons. *J Neurosci* 12: 2000–2006
28 Maura G, Raiteri M (1986) Cholinergic terminals in rat hippocampus possess 5-HT1B receptors mediating inhibition of acetylcholine release. *Eur J Pharmacol* 129: 333–337
29 Engel G, Gothert M, Hoyer D, Schlicker E, Hillenbrand K (1986) Identity of inhibitory presynaptic 5-hydroxytryptamine (5-HT) autoreceptors in the rat brain cortex with 5-HT1B binding sites. *Naunyn Schmiedebergs Arch Pharmacol* 332: 1–7
30 Buczek Y, Tomkins DM, Higgins GA, Sellers EM (1994) Dissociation of serotonergic regulation of anxiety and ethanol self-administration: a study with mCPP. *Behav Pharmacol* 5: 470–484
31 Maurel S, De Vry J, De Beun R, Schreiber R (1999a) 5-HT2A and 5-HT2C/5-HT1B receptors are differentially involved in alcohol preference and consummatory behavior in cAA rats. *Pharmacol Biochem Behav* 62: 89–96
32 Tomkins DM, O'Neill MF (2000) Effect of 5-HT(1B) receptor ligands on self-administration of ethanol in an operant procedure in rats. *Pharmacol Biochem Behav* 66: 129–136
33 Lappalainen J, Long JC, Eggert M, Ozaki N, Robin RW, Brown GL, Naukkarinen H, Virkkunen M, Linnoila M, Goldman D (1998) Linkage of antisocial alcoholism to the serotonin 5-HT1B receptor gene in 2 populations. *Arch Gen Psychiatry* 55: 989–994
34 Phillips TJ, Belknap JK, Buck KJ, Cunningham CL (1998) Genes on mouse chromosomes 2 and 9 determine variation in ethanol consumption. *Mamm Genome* 9: 936–941
35 Crabbe JC, Phillips TJ, Feller DJ, Hen R, Wenger CD, Lessov CN, Schafer GL (1996) Elevated alcohol consumption in null mutant mice lacking 5-HT1B serotonin receptors. *Nat Genet* 1996: 98–101
36 Crabbe JC, Wahlsten D, Dudek BC (1999) Genetics of mouse behavior: interactions with laboratory environment. *Science* 284: 1670–1672
37 Risinger FO, Doan AM, Vickrey AC (1999) Oral operant ethanol self-administration in 5-HT1b knockout mice. *Behav Brain Res* 102: 211–215

38 Cornea-Hebert V, Riad M, Wu C, Singh SK, Descarries L (1999) Cellular and subcellular distribution of the serotonin 5-HT2A receptor in the central nervous system of adult rat. *J Comp Neurol* 409: 187–209
39 Pompeiano M, Palacios JM, Mengod G (1994) Distribution of the serotonin 5-HT2 receptor family mRNAs: comparison between 5-HT2A and 5-HT2C receptors. *Brain Res Mol Brain Res* 23: 163–178
40 Maurel S, De Vry J, Schreiber R (1999) 5 HT receptor ligands differentially affect operant oral self-administration of ethanol in the rat. *Eur J Pharmacol* 370: 217–223
41 Pohorecky LA, Blakely CG, Southard H, Benjamin DA (1998) 5-HT2A receptors in the prefrontal cortex tonically inhibit alcohol intake in the rat. *Alcohol Clin Exp Res* 22: A85
42 Meert TF, Awouters F, Niemegeers CJ, Schellekens KH, Janssen PA (1991) Ritanserin reduces abuse of alcohol, cocaine, and fentanyl in rats. *Pharmacopsychiatry* 24: 159–163
43 Blakley GG, Pohorecky LA, Benjamin D (2001) Bidirectional changes in ethanol consumption in rats with site-specific antisense down-regulation of 5-hydroxytryptamine2A receptors in brain. *J Pharmacol Exp Ther* 299: 277–289
44 Abramowski D, Rigo M, Duc D, Hoyer D, Staufenbiel M (1995) Localization of the 5-hydroxytryptamine2C receptor protein in human and rat brain using specific antisera. *Neuropharmacology* 34: 1635–1645
45 Bloom F, Morales M (1998) The central 5-HT3 receptor in CNS disorders. *Neurochem Res* 23: 653–659
46 Campbell AD, McBride W (1995) Serotonin-3 receptor and ethanol-stimulated dopamine release in the nucleus accumbens. *Pharmacol Biochem Behav* 51: 835–842
47 Campbell AD, Korr KK, McBride WJ (1996) Serotonin-3 receptor and ethanol-stimulated somatodendritic dopamine release. *Alcohol* 13: 569–574
48 Lovinger DM (1999) 5-HT3 receptors and the neural actions of alcohol: an increasingly exciting topic. *Neurochem Int* 35: 125–130
49 Fadda F, Garrau B, Marchei F, Columbo G Gessa GL (1991) MDL 72222, a selective 5-HT3 receptor antagonist, suppresses voluntary alcohol consumption in alcohol-preferring rats. *Alcohol Alcoholism* 26: 107–110
50 Knapp DJ, Pohorecky LA (1992) Zacopride, a 5-HT_3 receptor antagonist, reduces voluntary ethanol consumption in rats. *Pharmacol Biochem Behav* 41: 847–850
51 Hodge CW, Samson HH, Lewis RS, Erickson HL (1993) Specific decreases in ethanol-but not water-reinforced responding produced by 5-HT3 antagonist ICS 205-930. *Alcohol* 10: 191–196
52 Dyr W, Kowtowski W (1995) Evidence that the amygdala is involved in the inhibitory effects of 5-HT3 receptor antagonists on alcohol drinking in rats. *Alcohol* 12: 387–391
53 Tomkins DM, Lê AD, Dsellers EM (1995) Effect of the 5-HT3 antagonist ondansetron on voluntary ethanol intake in rats and mice maintained on a limited access procedure. *Psychopharmacology* (Berl) 117: 479–485
54 Oakley NR, Jones BJ, Tyers MB, Costall B, Domeney AM (1988) The effects of GR 38032f on alcohol comsumption in the marmoset. *Br J Pharmacol* 95: 870P
55 Beardsley PM, Lopez OT, Gullikson G, Flynn D (1994) Serotonin 5-HT_3 antagonists fail to affect ethanol self-administration of rats. *Alcohol* 11: 389–395
56 McKinzie DL, McBride WJ, Murphy JM, Lumeng L, Li TK (2000) Effect of MDL 72222, a serotonin 3 antagonist on operant responding for ethanol by alcohol-preferring P rats. *Alcohol Clin Exp Res* 24: 1500–1504
57 Dyr W, Kostowski W (1997) Effects of 5-HT_3 receptor agonists on voluntary ethanol intake in rats maintained on a limited access procedure. *Alcohol Alcoholism* 2: 455–462
58 Engel SR, Lyons CR, Allan AM (1998) 5-HT_3 receptor over-expression decreases ethanol self administration in transgenic mice. *Psychopharmacology* 140: 243–248
59 Engel SR, Allan AM (1999) 5-HT_3 receptor over-expression enhances ethanol sensitivity in mice. *Psychopharmacology* 144: 411–415
60 Jankowska E, Bidzinski A, Kowstowski W (1995) Alcohol drinking in rats injectyed ICV with 6-OHDA: effect of 8-OHDPAT and tropisetron. *Alcohol* 12: 121–126
61 Rodd-Henricks ZA, McKinzie DL, Melendez RI, Berry N, Murphy JM, McBride WJ (2003) Effects of serotonin-3 receptor antagonists on the intracranial self-administration of ethanol within the ventral tegmental area of Wistar rats. *Psychopharmacology* 165: 252–259
62 Berke JD, Hyman SE (2000) Addiction, dopamine, and the molecular mechanisms of memory. *Neuron* 25: 515–532

63 Sinclair JD, Seneter RJ (1967) Increased preference for ethanol in rats following alcohol deprivation. *Psychonom Sci* 8: 11–16
64 Spanagel R, Holter SM (2000) Pharmacological validation of a new animal model for alcoholism. *J Neural Transm* 107: 669–680
65 Rodd-Henricks ZA, McKinzie DL, Edmundson DE, Dagon CL, Murphy JM, McBride WJ, Lumeng L, LiTK (2000) Effects of 5-HT3 receptor antagonists on daily ethanol intake under acquisition, maintenance, and relapse conditions in alcohol-preferring P rats. *Alcohol* 21: 73–85
66 Chiamulera C, Valerio E, Tessari M (1995) Resumption of ethanol-seeking behaviour in rats. *Behav Pharmacol* 6: 32–39
67 Katner SN, Weiss F (1999) Ethanol-associated olfactory stimuli reinstate ethanol-seeking behavior after extinction and modify extra-cellular dopamine levels in the nucleus accumbens. *Alcohol Clin Exp Res* 23: 1751–1760
68 Lê AD, Quan B, Juzytsch W, Joharchi N, Shaham Y (1998) Reinstatement of alcohol-seeking by priming injections of alcohol and exposure to stress in rats. *Psychopharmacology* 135: 168–174
69 Lê AD, Poulos CX, Harding S, Watchus J, Juzytsch W, Shaham Y (1999) Effects of naltrexone and fluoxetine on alcohol self-administration and reinstatement of alcohol seeking induced by priming injections with alcohol and exposure to stress in rats. *Neuropsychopharmacology* 21: 435–444
70 Lê AD, Harding S, Juzytsch W, Funk D, Shaham Y (2003) Stress-induced reinstatement of alcohol seeking: Involvement of 5-HT system. *Alcohol Clin Exp Res* 27 (suppl): Abst. 743
71 Haiser LK, Tescott LH (2000) A paradoxical locomotor response in serotonin 5-HT2C receptor mutant mice. *J Neurosci* 20 (RC71): 1–5

Drugs for Relapse Prevention of Alcoholism
Edited by R. Spanagel and K. Mann

Serotonergic compounds: clinical data

Bankole A. Johnson

South Texas Addiction Research & Technology Center, Departments of Psychiatry and Pharmacology, The University of Texas Health Science Center at San Antonio, 3939 Medical Drive, Suite 100, San Antonio, Texas 78229-3900, USA, and Departments of Psychiatric Medicine and Neuroscience, University of Virginia, P.O. Box 800623, Charlottesville, VA 22908-0623, USA

Introduction

In the last two decades, serotonergic agents have been studied intensely as potential treatments for alcoholism. Basic science studies have implicated the serotonin (5-HT) system, and especially its interaction with midbrain and cortico-mesolimbic dopamine pathways, as being central to the expression and appreciation of alcohol's reinforcing effects associated with its abuse liability. Such studies have increased our understanding of the bio-behavioral basis of alcoholism and provided the foundation for testing medications that interact with specific 5-HT receptor subtypes as treatment for particular types of alcoholic both with and without co-morbid psychiatric disorder.

The sections of this chapter are organized, first, to briefly provide an overview of the scientific rationale for testing a particular class of serotonergic medications for treating alcoholism and, subsequently, to evaluate the reported efficacy of these agents in clinical studies.

Selective serotonin reuptake inhibitors

Ethanol preference can be altered by lesions or pharmacological manipulations that result in 5-HT depletion and reduced 5-HT turnover [1, 2]. Augmentation of serotonergic function by administration of selective serotonin reuptake inhibitors (SSRIs) decreases voluntary ethanol consumption in preference paradigms [3–7]. SSRI-associated suppression of ethanol consumption may be due to a decrease in the reinforcing effects of alcohol as well as an effect of SSRIs to reduce consummatory behavior [8–10]. Operant studies show that the SSRI, fluoxetine, dose-dependently decreases ethanol responding [11, 12]. Hence, the results of basic science studies suggest that SSRIs would be efficacious in treating alcohol dependence.

Clinical data on the utility of SSRIs as a single agent for the treatment of alcoholics without major depressive disorder have been inconsistent. Early clinical trials of small sample size reported that SSRIs could produce short-term (1–4 weeks) reductions in alcohol consumption among problem drinkers [13–17]. Nevertheless, these studies had three additional limitations. First, the cohort was predominantly male [13–15], thereby reducing the generalizability of the results to women. Second, standardized psychosocial treatments, which have a tendency to diminish the effect size of the response to pharmacotherapy, were used infrequently; hence, non-specific factors might be responsible for at least some of the clinical improvement that was observed. Third, the short treatment period did not allow for determination of whether SSRI-related clinical improvement in problem drinkers was sustained. Studies that have employed longer SSRI treatment periods have not demonstrated efficacy. For example, Gorelick and Paredes [18] showed that an initial dramatic decrease in alcohol consumption (approx. 15%) during the first of four weeks in a clinical trial did not result in an overall therapeutic treatment response for fluoxetine compared with placebo over the entire trial period. Naranjo and coworkers [19] also were unable to show differences in alcohol consumption between the SSRI, citalopram (40 mg/day), and placebo after 12 weeks of treatment. Kabel and Petty [20] also did not find 12 weeks of fluoxetine (60 mg/day) treatment to be superior to placebo at reducing drinking among 28 alcohol-dependent men. Furthermore, Kranzler and colleagues [21], in a well-executed 12-week study, did not find fluoxetine 60 mg/day to be superior to placebo in treating alcoholism. Interestingly, predicated on the findings of human laboratory studies that alcoholics with an early onset of disease may have reduced 5-HT levels [22–25], Kranzler and colleagues [26] re-analyzed their data. Presumably, their premise was that if type B alcoholics, or those who develop the disease early in life, were deficient in 5-HT, then SSRIs would be expected to be particularly beneficial for this alcoholic subtype. Paradoxically, fluoxetine appeared to actually worsen rather than improve the clinical benefit of the adjunctive cognitive behavioral therapy, and it was certainly no better than placebo. Interestingly, Pettinati and colleagues [27] have shown that sertraline (another SSRI) appears to improve the drinking outcomes of type A alcoholics, or those who develop the disease later in life. Early-onset (i.e., type B-like) alcoholics differ from late-onset (i.e., type A-like) alcoholics by having greater family history and greater propensity toward impulse dyscontrol, serotonergic dysfunction, and antisocial behaviors [28, 29]. It is, therefore, tempting to speculate that the relationship between 5-HT dysfunction and the onset of alcoholism might not be a simple deficiency state but may be due to more fundamental biological differences within the 5-HT system [30, 31].

Although outside the scope of this chapter, SSRIs may be useful in treating alcoholics with severe co-morbid major depression [32]; however, further studies are needed to fully characterize these results, given that tricyclic antidepressants provided to similar populations in other studies have been shown to reduce the dysphoric symptoms with little effect on drinking behavior [33, 34].

In conclusion, SSRIs are not efficacious at improving the drinking outcomes of a heterogeneous alcoholic group. SSRIs may, however, be useful as treatment for alcoholics who develop the disease later in life, or alcoholics with co-morbid major depression.

5-HT_1 partial agonists

The 5-HT_{1A} partial agonist, buspirone, reduces ethanol consumption in a variety of animal paradigms. For instance, buspirone decreases volitional ethanol consumption by up to 60% in macaque monkeys [35]. Also, buspirone suppresses ethanol intake in Sprague-Dawley rats induced to drink by repeated brain-stem injection of tetrahydropapaveroline [36]. Dose-specific effects of buspirone to suppress ethanol intake also have been reported among medium alcohol-preferring rats [37]. Further, buspirone decreases stimulus-conditioned responding for ethanol [38]. Results of studies conducted with relatively 5-HT-deficient Fawn-Hooded rats show that they have a high propensity toward ethanol consumption [39]; correspondingly, alcohol-preferring rats, compared with their non-alcohol-preferring counterparts, have reduced cortical and subcortical gray matter reductions in 5-HT [40, 41 c.f. 42]. Hence, enhancing 5-HT neurotransmission should decrease ethanol consumption. 5-HT_{1A} receptors manifest differential sensitivity, the greater effect being seen at the postsynaptic receptor compared with the autoreceptor. Chronic buspirone administration, therefore, raises 5-HT function and down-regulates autoreceptor function [43]. Despite the paucity of operant dose-response studies examining buspirone's effects on ethanol consumption, there appears to be ample foundation for testing its efficacy as a treatment agent for alcoholism.

Generally, the results of clinical trials do not support a finding of efficacy for buspirone in treating alcoholics without concurrent co-morbid anxiety disorder. Indeed, Malcolm and colleagues [44] did not find that buspirone was superior to placebo at decreasing either the symptoms of anxiety or drinking among anxious alcoholics. In a similarly well-executed trial ($N = 61$), Kranzler and co-workers [45] did not find buspirone to be an efficacious treatment for alcoholism. An authoritative review of five published studies and a more recent re-evaluation of the data [46] also did not support the utility of buspirone in treating alcoholics; however, in contrast to the study of Kranzler and colleagues [45], alcoholics with co-morbid anxiety disorder appeared to derive therapeutic benefit [47, 48]. Finally, George and colleagues [49] showed that the younger the age of onset, the lower was the cerebrospinal fluid level of 5-HT's major metabolite, 5-hydroxyindole-acetic acid (5-HIAA). Despite the likelihood of 5-HT facilitation by chronic buspirone treatment, buspirone was not superior to placebo at improving the drinking outcomes of early-onset alcoholics. Again, this demonstrates that the apparent 5-HT dysfunction among alcoholics early in life may have complex bio-genetic underpinnings.

In sum, buspirone does not appear to be an effective treatment for alcoholics without co-morbid disease. Buspirone may, however, be useful in treating alcoholics with co-morbid anxiety disorder.

188–190
72 Costall B, Domeney AM, Naylor RJ, Tyers MB (1987) Effects of the 5-HT3 receptor antagonist, GR38032F, on raised dopaminergic activity in the mesolimbic system of the rat and marmoset brain. *Br J Pharmacol* 92: 881–894
73 Hodge CW, Samson HH, Lewis RS, Erickson HL (1993) Specific decreases in ethanol- but not water-reinforced responding produced by the 5-HT3 antagonist ICS 205-930. *Alcohol* 10: 191–196
74 Fadda F, Garau B, Marchei F, Colombo G, Gessa GL (1991) MDL 72222, a selective 5-HT3 receptor antagonist, suppresses voluntary ethanol consumption in alcohol-preferring rats. *Alcohol Alcoholism* 26: 107–110
75 Rodd-Henricks ZA, McKinzie DL, Li T-K, Crile RS, Murphy JM, McBride WJ (1999) Intracranial self-administration of ethanol into the posterior VTA of Wistar rats is mediated by 5-HT3 receptors [abstract]. *Alcohol Clin Exp Res* 23(suppl 5): 49A
76 McBride WJ, Li TK (1998) Animal models of alcoholism: neurobiology of high alcohol-drinking behavior in rodents. *Crit Rev Neurobiol* 12: 339–369
77 Tomkins DM, Le AD, Sellers EM (1995) Effect of the 5-HT3 antagonist ondansetron on voluntary ethanol intake in rats and mice maintained on a limited access procedure. *Psychopharmacology* 117: 479–485
78 Barnes NM, Sharp T (1999) A review of central 5-HT receptors and their function. *Neuropharmacology* 38: 1083–1152
79 Beardsley PM, Lopez OT, Gullikson G, Flynn D (1994) Serotonin 5-HT3 antagonists fail to affect ethanol self-administration of rats. *Alcohol* 11: 389–395
80 Johnson BA, Campling GM, Griffiths P, Cowen PJ (1993) Attenuation of some alcohol-induced mood changes and the desire to drink by 5-HT3 receptor blockade: a preliminary study in healthy male volunteers. *Psychopharmacology* 112: 142–144
81 Johnson BA, Cowen PJ (1993) Alcohol-induced reinforcement: dopamine and 5-HT3 receptor interactions in animals and humans. *Drug Dev Res* 30: 153–169
82 Doty P, Zacny JP, de Wit H (1994) Effects of ondansetron pretreatment on acute responses to ethanol in social drinkers. *Behav Pharmacol* 5: 461–469
83 Swift RM, Davidson D, Whelihan W, Kuznetsov O (1996) Ondansetron alters human alcohol intoxication. *Biol Psychiatry* 40: 514–521
84 Sellers EM, Toneatto T, Romach MK, Somer GR, Sobell LC, Sobell MB (1994) Clinical efficacy of the 5-HT3 antagonist ondansetron in alcohol abuse and dependence. *Alcohol Clin Exp Res* 18: 879–885
85 Johnson BA, Roache JD, Javors MA, DiClemente CC, Cloninger CR, Prihoda TJ, Bordnick PS, Ait-Daoud N, Hensler J (2000) Ondansetron for reduction of drinking among biologically predisposed alcoholic patients: a randomized controlled trial. *JAMA* 284: 963–971
86 Johnson BA, Roache JD, Ait-Daoud N, Zanca NA, Velazquez M (2002) Ondansetron reduces the craving of biologically predisposed alcoholics. *Psychopharmacology* 160: 408–413
87 Kranzler HR, Pierucci-Lagha A, Feinn R, Hernandez-Avila C (2003) Effects of ondansetron in early- *versus* late-onset alcoholics: a prospective, open-label study. *Alcohol Clin Exp Res* 27: 1150–1155
88 Bisaga A, Sikora J, Kostowski W (1993) The effect of drugs interacting with serotonergic 5HT3 and 5HT4 receptors on morphine place conditioning. *Pol J Pharmacol* 45: 513–519
89 Panocka I, Ciccocioppo R, Polidori C, Pompei P, Massi M (1995) The 5-HT4 receptor antagonist, GR113808, reduces ethanol intake in alcohol-preferring rats. *Pharmacol Biochem Behav* 52: 255–259

Drugs for Relapse Prevention of Alcoholism
Edited by R. Spanagel and K. Mann

Opioidergic compounds: preclinical data

Petri Hyytiä

Department of Mental Health and Alcohol Research, National Public Health Institute, POB 33, FI-00251 Helsinki, Finland

Introduction

Ethanol-induced activation of the endogenous opioid system has been hypothesized to be one of the mechanisms mediating ethanol reinforcement and enhanced ethanol drinking. This hypothesis is supported by many lines of evidence, including ethanol-induced elevation of extracellular levels of β-endorphin in the nucleus accumbens [1], and genetically determined differences in the basal levels of β-endorphin, enkephalins, and dynorphins in distinct brain areas of rodent lines that differ in ethanol self-administration behavior [2–4]. In accordance with this hypothesis, the non-selective opioid receptor antagonists naltrexone, naloxone and nalmefene suppress ethanol-reinforced behavior in widely different experimental conditions.

Opioid receptors are divided into three major classes, the μ-, δ- and κ-opioid receptors, which have all been cloned and sequenced [5–8]. Furthermore, on the basis of pharmacological evidence, subtypes of these receptors have been proposed. β-endorphin recognizes both μ- and δ-receptors with almost equal potency. Also, enkephalins and dynorphins interact with μ-receptors but with a lower affinity as compared to δ- and κ- receptors, respectively [9]. Increasing evidence for the importance of ethanol-induced activation of both β-endorphin and enkephalin systems in ethanol reward has further prompted research on the contribution of the different opioid receptor types, especially the μ- and δ- receptors, in the reinforcing effect of ethanol. Because the commonly used non-selective antagonists (naltrexone, naloxone and nalmefene) bind to all opioid receptor types as a function of the dose administered [10–12], the roles of the opioid receptor types in ethanol reinforcement, in recent studies, have been studied with antagonists selective for these receptors.

In the majority of these studies, the selective antagonists have been tested using behavioral models that measure the direct reinforcing effects of ethanol, including various free-choice drinking and operant self-administration models. These models do not permit assessment of the conditioned ethanol effects, which underlie craving and relapse. However, there is evidence showing that the modulation of the reinforcing effects of ethanol by opioid antagonists may also predict their effects on the appetitive conditioned aspects of ethanol consumption [13, 14].

Effects of selective μ-receptor antagonists on ethanol consumption

Selective μ-opioid receptor antagonists β-funaltrexamine, D-Pen-Cys-Tyr-D-Trp-Orn-Thr-Pen-Thr-NH_2 (CTOP), and naloxonazine have been shown to reliably suppress ethanol intake in many experimental paradigms. Systemically administered β-funaltrexamine (5–20 mg/kg) produced a dose-dependent decrease in ethanol consumption both in the high alcohol drinking (HAD) rat line on a fluid-deprivation schedule with a 2-h daily access to ethanol and water, as well as in Wistar rats, given limited access 1-h to ethanol with *ad libitum* water and food [15, 16]. Injections of β-funaltrexamine were administered 16–20 h before the opportunity to drink, because this antagonist has an initial κ-agonist effect lasting 2–3 h, followed by a long (2–4 days) μ-antagonist effect. In both experiments, the highest dose, 20 mg/kg β-funaltrexamine, decreased ethanol intake even during the second post-injection session, probably reflecting the long-lasting μ-antagonist action. Although β-funaltrexamine did not attenuate the 2-h scheduled saccharin intake, decrease in 24-h water consumption suggests that this antagonist displays a general suppressive effect on ingestive behavior [15].

Another μ-opioid receptor antagonist, CTOP, decreased both limited access ethanol drinking and operant responding for ethanol in alcohol-preferring AA (Alko, Alcohol) and Wistar rats after intracerebroventricular (0.3–3 μg) administration [17] (Fig. 1). Moreover, systemic administration of this com-

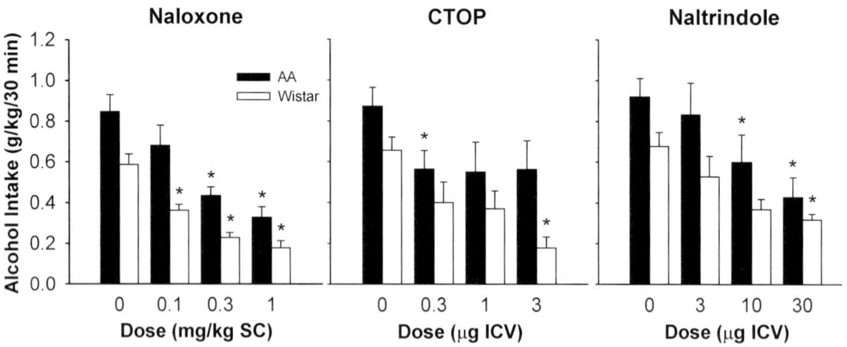

Figure 1. Effects of subcutaneous naloxone, and intracerebroventricular CTOP and naltrindole injections on ethanol consumption in alcohol-preferring AA and heterogeneous Wistar rats. The animals were allowed to respond for a 0.1 ml drop of 10% ethanol solution on a FR1 schedule during daily 30-min sessions. Data are expressed as mean (±SEM) ethanol intake (g/kg) during the 30-min session. (* $p < 0.05$, significantly different from the vehicle conditions). Adapted with permission of Lippincott Williams & Wilkins, Baltimore, from Hyytiä P & Kiianmaa K (2001) Suppression of ethanol responding by centrally administered CTOP and naltrindole in AA and Wistar rats. *Alcohol Clin Exp Res* 25: 25–33.

pound has been reported to suppress ethanol drinking in mice as well [18]. With repeated administration of CTOP across three sessions, a progressive decline in limited access ethanol drinking was observed, with a transient decrease in 24-h water and food intake [19]. It is not clear, however, whether this finding could be interpreted as an extinction-like decrease caused by an antagonist-induced attenuation of ethanol's reinforcing effects or whether it is due to cumulative drug effects. Moreover, the highest dose, 3-μg, had a tendency of slowing the initiation of operant responding for ethanol, suggestive of aversive effects produced by CTOP [17].

Systemic naloxonazine blocks central μ_1-opioid receptors irreversibly, well over 24 h, but can also block μ_2-receptors in a reversible manner during the first hours after administration. Thus, when administered 20 h prior to ethanol and concomitant food access, the dose-dependent (1–15 mg/kg) decrease in both ethanol and food intake by naloxonazine could be attributed to a μ_1-blockade [20]. However, naloxonazine suppressed ethanol and food consumption as well, when given 15 min before the session, suggesting that both μ_1- and μ_2-receptor blockade could modulate ethanol reinforcement [21]. When the 15 mg naloxonazine was administered before three successive ethanol sessions, a decrease was seen only during the first session, after which tolerance to the drug's effect developed. This was probably due to a μ-receptor up-regulation, produced by the prolonged receptor blockade [20].

Effects of selective δ-receptor antagonists on ethanol consumption

Although δ-opioid receptor antagonists have been shown to attenuate ethanol reinforcement in many studies, the results are generally more inconsistent when compared to those from experiments with μ-antagonists. The first δ-antagonist tested for its effects on ethanol drinking was ICI 174864. When both ethanol and water availability was limited to one 30-min daily access period, ICI 174864 (0.5–3 mg/kg) dose-dependently decreased ethanol but not water intake in the alcohol-preferring HAD rats [22]. Similarly, single injections of this compound (3–8 mg/kg) suppressed ethanol drinking during the first of the three 1-h access periods every 4 h in alcohol-preferring P rats without effects on 24-h water drinking [23]. The effect of naltrindole, another δ-receptor antagonist (5–20 mg/kg) tested in the same model, lasted for the first two access periods, and no effects were seen on water consumption [23]. However, separate control experiments indicated that naltrindole also affected ethanol-saccharin and saccharin solution intake, suggesting that the suppressive effects were not specific for ethanol.

Attenuation of ethanol consumption by naltrindole in rats and mice has been reported in several other studies [17, 18, 24, 25], but these positive findings have not been uniformly replicated [16, 20, 26, 27]. The reasons for the discrepant findings are not clear. The failure to see decreases in ethanol drinking by naltrindole was probably not always caused by insufficient dosing, as negative findings were also reported from experiments where the naltrindole doses matched those employed in positive reports. The genetic background of the animals cannot easily explain the differences either, as indicated by similar naltrindole-induced decreases in operant responding for ethanol both in the alcohol-preferring AA as well as the heterogeneous Wistar rats [17] (Fig. 1). Finally, a meaningful comparison of the various behavioral models used for measuring ethanol consumption is complicated by different periods of ethanol and/or fluid deprivation affecting the motivational state of the animals and by the various combinations of conditioned factors present at the time of ethanol access.

Based on pharmacological and behavioral evidence, the existence of two δ-opioid receptor subtypes in the rodent brain, δ_1- and δ_2-receptors, has been proposed [28, 29]. Naltrindole blocks both subtypes but naltriben is a putative δ_2-receptor antagonist. Systemic naltriben has been shown to decrease ethanol drinking in a daily 8-h limited access situation and operant responding for ethanol [30, 31]. The effect of naltriben on ethanol reinforcement was relatively specific, as it did not affect intake of saccharin/ethanol or quinine/ethanol solutions in the limited access paradigm nor did it affect responses for the concomitantly available saccharin solution in the operant model.

Effects of selective opioid receptor antagonists on ethanol seeking

Drug-paired environmental stimuli may acquire powerful incentive-motivational properties through classical conditioning and elicit drug craving and relapse in drug abusers as well as in laboratory animals trained to self-administer drugs [32, 33]. There is both clinical and preclinical evidence showing that the endogenous opioid system is not only involved in the direct reinforcing effects of ethanol, but may also partly mediate the effects of conditioned contextual cues on ethanol seeking. For example, naltrexone was demonstrated to attenuate the efficacy of ethanol-associated environmental stimuli to reinstate extinguished responding for ethanol in laboratory rats [13]. Using the same behavioral model, naltrindole and naloxonazine were used for assessing the contribution of μ- and δ-opioid receptors in ethanol seeking [14] (Fig. 2). Naltrindole decreased ethanol-seeking behavior at the highest dose (5 mg/kg) under stimuli predictive of ethanol reward, but did not affect responding under the stimulus condition associated with non-reward. In contrast, the effect of naloxonazine was not specific, because the effective dose (15 mg/kg) that suppressed ethanol-seeking behavior also decreased responding under the non-reward stimulus condition.

The conditioned place preference methods have long been used to measure the motivating effects of drug-paired environmental stimuli. The non-selective opioid antagonist naloxone has been shown to attenuate ethanol-induced conditioned place preference in mice [34, 35]. In rats, however, it has been difficult to observe ethanol-induced place preference without special conditioning procedures, including exposure to stress. For example, ethanol produced con-

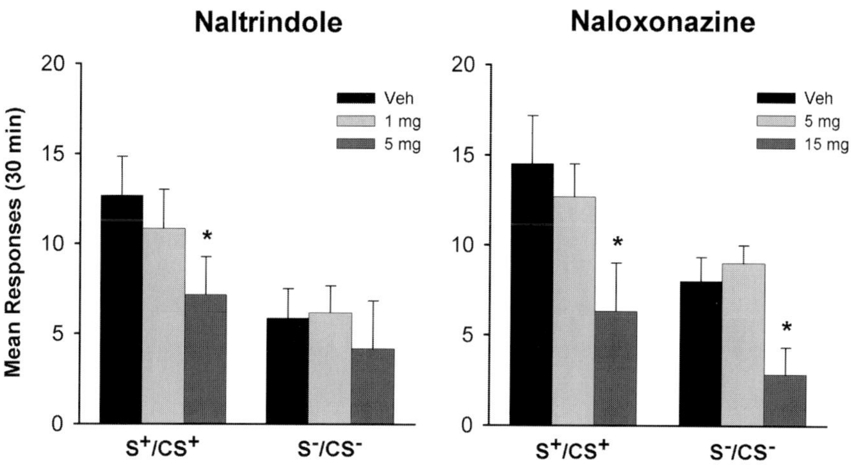

Figure 2. Effects of naltrindole and naloxonazine on cue-induced reinstatement of ethanol-seeking behavior. Both drugs were tested under the stimulus conditions previously associated with ethanol (S^+/CS^+) or non-reward/water (S^-/CS^-). Data are expressed as mean (±SEM) responses at the active lever during 30-min reinstatement sessions. (* $p < 0.05$, significantly different from the vehicle conditions). Adapted with permission of Nature Publishing Group, London, from Ciccocioppo R, Martin-Fardon R, Weiss F (2002) Effect of selective blockade of μ_1 or δ opioid receptors on reinstatement of alcohol-seeking behavior by drug-associated stimuli in rats. *Neuropsychopharmacology* 27: 391–399.

ditioned place preference in rats that were given electric foot shocks prior to ethanol injections (0.3 g/kg) and conditioning. In these animals, pre-treatment with the μ-antagonist β-funaltrexamine or the δ-antagonist naltrindole dose-dependently reduced preference for the ethanol-paired compartment at doses that have been shown to decrease ethanol drinking [36, 37].

Opioidergic mechanisms of ethanol reinforcement and conditioned ethanol effects

Reductions produced in ethanol consumption and ethanol-seeking behavior by the selective opioid receptor antagonists, generally suggest a role for both the μ- and δ-receptors in these phenomena. However, there are also discrepant findings, especially with respect to the involvement of the δ-receptors in ethanol drinking behavior. Moreover, the experimental data suggest that in most experimental situations, the suppressive effects of opioid receptor antagonists on ethanol consumption are not specific, but reflect the well-known involvement of opioid receptors with regard to ingestive behavior [38].

The precise roles of the opioid receptors in ethanol reinforcement and conditioned reinforcement processes are not yet very well known. There is evidence that systemically administered ethanol increases the level of extracellular β-endorphin in the nucleus accumbens [1]. Moreover, intra-accumbal infusions of methylnaloxonium and naltrindole suppressed ethanol self-administration, suggesting that reductions in ethanol reinforcement by opioid antagonists could be related to their inhibition of the endogenous opioid peptide action in the nucleus accumbens [17, 39]. Infusion of these antagonists into the amygdala also attenuated ethanol reinforcement, which could be related to the role of the amygdala in stimulus-reward associations [17]. Another line of evidence suggests that the suppressive effect of opioid antagonists on ethanol reinforcement could involve interaction with mesolimbic dopamine transmission. For example, ethanol-induced increase in extracellular dopamine level in the nucleus accumbens was attenuated by systemic naltrexone and focal naltrindole administration [40–42].

In addition to the direct pharmacological actions of ethanol, ethanol-associated contextual stimuli can increase dopamine levels in the nucleus accumbens as well. This is consistent with the view of the role of the midbrain dopamine neurons in the processing of motivational signals [43, 44]. So far, there are no data on the effects of selective opioid antagonists on cue-induced enhancement in dopamine transmission. Both the μ- and δ-receptors are involved in tonic modulation of mesolimbic dopamine transmission [45], and blockade of these receptors could therefore blunt the efficacy of the contextual cues in enhancing dopamine transmission and reinstating ethanol seeking.

References

1 Olive MF, Koenig HN, Nannini MA, Hodge CW (2001) Stimulation of endorphin neurotransmission in the nucleus accumbens by ethanol, cocaine, and amphetamine. *J Neurosci* 21: 184RC

2 de Waele J-P, Papachristou DN, Gianoulakis C (1992) The alcohol-preferring C57BL/6 mice present an enhanced sensitivity of the hypothalamic β-endorphin system to ethanol than the alcohol-avoiding DBA/2 mice. *J Pharmacol Exp Ther* 261: 788–794

3 Gianoulakis C, de Waele J-P, Kiianmaa K (1992) Differences in the brain and pituitary β-endorphin system between the alcohol-preferring AA and alcohol-avoiding ANA rats. *Alcohol Clin Exp Res* 16: 453–459

4 Nylander I, Hyytiä P, Forsander O, Terenius L (1994) Differences between alcohol-preferring (AA) and alcohol-avoiding (ANA) rats in the prodynorphin and proenkephalin systems. *Alcohol Clin Exp Res* 18: 1272–1279

5 Chen Y, Mestek A, Liu J, Hurley JA, Yu L (1993) Molecular cloning and functional expression of a mu-opioid receptor from rat brain. *Mol Pharmacol* 44: 8–12

6 Evans CJ, Keith DE, Jr, Morrison H, Magendzo K, Edwards RH (1992) Cloning of a delta opioid receptor by functional expression. *Science* 258: 1952–1955

7 Kieffer BL, Befort K, Gaveriaux-Ruff C, Hirth CG (1992) The delta-opioid receptor: isolation of a cDNA by expression cloning and pharmacological characterization. *Proc Natl Acad Sci USA* 89: 12048–12052

8 Minami M, Toya T, Katao Y, Maekawa K, Nakamura S, Onogi T, Kaneko S, Satoh M (1993) Cloning and expression of a cDNA for the rat kappa-opioid receptor. *FEBS Lett* 329: 291–295

9 Akil H, Owens C, Gutstein H, Taylor L, Curran E, Watson S (1998) Endogenous opioids: overview and current issues. *Drug Alcohol Depend* 51: 127–140

10 Chang KJ, Cooper BR, Hazum E, Cuatrecasas P (1979) Multiple opiate receptors: different regional distribution in the brain and differential binding of opiates and opioid peptides. *Mol Pharmacol* 16: 91–104

11 Chang KJ, Cuatrecasas P (1981) Heterogeneity and properties of opiate receptors. *Fed Proc* 40: 2729–2734

12 Childers SR, Creese I, Snowman AM, Synder SH (1979) Opiate receptor binding affected differentially by opiates and opioid peptides. *Eur J Pharmacol* 55: 11–18
13 Katner SN, Magalong JG, Weiss F (1999) Reinstatement of alcohol-seeking behavior by drug-associated discriminative stimuli after prolonged extinction in the rat. *Neuropsychopharmacology* 20: 471–479
14 Ciccocioppo R, Martin-Fardon R, Weiss F (2002) Effect of selective blockade of μ_1 or δ opioid receptors on reinstatement of alcohol-seeking behavior by drug-associated stimuli in rats. *Neuropsychopharmacology* 27: 391–399
15 Krishnan-Sarin S, Wand GS, Li XW, Portoghese PS, Froehlich JC (1998) Effect of mu opioid receptor blockade on alcohol intake in rats bred for high alcohol drinking. *Pharmacol Biochem Behav* 59: 627–635
16 Stromberg MF, Casale M, Volpicelli L, Volpicelli JR, O'Brien CP (1998) A comparison of the effects of the opioid antagonists naltrexone, naltrindole, and beta-funaltrexamine on ethanol consumption in the rat. *Alcohol* 15: 281–289
17 Hyytiä P, Kiianmaa K (2001) Suppression of ethanol responding by centrally administered CTOP and naltrindole in AA and Wistar rats. *Alcohol Clin Exp Res* 25: 25–33
18 Kim SG, Stromberg MF, Kim MJ, Volpicelli JR, Park JM (2000) The effect of antagonists selective for mu- and delta-opioid receptor subtypes on alcohol consumption in C57BL/6 mice. *Alcohol* 22: 85–90
19 Hyytiä P (1993) Involvement of mu-opioid receptors in alcohol drinking by alcohol-preferring AA rats. *Pharmacol Biochem Behav* 45: 697–701
20 Honkanen A, Vilamo L, Wegelius K, Sarviharju M, Hyytiä P, Korpi ER (1996) Alcohol drinking is reduced by a μ_1- but not by a δ-opioid receptor antagonist in alcohol-preferring rats. *Eur J Pharmacol* 304: 7–13
21 Mhatre M, Holloway F (2003) Micro1-opioid antagonist naloxonazine alters ethanol discrimination and consumption. *Alcohol* 29: 109–116
22 Froehlich JC, Zweifel M, Harts J, Lumeng L, Li TK (1991) Importance of delta opioid receptors in maintaining high alcohol drinking. *Psychopharmacology (Berl)* 103: 467–472
23 Krishnan-Sarin S, Jing SL, Kurtz DL, Zweifel M, Portoghese PS, Li TK, Froehlich JC (1995) The delta opioid receptor antagonist naltrindole attenuates both alcohol and saccharin intake in rats selectively bred for alcohol preference. *Psychopharmacology (Berl)* 120: 177–185
24 Franck J, Lindholm S, Raaschou P (1998) Modulation of volitional ethanol intake in the rat by central delta-opioid receptors. *Alcohol Clin Exp Res* 22: 1185–1189
25 Le AD, Poulos CX, Quan B, Chow S (1993) The effects of selective blockade of delta and mu opiate receptors on ethanol consumption by C57BL/6 mice in a restricted access paradigm. *Brain Res* 630: 330–332
26 Williams KL, Woods JH (1998) Oral ethanol-reinforced responding in rhesus monkeys: effects of opioid antagonists selective for the mu-, kappa-, or delta-receptor. *Alcohol Clin Exp Res* 22: 1634–1639
27 Middaugh LD, Kelley BM, Groseclose CH, Cuison ER Jr (2000) Delta-opioid and 5-HT3 receptor antagonist effects on ethanol reward and discrimination in C57BL/6 mice. *Pharmacol Biochem Behav* 65: 145–154
28 Negri L, Potenza RL, Corsi R, Melchiorri P (1991) Evidence for two subtypes of delta opioid receptors in rat brain. *Eur J Pharmacol* 196: 335–336
29 Sofuoglu M, Portoghese PS, Takemori AE (1991) Differential antagonism of delta opioid agonists by naltrindole and its benzofuran analog (NTB) in mice: evidence for delta opioid receptor subtypes. *J Pharmacol Exp Ther* 257: 676–680
30 Krishnan-Sarin S, Portoghese PS, Li TK, Froehlich JC (1995) The delta 2-opioid receptor antagonist naltriben selectively attenuates alcohol intake in rats bred for alcohol preference. *Pharmacol Biochem Behav* 52: 153–159
31 June HL, McCane SR, Zink RW, Portoghese PS, Li TK, Froehlich JC (1999) The delta 2-opioid receptor antagonist naltriben reduces motivated responding for ethanol. *Psychopharmacology (Berl)* 147: 81–89
32 O'Brien CP, Childress AR, Ehrman R, Robbins SJ (1998) Conditioning factors in drug abuse: can they explain compulsion? *J Psychopharmacol* 12: 15–22
33 See RE (2002) Neural substrates of conditioned-cued relapse to drug-seeking behavior. *Pharmacol Biochem Behav* 71: 517–529
34 Cunningham CL, Henderson CM, Bormann NM (1998) Extinction of ethanol-induced condi-

tioned place preference and conditioned place aversion: effects of naloxone. *Psychopharmacology (Berl)* 139: 62–70
35 Kuzmin A, Sandin J, Terenius L, Ogren SO (2003) Acquisition, expression, and reinstatement of ethanol-induced conditioned place preference in mice: effects of opioid receptor-like 1 receptor agonists and naloxone. *J Pharmacol Exp Ther* 304: 310–318
36 Matsuzawa S, Suzuki T, Misawa M, Nagase H (1998) Involvement of mu- and delta-opioid receptors in the ethanol-associated place preference in rats exposed to foot shock stress. *Brain Res* 803: 169–177
37 Matsuzawa S, Suzuki T, Misawa M, Nagase H (1999) Different roles of mu-, delta- and kappa-opioid receptors in ethanol-associated place preference in rats exposed to conditioned fear stress. *Eur J Pharmacol* 368: 9–16
38 Cooper SJ (1983) Benzodiazepine-opiate antagonist interactions in relation to feeding and drinking behavior. *Life Sci* 32: 1043–1051
39 Heyser CJ, Roberts AJ, Schulteis G, Koob GF (1999) Central administration of an opiate antagonist decreases oral ethanol self-administration in rats. *Alcohol Clin Exp Res* 23: 1468–1476
40 Benjamin D, Grant ER, Pohorecky LA (1993) Naltrexone reverses ethanol-induced dopamine release in the nucleus accumbens in awake, freely moving rats. *Brain Res* 621: 137–140
41 Acquas E, Meloni M, Di Chiara G (1993) Blockade of delta-opioid receptors in the nucleus accumbens prevents ethanol-induced stimulation of dopamine release. *Eur J Pharmacol* 230: 239–241
42 Gonzales RA, Weiss F (1998) Suppression of ethanol-reinforced behavior by naltrexone is associated with attenuation of the ethanol-induced increase in dialysate dopamine levels in the nucleus accumbens. *J Neurosci* 18: 10663–10671
43 Weiss F, Lorang MT, Bloom FE, Koob GF (1993) Oral alcohol self-administration stimulates dopamine release in the rat nucleus accumbens: genetic and motivational determinants. *J Pharmacol Exp Ther* 267: 250–258
44 Katner SN, Weiss F (1999) Ethanol-associated olfactory stimuli reinstate ethanol-seeking behavior after extinction and modify extracellular dopamine levels in the nucleus accumbens. *Alcohol Clin Exp Res* 23: 1751–1760
45 Devine DP, Leone P, Pocock D, Wise RA (1993) Differential involvement of ventral tegmental mu, delta and kappa opioid receptors in modulation of basal mesolimbic dopamine release: *in vivo* microdialysis study. *J Pharmacol Exp Ther* 266: 1236–1246

Drugs for Relapse Prevention of Alcoholism
Edited by R. Spanagel and K. Mann

Second generation opioidergic compounds: clinical data

John David Sinclair

POB 33, Department of Mental Health and Alcohol Research, National Public Health Institute (KTL), Helsinki 00251, Finland

Introduction

The only second generation opioidergic compound for which there is substantial clinical information is nalmefene. Nalmefene, $C_{21}H_{25}NO_3$, with a molecular weight of 339.43, is a pure opioid antagonist with no abuse potential [1].

Like naltrexone and naloxone, nalmefene was originated by Jack Fishman and co-workers. The three antagonists have rather similar structures and properties. The molecular structure of nalmefene is shown in the Appendix for Chemical Structures. Nalmefene is an analog of naltrexone, while naltrexone is a congener of naloxone.

Nalmefene, like naltrexone and naloxone, is a relatively non-specific opioid antagonist, i.e., binding to mu, delta, and kappa opioid receptors. The binding abilities of the three antagonists [2] are shown in Figure 1. Nalmefene is similar to naltrexone in its binding of mu receptors, but somewhat more potent than naltrexone for kappa and delta binding. Both are more potent than naloxone in binding all three opioid receptor types. Thus, naltrexone has been characterized as being relatively selective for mu receptors, whereas nalmefene can be seen as a more universal opioid blocker. There is, however, currently no evidence that these three opioid antagonists differ in their basic actions once they reach the central nervous system.

Pharmacokinetics

Oral nalmefene is rapidly absorbed [3]. It has a mean elimination half-life of 10.7 h (range 7–15 h), which is substantially longer than for naloxone (1.1 h) and naltrexone (4 h) [4]. About 4% of nalmefene is excreted in the urine as unchanged nalmefene, and up to 60% is excreted as inactive glucuronide conjugates. Naltrexone, in contrast, is metabolized to an active compound, 6-beta-naltrexol (half-life: 12.9 h) which may be important for the efficacy of naltrexone in treating alcoholism [5].

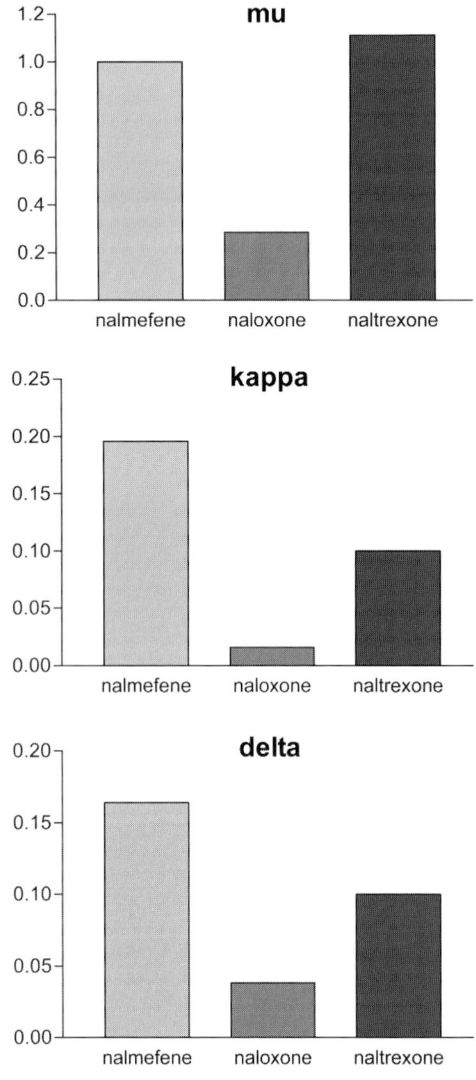

Figure 1. Binding of nalmefene, naloxone, and naltrexone to the three types of opioid receptors. Derived from Michel et al. [2]. Shown are the reciprocals of the IC_{50}s (nM), in order to represent stronger bindings as larger values. The specific binding compounds displaced were [^{3}H]-dihydromorphine for mu receptors, [^{3}H]-ethylketocyclazocine for kappa, and [^{3}H]-D-ala-D-leu enkephalin for delta receptors.

Oral dosing with nalmefene is subject to less variability than with naltrexone. Oral nalmefene has a bioavailability of 40–50% with relatively small differences between individuals [6]. In contrast, oral naltrexone is subject to high but variable amounts being removed by first-pass metabolism resulting in a bioavailability ranging from only 5% up to 40% [4, 7]. The 3- to 4-fold variation between individuals in first-pass metabolism complicates naltrexone dosing [8]. Furthermore, the metabolism of naltrexone is altered in patients with liver cirrhosis [9].

Naloxone is metabolized so rapidly in the liver that the first-pass metabolism prevents it from being used orally; i.e., if naloxone is ingested, essentially all of it is metabolized on the first pass through the liver and none reaches the brain.

Nalmefene continues to block opiate effects longer than naltrexone and naloxone [10]. Oral doses of both 50 mg and 100 mg nalmefene blocked the effects of morphine for up to 48 h [11], whereas only a 100 mg naltrexone dose is still effective at 48 h [4]. Nalmefene also maintains occupancy of opioid receptors longer than naloxone does [12]. Renal insufficiency prolonged the half-life, but hemodialysis did not change the basic pharmacokinetics of nalmefene [13]. The kinetics of nalmefene are similar in children (ages 2 to 12 years) to those reported for adults [14]. The terminal half-life in children was 8.7 ± 2.3 h.

Hepatic toxicity

Theoretically, effective usage of an opioid antagonist should not depend upon the ability to deliver a specific dose to the brain. The goal is to block all the opioid receptors (or perhaps all the mu opioid receptors); it should not matter if too much antagonist is used and some is left over. In practice, however, the naltrexone dose cannot be raised too high because of hepatic toxicity reported with 300 mg daily [15, 16]. Naltrexone is marketed with a black box warning of liver toxicity. Naltrexone is contraindicated for cirrhotic patients and should not be started in alcohol-dependent patients until after a blood test for liver cirrhosis.

In contrast, there is no evidence of nalmefene being detrimental to the liver [17]. No dose-dependent liver toxicity has been reported in over 1300 patients, including patients already having liver disorders [18]. Oral nalmefene was well tolerated in doses up to 300 mg [3]. Chronic administration is also well tolerated with no clinically significant adverse effects.

There has been discussion that the hepatic toxicity warnings for naltrexone are not warranted. Treatment of alcoholism with naltrexone usually results in reductions in the markers of liver damage, since alcohol itself is more toxic than naltrexone. In practice, only a small percentage of alcoholics are rejected initially for naltrexone administration because of cirrhosis. Nevertheless, if nalmefene turns out to be at least equivalent to naltrexone in efficacy for treating alcoholism, as seems likely, nalmefene without the toxicity would seem to be the preferred medication.

Blocking opioid effects

The Food and Drug Administration (FDA) has approved nalmefene, packaged for administration by injection, for complete or partial reversal of opioid drug effects, including respiratory depression, induced by either natural or synthetic opioids. It is marketed under the brand name Revex® by Ohmeda. Much of the information concerning nalmefene HCl injections [17] is applicable also for the oral forms used in the treatment of alcoholism.

Hormonal effects

Like naltrexone and naloxone, acute nalmefene increases levels of testosterone and luteinizing hormone [19, 20]. In older impotent men it also increased follicle-stimulating hormone and cortisol levels [20]. Nalmefene caused more activation of the hypothalamic-pituitary-adrenal axis than naloxone [21].

Eating

Chronically injected nalmefene significantly suppressed caloric intake in the study on older impotent men by decreasing specifically their eating of highly palatable foods that are high in fat content [20]. The effect was replicated in a double-blind placebo-controlled clinical trial [22]. Nalmefene reduced caloric intake by 22%, without causing a significant change in hunger or satiety rating. The reduction was produced by a decrease in high fat and protein foods rated to be highly palatable. Nalmefene was given in grapefruit juice in doses of either 2.5 mg or 5.0 mg. Subjects were able to identify the higher dose from placebo but not the lower dose, whereas the effect on eating was seen with both doses. It might be noted that these are much lower than the oral doses found to be effective with alcoholism treatment.

Flushing and acetaldehyde

Nalmefene was found to inhibit the flushing reaction that some Orientals display after drinking alcohol [23]. The flushing reaction is caused by an accumulation of acetaldehyde resulting from a lower rate of aldehyde dehydrogenase activity. Flushers given 2 mg i.v. nalmefene plus alcohol had significantly less flushing (measured as facial skin temperature) than those given placebo plus alcohol. Nalmefene tended to decrease acetaldehyde levels found in flushers after drinking alcohol, although the effect did not reach significance. The fact, however, that nalmefene at the very least did not increase acetaldehyde levels, demonstrates that it is not acting like disulfiram (Antabuse®) and other sensitizing drugs that cause higher acetaldehyde levels and thus adverse reactions when alcohol is drunk.

Clinical trials of nalmefene in alcoholism treatment

Our pre-clinical experiments showed that when nalmefene was administered to rats before several sessions in which they had access to 10% ethanol solution as well as food and water, the alcohol intake decreased progressively with each session and remained significantly lower than baseline on subsequent sessions, when no nalmefene was administered [6, 24]. Nalmefene administered similarly but to rats without concomitant access to alcohol, however, did not reduce subsequent alcohol drinking but instead tended to increase it. These results were virtually the same as we observed in our naltrexone clinical trial: naltrexone given to patients instructed to control drinking was significantly better than placebo but naltrexone given with instructions to abstain completely, tended to be worse than placebo and was significantly worse than naltrexone with controlled drinking [25]. The findings are consistent with Wikler's hypothesis that opioid antagonists work by extinguishing opiate self-administration when the drug-taking response is made while the antagonist blocks reinforcement [26]. Nearly all of the naltrexone trials, not only with alcohol, but also with heroin and cocaine, have also supported this hypothesis (see review [24]), as have the results from the clinical trials with nalmefene.

A small double-blind placebo-controlled clinical trial of two doses of nalmefene was conducted by Barbara Mason and coworkers at the University of Miami [27–29]. In order to examine the safety and efficacy of different doses, 21 patients were given either 10 mg (5 mg twice daily) or 40 mg (20 mg twice daily) of oral nalmefene or placebo (twice daily) for 12 weeks. The 40 mg nalmefene patients had significantly fewer relapses to heavy drinking than either of the other two groups. Relapse to heavy drinking was defined as five or more drinks per day, or 5 or more drinking days per week. They also had highly significant reductions from baseline in the mean number of abstinent days and in the mean number of drinks per drinking day. The later measure was also highly significant in the 10 mg nalmefene group but not in the control group. The 40 mg nalmefene had the greatest reduction in SGOT levels (validated marker for drinking), while the placebo group had the least reduction. Nalmefene had no significant effect on the number of days to first sampling of alcohol. There was a tendency for reduced depression, indicating nalmefene was not causing dysphoria. Nalmefene in both doses was well tolerated. The most common drug reactions were headache, insomnia, and nausea. Patients could choose psychosocial therapy as they wished. None of the 40 mg nalmefene patients were on any other therapy. One of the 10 mg nalmefene patients chose to attend group therapy meetings. Two of the placebo patients attended AA meetings and these were the only placebo patients who did not relapse to heavy drinking.

Mason's group subsequently tested 16 more alcoholics with nalmefene doses of either 20 mg (10 mg twice daily) or 80 mg (40 mg twice daily) [18] and compared the results with those previously tested with 10 mg or 40 mg [29]. The best results were obtained with the 80 mg dose. All of them remained in the study for the entire 12 weeks. 62% of the 80 mg patients had no more than two days of heavy drinking in the 84-day trial, while none of the 10 mg patients and only about a third of the 20 mg and 40 mg patients met this criterion.

A larger double-blind placebo-controlled trial with doses of 20 mg and 80 mg was later reported [30, 31]. 105 outpatient alcoholics who had been abstinent for at least two weeks, were studied for 12 weeks, with Coping therapy (i.e., cognitive behavioral skills taught for coping with drinking situations) also being given to all subjects weekly. This is the protocol with which naltrexone has consistently been successful [24], and nalmefene with Coping therapy also was effective here. During the last 2 weeks, the 80 mg nalmefene group showed significantly fewer drinks per drinking day [30]. Otherwise, the results with the two nalmefene doses here were similar and thus were combined. The double-blind procedure was demonstrated to be effective by the inability of subjects to guess their group significantly better than chance. Nalmefene significantly reduced the rate of relapsing to heavy drinking (here, >4 drinks/day for men, >3 drinks/day for women). Although this difference between nalmefene and placebo was already significant in the first week, the difference between the groups expanded progressively during the first half of the study and thereafter remained essentially constant. Nalmefene was significantly better than placebo, not only in the percentage of patients relapsing but also in the number of relapses. Nalmefene tended ($p = 0.06$) to be better than placebo for the duration of relapses and for the number of drinks per drinking day. The latter measure decreased significantly from baseline in both groups. These results in turn were significantly correlated with final GGT levels, which decreased significantly over the 12 weeks. The group differences were still clearer in the subgroups of patients who had at least one drink during the entire study: 80% of those on placebo relapsed to heavy drinking while only 56% of those on nalmefene did ($p < 0.03$). The percentage of abstinent days did not differ significantly between nalmefene and placebo groups, with both groups increasing significantly from baseline. Both groups had significant reductions in craving but there was no significant difference between groups. Patients on nalmefene showed no medically serious adverse drug experiences and had high rates of medication compliance and treatment completion.

Newer clinical results

The 1999 study by Mason et al. [31] is the latest nalmefene trial to be published in a full-length paper. Several additional nalmefene trials have been conducted and reported at scientific conferences since then, but the published references for them are at best only abstracts. Consequently, they will be covered only briefly here, with the reader being cautioned that the results have not been subjected to peer review.

The clinical results with naltrexone have been shown to be highly dependent upon the protocol with which the medicine is given [24, 32]. This is best shown in the three double trials comparing naltrexone and placebo with two protocols: Coping (training patients how to cope with minor slips so they do not turn into binges) and Supportive (traditional procedure supporting complete abstinence) [25, 33, 34]. All three found several measures for which naltrexone was significantly better than placebo when used with Coping therapy. None of the three double clinical trials found a single measure for which naltrexone plus Supportive therapy was significantly better than placebo plus Supportive therapy. Indeed, there were highly significant interactions demonstrating that the efficacy of naltrexone was significantly better with Coping than Supportive therapy. A double clinical trial of naltrexone for treating cocaine addiction has reported the same results: naltrexone with a Coping form of treatment ("relapse prevention") was significantly better than placebo but naltrexone with Support of abstinence tended to be worse than placebo [35].

The efficacy of nalmefene also appears to be dependent upon protocol. Mason and coworkers [31] had shown that nalmefene, like naltrexone, was effective when combined with Coping therapy. In 2002, Anton reported negative results from a multisite trial of nalmefene with Motivational Enhancement Therapy (MET) [36], a form of psychosocial therapy that had not previously been used in tests with opioid antagonists. The next year Anton reported results from a double trial comparing naltrexone with Coping to naltrexone with MET [37]. Naltrexone with Coping once again provided significant benefits, but naltrexone with MET, like nalmefene with MET, produced negative results. It seems likely that MET is similar to Supportive therapy and not suited for use with opioid antagonists.

Meanwhile, Mäkelä reported results from a randomized double-blind placebo-controlled nalmefene study of 150 outpatients with impaired control over their drinking at 6 Finnish sites [38, 39]. The study used no structured psychosocial therapy. Nalmefene (10 mg or 40 mg) or placebo was taken once daily for 16 weeks. The smaller nalmefene dose produced only transient benefits, but the larger nalmefene dose was significantly better than placebo in the reduction of heavy drinking days. The increase in abstinence days and the decrease in mean weekly consumption were best in the 40 mg nalmefene group. Nalmefene was well tolerated with only minor adverse events, primarily at the beginning of treatment. It was concluded that "nalmefene 40 mg once daily is safe and appears effective in the reduction of heavy alcohol consumption without structured psychosocial treatment."

Subsequently, a Phase III double-blind placebo-controlled nalmefene trial in Finland and the UK was reported [40]. Approximately 570 patients took nalmefene or placebo for 28 weeks only before alcohol drinking. Patients on nalmefene had a significantly greater reduction (nearly 50%) in heavy drinking days than did placebo patients. This was significant also in the Finnish patients separately but not in the British subgroup alone due to higher dropout rates. Significantly more nalmefene patients than controls were rated as "much improved" or "very much improved". This was significant in both Finnish and British patients separately, too. No serious adverse effects were observed.

Finally, a double-blind placebo-controlled Phase II clinical trial with 200 subjects found that nalmefene was significantly better than placebo in the treatment of compulsive gambling [41]. Nalmefene reduces craving and thoughts about gambling to a level about half that in the placebo group.

Relapse: definition of terms

The results with nalmefene, and with naltrexone, emphasize the importance of precisely defining the word "relapse". With traditional treatments, breaking abstinence by sampling alcohol often leads to a resumption of heavy drinking, so "relapse" could be used indiscriminately to mean both the first sip and also the return to heavy drinking. When opioid antagonists are used, however, there is a great difference between the two behaviors. The resumption of sampling has almost never been found to be affected by treatment with naltrexone or nalmefene [24], but resumption of heavy drinking has been the measure showing the greatest benefits. Probably for the same reason, naltrexone has been found to be helpful for alcoholics who were actively drinking at the onset of treatment but not for ones who had already been abstinent for weeks [42].

It would be advantageous if the word "relapse" were used exclusively for the resumption of heavy drinking, and if agreement could be reached as to what precisely constitutes heavy drinking. Unfortunately, we then need some other word for resumption of sampling and moderate consumption.

Conclusions

- The clinical results show that nalmefene is safe and well tolerated by heavy drinkers and alcohol-dependent patients, regardless of whether they have been detoxified or are still actively drinking.
- Nalmefene has inherent advantages over naltrexone in its dosing, duration of action, lack of active metabolite, and lack of liver toxicity.
- Nalmefene can be effective, especially in reducing relapses to heavy drinking, in the treatment of alcoholism.
- The efficacy of nalmefene in alcoholism treatment, like that of naltrexone, is dependent upon the protocol with which it is used:
  - Nalmefene is effective with Coping therapy.
  - Nalmefene is effective without any structured psychosocial therapy.
  - Nalmefene is not more effective than placebo with Motivational Enhancement Therapy and probably, like naltrexone, not with programs strongly supporting complete abstinence.
- Nalmefene is effective also in treating compulsive gambling.

References

1 Fudala PJ, Heishman SJ, Henningfield JE, Johnson RE (1991) Human pharmacology and abuse potential of nalmefene. *Clin Pharmacol Ther* 49: 300–306
2 Michel ME, Bolgan G, Weisman B-A (1985) Binding of a new opiate antagonist, nalmefene, to rat brain membranes. *Meth Find Exp Clin Pharmacol* 7: 175–177
3 Dixon R, Gentile J, Hsu HB, Hsiao J, Howes J, Garg D, Weidler D (1987) Nalmefene: safety and kinetics after single and multiple oral doses of a new opioid antagonist. *J Clin Pharmacol* 27: 233–239
4 Dupont Pharma (1995) package insert for ReVia® available at *http://www.fda.gov/cder/ogd/rld/18932s14.pdf*
5 McCaul ME, Wand GS, Sullivan J, Mummford G, Quigley J (1997) Beta-naltrexol level predicts alcohol relapse. *Alcohol Clin Exp Res* 21: 32A
6 Sinclair JD, Scheinin H, Lammintausta R (1992) Method for treating alcoholism with nalmefene. *USA Patent* 5,086,058 Feb. 4
7 Gonzalez JP, Brogden RN (1988) Naltrexone: A review of its pharmacodynamic and pharmacokinetic properties and therapeutic efficacy in the management of opioid dependence. *Drugs* 35: 192–213
8 Verebey K, De Pace A, Jukofsky D, Volabka JV, Mule SJ (1980) Quantitative determination of 2-hydroxy-3-methoxy-6 beta-naltrexol (HMN), naltrexone, and 6 beta-naltrexol in human plasma, red blood cells, saliva, and urine by liquid chromatography. *J Anal Toxicol* 4: 33–37
9 Bertolotti M, Ferrari A, Vitale G, Stefani M, Trenti T, Loria P, Carubbi F, Carulli N, Sternieri E (1997) Effect of liver cirrhosis on the systemic availability of naltrexone in humans. *J Hepatol* 27: 505–511
10 Gal TJ, DiFazio CA, Dixone R (1986) Prolonged blockage of opioid effects with oral nalmefene. *Clin Pharmacol Ther* 40: 537–542
11 Jones HE, Johnson RE, Fudala PJ, Henningfield JE, Heishman SJ (2000) Nalmefene: blockade of intravenous morphine challenge effects in opioid abusing humans. *Drug Alcohol Depend* 60: 29–37
12 Kim S, Wagner KN, Villemagne VL, Kao PF, Dannals RF, Ravert HT, Job T, Dixon RB, Civelek AC (1997) Longer occupancy of opioid receptors by nalmefene compared to naloxone as measured *in vivo* by a dual-detector system. *J Nucl Med* 38: 1726–1731
13 Matzke GR, Frye RF, Alexander AC, Reynolds R, Dixon R, Johnston J, Rault RM (1996) The effect of renal insufficiency and hemodialysis on the pharmacokinetics of nalmefene. *J Clin Pharmacol* 36: 144–151
14 Chumpa A, Kaplan RL, Burns MM, Shannon MW (2001) Nalmefene for elective reversal of procedural sedation in children. *Am J Emerg Med* 19: 545–548
15 Mitchell JE, Morley JE, Levine AS, Hatsukami D, Gannon M, Pfohl D (1987) High-dose naltrexone therapy and dietary counseling for obesity. *Biol Psychiatry* 22: 35–42
16 Morgan CJ, Kosten TR (1990) Potential toxicities of high doses of naltrexone in patients with appetitive disorders. In: LD Reid (ed.): *Opioids, bulimia, and alcohol abuse and alcoholism.* Springer-Verlag, New York, 261–273
17 Ohmeda (2000) Revex® package insert, available at *http://www.fda.gov/cder/foi/label/2000/20459S2lbl.pdf*
18 Mason B (1996) Dosing issues in the pharmacotherapy of alcoholism. *Alcohol Clin Exp Res* 20 (Suppl): 10A–16A
19 Graves GR, Kennedy TG, Weick RF, Casper RF (1993) The effect of nalmefene on pulsatile secretion of luteinizing hormone and prolactin in men. *Hum Reprod* 8: 1598–1603
20 Billington CJ, Shafer RB, Morley JE (1990) Effects of opioid blockade with nalmefene in older impotent men. *Life Sci* 47: 799–805
21 Schluger JH, Ho A, Borg L, Porter M, Maniar S, Gunduz M, Perret G, King A, Kreek MJ (1998) Nalmefene causes greater hypothalamic-pituitary-adrenal axis activation than naloxone in normal volunteers: Implications for the treatment of alcoholism. *Alcohol Clin Exp Res* 22: 1430–1436
22 Yeomans MR, Wright P, Macleod HA, Critchley JAJH (1988) Effects of nalmefene on feeding in humans. *Psychopharmacology* 100: 426–432
23 Ho SB, DeMaster EG, Shafer RB, Levine AS, Morley JE, Go VLW, Allen JI (1988) Opiate antagonist nalmefene inhibits ethanol-induced flushing in Asians: A preliminary study. *Alcohol Clin Exp Res* 12: 705–712

24 Sinclair JD (2001) Evidence about the use of naltrexone and for different ways of using it in the treatment of alcoholism. *Alcohol Alcoholism* 36: 2–10
25 Heinälä P, Alho H, Kiianmaa K, Lönnqvist J, Kuoppasalmi K, Sinclair JD (2001) Targeted use of naltrexone without prior detoxification in the treatment of alcohol dependence: A factorial double-blind placebo-controlled trial. *J Clin Psychopharmacol* 21: 287–292
26 Wikler A (1976) The theoretical basis of narcotic addiction treatment with narcotic antagonists. In: D Julius, P Renault (eds): *Narcotic Antagonists: Naltrexone*. National Institute of Drug Abuse, Rockville, MD, 119–122
27 Mason BJ, Ritvo EC, Salvato F, Zimmer E, Goldberg G, Welch B (1993) Nalmefene modification of alcohol dependence: A pilot study. *Proceedings of American Psychiatric Association Annual Meeting*, San Francisco, CA, May 1993, 170 (abstract NR442)
28 Mason BJ, Ritvo EC, Salvato FR, Goldberg G (1994) Preliminary dose finding for nalmefene treatment of alcoholism. *Alcohol Clin Exp Res* 18: 464 (abstract 270)
29 Mason BJ, Ritvo EC, Morgan RO, Salvato FR, Goldberg G, Welch B, Mantero-Atienza E (1994) A double-blind, placebo-controlled pilot study to evaluate the efficacy and safety of oral Nalmefene HCL for alcohol dependence. *Alcohol Clin Exp Res* 18: 1162–1167
30 Litten RZ, Croop RS, Chick J, McCaul ME, Mason B, Sass H (1996) International update: New findings on promising medications. *Alcohol Clin Exp Res* 20: 216A–218A
31 Mason BJ, Salvato FR, Williams LD, Ritvo EC, Cutler RB (1999) A double-blind, placebo-controlled study of oral Nalmefene for alcohol dependence. *Arch Gen Psychiatry* 56: 719–725
32 Sinclair JD, Alho H, Shinderman M (2002) Treatment of alcoholism: Efficacious use of naltrexone. *N Engl J Med* 17: 1329–1331
33 O'Malley SS, Jaffe AJ, Chang G, Schottenfeld RS, Meyer RE, Rounsaville B (1992) Naltrexone and coping skills therapy for alcohol dependence. *Arch Gen Psychiatry* 49: 881–887
34 Balldin J, Berglund M, Borg S, Månsson M, Berndtsen P, Franck J, Gustafsson L, Halldin J, Hollstedt C, Nilsson L-H et al. (1997) A randomized 6 month double-blind placebo-controlled study of Naltrexone and coping skills education programme. *Alcohol Alcoholism* 32: 325
35 Schmitz JM, Stotts AL, Rhoades HM, Grabowski J (2001) Naltrexone and relapse prevention treatment for cocaine-dependent patients. *Addict Behav* 26: 167–180
36 Anton R (2002) Multisite study of Nalmefene combined with modified motivational enhancement therapy in the treatment of outpatient alcoholics *Presented at the 25th Annual Scientific Meeting of the Research Society on Alcoholism,* June 28-July 3, 2002, San Francisco, CA, USA
37 Anton RF, Moak DM, Latham PK, Myrick DL, Waid LR (2003) A double blind comparison of naltrexone combined with CBT or MET in the treatment of alcohol dependence. Presented at the 26th Annual Scientific Meeting of the Research Society on Alcoholism, June 21–25, 2003, Fort Lauderdale, Florida. *Alcohol Clin Exp Res* 27 (suppl): 191A (abstract S170)
38 Mäkelä R, Kallio A, Karhuvaara S (2001) Nalmefene in the treatment of heavy drinking. *Programme and Abstracts of the 2001 ISAM Meeting, Trieste, Italy, September 12–14* (no page numbers)
39 Mäkelä R (2002) Multisite study of Nalmefene for the treatment of heavy alcohol drinkers with impaired control. *Presented at the 25th Annual Scientific Meeting of the Research Society on Alcoholism,* June 28–July 3, 2002, San Francisco, CA, USA
40 BioTie Therapies Corp., press release, April 24 (2003) Phase III clinical studies in alcoholism and alcohol abuse. Available at *http://www.biotie.com/*
41 BioTie Therapies Corp, press release, May 30 (2003) Results from a Phase II clinical study suggest Nalmefene effective in the treatment of pathological gambling. Available at *http://www.biotie.com/*
42 Killeen T, Brady K, Faldowski R, Gold P, Simpson K, Anton, R (2003) The efficacy of naltrexone in a community treatment program. *Alcohol Clin Exp Res* 27 (suppl): 146a (abstract 846)

Drugs for Relapse Prevention of Alcoholism
Edited by R. Spanagel and K. Mann

Dopaminergic compounds: preclinical data

Friedbert Weiss

Department of Neuropharmacology, The Scripps Research Institute (CVN 15), 10550 North Torrey Pines Road, La Jolla, CA 92037, USA

Introduction

Like other addictive disorders, alcoholism is characterized by long-lasting vulnerability to relapse after cessation of drinking. Increasing attention has, therefore, been directed toward identifying specific risk factors for relapse and to establish the neurobiological mechanisms by which these factors convey vulnerability to relapse. At the same time, relapse prevention has emerged as an important focus of treatment and medication development efforts.

An important neuropharmacological substrate for many of the neurobehavioral effects of ethanol is the mesocorticolimbic dopamine (DA) system. Evidence has accumulated in recent years implicating this system also in a range of behavioral and neurobiological processes relevant for alcohol craving and relapse. These findings identify DA transmission as a possible pharmacotherapeutic target for relapse prevention.

Evidence implicating dopamine neurotransmission as a target for relapse prevention

Craving and relapse associated with ethanol cue exposure

Ethanol acutely increases the activity of the mesolimbic DA reward pathway [1–5] and this effect has been widely implicated as a mechanism by which ethanol exerts its reinforcing actions [6–9]. However, not only ethanol consumption [5, 10], but also exposure to environmental stimuli associated with ethanol can activate DA transmission in the nucleus accumbens [5, 10–12]. Thus, in addition to its role in the reinforcing effects of ethanol, mesolimbic DA transmission may mediate the incentive-motivational effects of ethanol-related environmental stimuli that are thought to be relevant for ethanol-seeking, craving and relapse. The conditioning of ethanol's pharmacological actions with discrete environmental stimuli is a major factor in the abuse potential of this drug [13]. Ethanol cues can evoke drug desire that may lead to the resumption of drinking in abstinent alcoholics [14–21]. Drug-related stimuli may perhaps also elicit automatic responses that lead to drug-seeking behavior and relapse without the intervention of distinct feelings of craving [22, 23]. A role of DA in such conditioned responses to ethanol cues would be consistent with the established general role of mesocorticolimbic DA transmission in incentive learning and conditioned reinforcement associated with both natural and drug rewards (e.g., [24–26]). In support of this possibility, alcohol-related stimuli not only increase accumbal DA release in rats [5, 10–12], but activate the ventral striatum (i.e., a DA-rich brain region) in abstinent alcoholics [27]. Consistent with this hypothesis as well, pharmacological activation of DA transmission in the nucleus accumbens enhances the ability of ethanol-paired stimuli to function as conditioned reinforcers [28].

Craving elicited by "priming" doses of ethanol

It is well established that small doses of drugs of abuse including ethanol, rather than reducing drug desire, elicit further drug craving (e.g., [29, 30]). Moreover, in alcoholics, the first drink after abstinence is often associated with "loss of control" leading to severe intoxication and return to continued alcohol abuse [30]. A role for DA in this aspect of relapse has emerged in controlled laboratory studies showing that the D_2-preferring DA receptor antagonist haloperidol reduces alcohol craving in response to a "priming" dose of ethanol in alcoholic patients [31]. Haloperidol also blocked the stimulant and euphorigenic effects of ethanol in social drinkers [32]. The attenuation of ethanol's priming effects supports the view that craving and loss of control following resumption of drinking are linked to DA-activating effects of ethanol. More generally, these findings support the wealth of preclinical evidence that has implicated DA as a major neuropharmacological substrate for ethanol reward.

Neuroadaptation induced by chronic ethanol

In contrast to the acute effects of ethanol that activate mesolimbic DA neurotransmission, chronic dependence-inducing ethanol treatments lead to the development of DA hypofunction. Animals chronically exposed to ethanol show impaired DA synthesis (e.g., [33–35]) paired with elevated DA transporter levels and, thus, presumably enhanced clearance of synaptic DA [34]. Impairments in mesolimbic DA function are particularly evident during withdrawal with marked decrements in the activity of ventral tegmental DA neurons [36, 37] and reduced extracellular DA levels in the nucleus accumbens [38, 39]. Renewed ethanol administration reverses these functional deficits as well as behavioral signs of withdrawal in rats [39, 40]. Consistent with these findings in animals, the DA D_2 agonist bromocriptine ameliorates ethanol withdrawal symptoms, including anxiety, depression, and restlessness in alcoholics [41]. Deficient DA function, therefore, is likely to represent a neural basis for the state of dysphoria and negative affect that accompanies ethanol withdrawal such that restoration of normal DA function by appropriate pharmacological agents may be a suitable strategy for relapse prevention.

Deficits in DA neurotransmission induced by chronic ethanol appear to be long lasting. Midbrain DA neural activity still shows suppression three days after withdrawal [42]. Measures of accumbal DA synthesis and turnover indicate that the release of DA is deficient as late as two months after withdrawal [43], and impairments in DA D_1-mediated signal transduction have been observed as late as four months following withdrawal from long-term ethanol exposure [44]. The persistent impairment in the functioning of mesolimbic DA transmission likely has implications for vulnerability to relapse during protracted ethanol withdrawal. Indeed, clinical studies suggest that retardation in the recovery of impairments in DA receptor function or DA-dependent signal transduction is associated with the risk of early relapse [45, 46]. Relapse risk linked to delayed recovery of DA receptor function may not, however, extend to heightened ethanol cue reactivity or cue-induced craving, since a more recent study failed to demonstrate a relationship between DA receptor sensitivity and craving induced by the smell of alcoholic beverages [47]. A role of deficient DA transmission in the resumption of drinking has also found important indirect support by findings that low endogenous DA function is linked to predisposition for increased alcohol preference and intake [48–50], whereas pharmacological enhancement of DA synthesis, inhibition of DA degradation, or overexpression of D_2 receptors by adenoviral DRD2 gene delivery exert protective effects reducing ethanol preference and intake [58, 51].

Exacerbation of drinking following ethanol deprivation

A well-described phenomenon in the alcohol literature is a marked increase in ethanol consumption that follows periods of alcohol deprivation [52, 53]. This "alcohol deprivation effect (ADE)" is considered a measure of motivation for alcohol [52, 54], loss of control [55], or relapse [50, 56]. Similarities exist between the alcohol deprivation effect in animals and human alcohol abuse such as enhanced ethanol consumption after abstinence in social drinkers [57], and aspects of the "loss of control" phenomenon surrounding the first drink after abstinence in alcoholics [30, 58, 59]. With repeated deprivation and increased length of deprivation periods, this phenomenon becomes resistant to manipulations of ethanol concentration, taste, and environmental factors [55, 60, 61]. Moreover, increases in ethanol intake produced by repeated deprivation outlast long abstinence phases [60] and may become irreversible [55]. The ADE, therefore, provides a model to study compulsive alcohol-seeking behavior and loss of control following resumption of drinking.

Neurochemical studies implicate dysregulation of mesolimbic DA transmission as a factor contributing to enhanced drinking following alcohol deprivation. DA release in the nucleus accumbens as measured by microdialysis [62] as well as electrically evoked release of [^{3}H]DA from accumbal tissue [63] was increased after 2–3 weeks of ethanol deprivation in alcohol-preferring P and Wistar rats following several weeks of free access to ethanol. In addition, under these conditions the ADE was associated with a reduction of autoregulatory increases on accumbal DA overflow induced by D_2 receptor stimulation [62]. The latter finding suggests that increased DA activity resulting from compromised D_2 autoreceptor function may play a role in the ADE [62], although this interpretation is difficult to reconcile with the wealth of data showing that increased DA function is associated with decreased ethanol preference and intake whereas preference and intake are increased in animals with low DA function (e.g., [48–51]). Data from a model that utilized repeated deprivation to induce high ethanol drinking also implicate impaired D_1-mediated signal transduction—as reflected by a decrease in the efficacy of DA to stimulate striatal adenylyl cyclase activity—in deprivation-induced exacerbation of ethanol intake [44]. Thus, persistent ethanol-induced changes in DA receptor function may be responsible for increased ethanol drinking associated with the ADE.

In summary, the literature points toward a prominent role for mesolimbic DA transmission in behavioral and neurobiological processes relevant for alcohol relapse risk. Despite this evidence, only few preclinical studies have explicitly tested the effects of pharmacological manipulation of DA neurotransmission on "relapse." The following review will therefore extend beyond data from strict animal models of relapse to include a wider spectrum of experimental approaches relevant for evaluating the potential of DA transmission as a treatment target for relapse prevention. Findings covered by this review are summarized in Figure 1.

Manipulation of dopamine receptors and ethanol-seeking behavior

Conditioned ethanol-seeking behavior

A dopaminergic role in drug-related learning and the potential of DA antagonists to reverse ethanol-seeking induced by alcohol cues has been studied in several animal models. These models have in common that they measure behavior controlled by the incentive-motivational or conditioned reinforcing effects of alcohol-associated environmental stimuli. These models differ, however, in terms of whether or not abstinence or extinction of ethanol-reinforced behavior is imposed before testing for conditioned ethanol-seeking, and whether or not conditioning occurs in conjunction with involuntary or self-administration of ethanol.

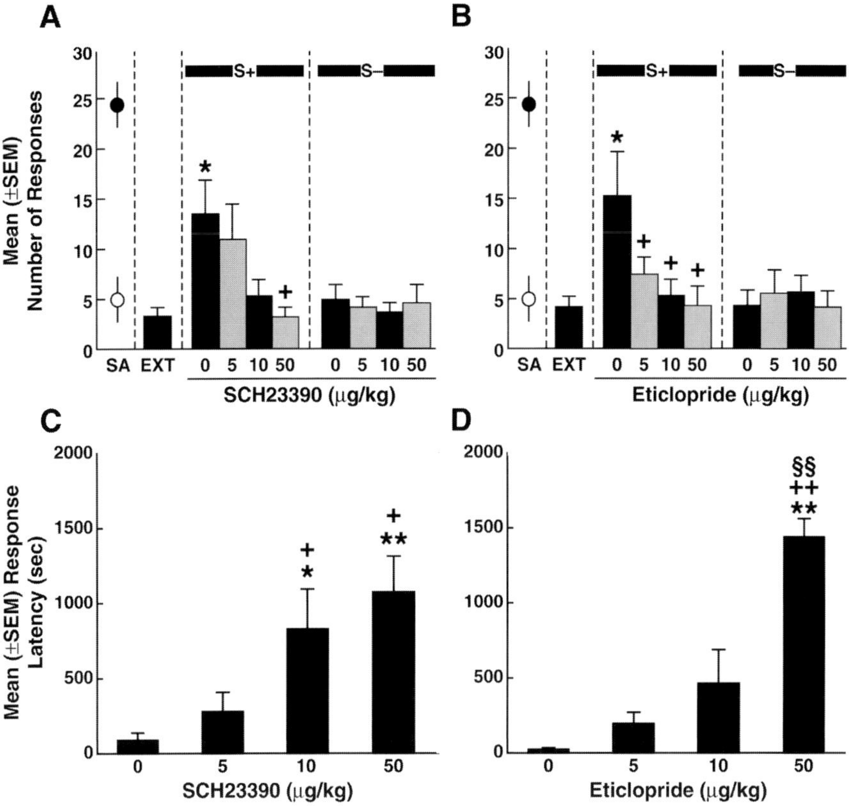

Figure 1. Conditioned reinstatement by olfactory discriminative stimuli associated with ethanol availability (S^+) *versus* nonreward (S^-), and modification of this effect by the D_1 antagonist SCH23390 and the D_2 antagonist eticlopride. Rats were treated with DA antagonists 30 min before reinstatement tests. Panels A, C: Total number of responses. For comparison, the figures show also the mean (±SEM) number of responses during ethanol self-administration (●) and nonreward (○) conditioning sessions collapsed across the last three days of the conditioning phase (SA), as well as the mean (±SEM) number of responses across the last three days of the extinction phase (EXT). * $p < 0.05$, ** $p < 0.01$ different from extinction baseline; + $p < 0.05$, different from vehicle. Panels B, D: Latency to initiate responding. * $p < 0.05$, ** $p < 0.01$ different from vehicle. + $p < 0.05$, ++ $p < 0.01$ different from 5 μg/kg. §§ $p < 0.01$ different from 10 μg/kg. Modified from Ref. 54 (with permission)

Conditioned reinstatement

In the context of the drug addiction literature, conditioned reinstatement refers to the resumption of extinguished instrumental responding induced by noncontingent exposure to a drug-related cue (for review see [64, 65]). Ethanol-associated contextual stimuli reliably elicit recovery of responding at a previously ethanol-paired lever following extinction without further alcohol availability [66–72]. Moreover, the behavioral effects of these stimuli are remarkably resistant to extinction in that recovery of ethanol-seeking does not diminish when these cues are presented repeatedly under non-reinforced conditions [73].

Reinstatement of ethanol-seeking by ethanol-associated contextual stimuli is sensitive to antagonism of DA transmission [74]. Rats operantly self-administering 10% ethanol in daily 30-min sessions were trained to associate distinct olfactory discriminative stimuli with the availability of ethanol or absence of reward. Following subsequent extinction of ethanol-reinforced responding the animals were exposed to the discriminative stimuli previously predictive of ethanol or no reward. Presentation of the ethanol-associated stimulus reinstated responding at the previously active lever. Both the DA D_1-selective antagonist SCH 23390, and the DA D_2-selective antagonist eticlopride, dose-dependently attenuated this effect by increasing the latency to initiate responding and decreasing the number of responses (Fig. 1). Similar effects were obtained in rats with a history of ethanol dependence, tested three weeks following withdrawal from a 12-day ethanol vapor inhalation procedure. Interestingly, the dose-effect curve for SCH 23390 and eticlopride in these animals was shifted to the left, resulting in an overall greater inhibition of reinstatement responses at lower doses of these agents (Fig. 2). These findings suggest that in both non-dependent and previously dependent rats, blockade of D_1 or D_2 receptors attenuates the conditioned incentive effects of ethanol-related contextual stimuli. A likely explanation for the increased DA antagonist potency in previously dependent rats involves the DA hypoactivity produced by chronic ethanol as discussed above. Specifically, reduced synaptic availability of DA at the time of testing may have been a factor contributing to the increased DA antagonist potency in rats with a history of ethanol dependence.

Appetitive ethanol-seeking behavior

Recently, an ethanol self-administration model has been developed [75–77] that dissociates ethanol-reinforced consummatory behavior (i.e., ethanol drinking) from appetitive ethanol-seeking responses (i.e., behavior induced and maintained by the incentive-motivational effects of ethanol-associated contextual cues present in the self-administration environment). In this procedure, rats must complete a set of responses at a lever during which time ethanol is not available (appetitive phase). Completion of a given response requirement within a specified time results in retraction of the lever and presentation of a sipper tube containing 10% ethanol from which rats are then allowed to freely drink (consummatory phase). Although this procedure is not, strictly speaking, a model of "relapse" because no significant period of abstinence is imposed before tests, responding during the appetitive phase provides a measure of the day-to-day strength of animals' motivation to initiate and engage in ethanol-seeking behavior when exposed to the ethanol-predictive stimulus environment of the operant conditioning chamber. Thus, pharmaco-

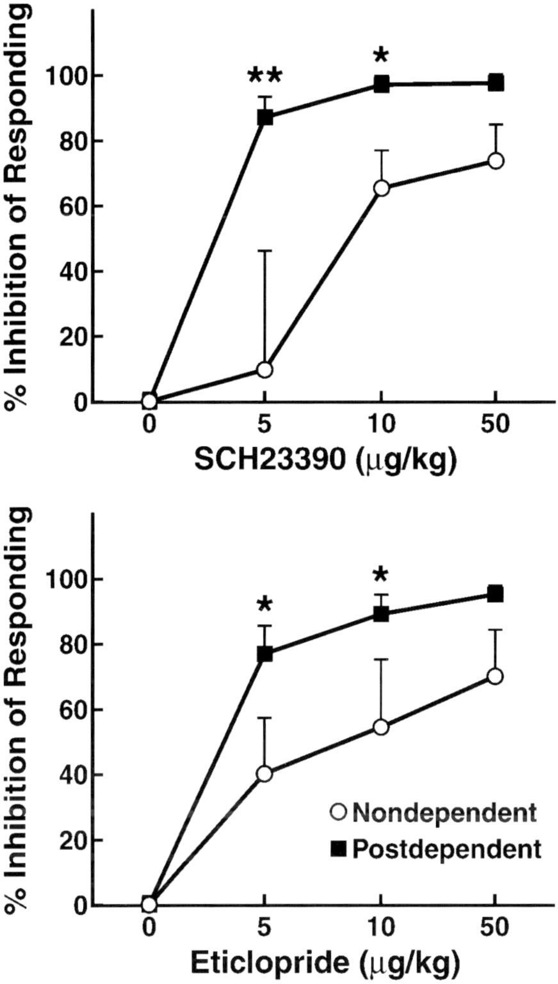

Figure 2. Dose-effect curves for the attenuation of conditioned reinstatement by SCH 23390 and eticlopride in nondependent rats and rats with a history of prior ethanol dependence. Tests were conducted in nondependent rats prior to a 12-day ethanol vapor inhalation procedure and, again, beginning on day 21 after termination of ethanol vapor exposure (Postdependent). Drug effects are expressed as mean (±SEM) percent inhibition of S^+-induced responding compared to the effects of vehicle injection. * $p < 0.05$ ** $p < 0.01$ different from Nondependent. Reproduced from Ref. 51 (with permission).

logical interventions that attenuate appetitive responding in this model may provide relevant leads as to potential treatment targets for conditioned ethanol-seeking and craving.

Data generated with this model have revealed that appetitively motivated responding preceding the availability of ethanol is more sensitive to reversal by dopamine D_2-selective antagonists than actual ethanol intake [78, 79].

Raclopride, microinjected into the nucleus accumbens, delayed the onset of appetitive responding and decreased the total number of responses. In contrast, only a single high dose of raclopride reduced the consummatory response. Systemic administration of another D_2 antagonist, remoxipride, produced even more selective effects on appetitive ethanol-seeking by dose-dependently decreasing the number of appetitive responses, while having no effect on ethanol consumption.

The findings above demonstrate that both conditioned reinstatement and appetitive ethanol-seeking behavior are sensitive to reversal by DA antagonists. A role of DA in cue-controlled ethanol-seeking is supported further by a study that examined the effects of the D_2 antagonist haloperidol on responding maintained by a conditioned stimulus (CS) that had been discretely paired with ethanol-reinforced responses [80]. Haloperidol significantly reduced conditioned responding at a previously active lever. Only a single dose was tested. However, the effects of this dose were selective for responding at the lever paired with the ethanol cue and responding at a control lever remained unaltered. This observation, therefore, lends additional support for the conclusion that DA transmission participates in mediating ethanol-seeking behavior initiated and maintained by ethanol-related environmental stimuli.

Expression of conditioned place preference

Place conditioning procedures permit examination of neuropharmacological substrates involved in the acquisition and expression of the conditioned reinforcing effects of ethanol. In this model, manipulations that interfere specifically with the expression of conditioned place preference (CPP), once acquired, provide information on the neuropharmacological basis of conditioned ethanol-seeking behavior, whereas interference with the acquisition of CPP is relevant for the understanding of mechanisms mediating the acute reinforcing effects of ethanol or the learning of Pavlovian associations. Like the model of appetitive ethanol-seeking behavior, studies on the expression of ethanol CPP have not involved imposition of abstinence and, thus, have limitations with respect to providing a measure of relapse. Nonetheless, the degree of preference for a previously ethanol-paired environment provides an index of the strength of ethanol-seeking associated with the incentive-motivational effects of an alcohol-associated stimulus context.

The strongest evidence in favor of a dopaminergic role in ethanol CPP comes from knockout studies. Mice deficient in either D_1 or D_2 receptors as well as mice lacking DARPP-32 (a phosphoprotein regulating D_1 receptor function) show reduced ethanol-induced conditioned place preference [81–84]. In contrast to these findings, antagonists of the D_1 (SCH 23390) or D_2 (raclopride, haloperidol) receptor failed to alter the expression of ethanol conditioned place preference (CPP) [85, 86]. The D_3-preferring antagonist U99194A was shown to enhance the acquisition of ethanol CPP [87, 88], an effect that may depend on an additive interaction with ethanol because this D_3 antagonist alone can produce CPP in rats [89]. The facilitating effects of U99194A in pharmacological studies are difficult to reconcile with data from D_3 knockout mice that showed no change in ethanol CPP [90]. More importantly, U99194A failed to prevent the expression of ethanol CPP [30] indicating that blockade of D_3 receptors does not interfere with the conditioned reinforcing effects of ethanol. In interpreting these CPP data it is important to consider that effects in knockout models may depend on life-long loss of DA receptor function, resulting in behavioral consequences different from those produced by acute DA receptor blockade. Also, use of knockout preparations leaves unclear whether the deficit in the targeted DA receptor population compromises the acquisition of ethanol CPP because ethanol is not reinforcing in these animals, whether these animals are unable to learn Pavlovian associations, or whether deficiency in a particular DA receptor population interferes with the expression of CPP.

As outlined earlier, the CPP data that pertain most directly to the question as to whether DA plays a role in the conditioned incentive effects of ethanol-paired environments are those that have examined whether the expression of CPP is sensitive to pharmacological manipulation of DA neurotransmission. In the studies that have taken this approach, neither D_1 nor D_2-selective antagonists altered ethanol CPP [85, 86], findings that are in clear contradiction with the attenuation of ethanol-seeking behavior by D_1 or D_2 antagonists in models of conditioned reinstatement [74], appetitively-motivated responding [78, 79], and conditioned reinforcement [80]. Several factors may account for these discrepancies. First, CPP studies typically employ involuntary administration of ethanol. The reinforcing actions of ethanol under these conditions may differ from those associated with voluntary oral self-administration. As a result, the strength or nature of associations that are formed between ethanol and environmental stimuli may differ in CPP *versus* self-administration procedures. Importantly, as well, the number of learning trials in models of ethanol-seeking that involve the conditioning of the effects of self-administered ethanol with environmental stimuli is typically greater than in the CPP procedure. Associations that are produced between specific environmental stimuli and ethanol therefore presumably are weaker in the CPP model. Due to these differences, the expression of conditioned ethanol-seeking responses may be differentially sensitive to DA antagonists in CPP *vs.* self-administration models. Lastly, it cannot be ruled out that different neural substrates are involved in contextual conditioning in the case of CPP as opposed to stimuli associated with active self-administration.

With these considerations in mind, data from behavioral models that involve the conditioning of self-administered ethanol with environmental stimuli consistently confirm that DA antagonists attenuate the motivating effects of ethanol-related environmental stimuli. What limits this information with respect to the potential of DA transmission to serve as a treatment target for the prevention of cue-induced craving and relapse is that the great majority of data has been generated in ethanol nondependent animals. In addition to its positive reinforcing effects, alcohol serves as a negative reinforcer during the development of dependence by alleviating aversive withdrawal symptoms (e.g., [91]). This type of drug-related learning may increase the salience of ethanol as a reinforcer and consequently craving or drug-seeking produced by ethanol cues compared to nondependent subjects for which the drug serves as a positive reinforcer only. Additionally, ethanol-seeking in individuals with a history of dependence is likely to be modified by neuroadaptive changes. Chronic ethanol exposure alters DA function and these changes may modify the effects of DA antagonist treatments on ethanol-seeking behavior as illustrated, for example, by the finding that the potency with which D_1 and D_2 antagonists suppress conditioned reinstatement is increased in previously ethanol-dependent rats [74]. Thus, it will be important to more systematically evaluate the role of DA in conditioned ethanol-seeking using animals with a history of ethanol dependence and withdrawal.

The alcohol deprivation effect

Despite growing evidence of a dopaminergic role in the exacerbation of ethanol intake following deprivation, only few studies have examined whether the ADE is sensitive to pharmacological manipulation of DA receptors. In one study [92], haloperidol administered daily during 14 days of deprivation following six weeks of free access to ethanol reversed the ADE measured on the first post-treatment day in mice. The reduction in drinking was most pronounced during the first 1.5 h of renewed access, a time period during which ethanol intake was greatest in vehicle-treated controls. A proportionally smaller, but significant reduction was observed during the remaining 22.5 h of the first post-deprivation day, suggesting that drinking during the early and late phases of the ADE is differentially sensitive to chronic haloperidol. Nonetheless, the reversal of the ADE during the first hour of renewed access may be indicative of protective effects of chronic D_2 antagonist treatments against “loss of control.” This effect may be mediated via haloperidol-induced upregulation of postsynaptic D_2 receptors and consequently enhanced DA transmission, in agreement with the literature implicating D_2 deficits in alcohol preference (e.g., [48–50]). In a second study, lisuride, an ergot derivative acting as a D_2 agonist, failed to reduce ethanol intake in rats tested in a long-term free-access model that utilizes repeated brief deprivation periods to induce high ethanol intake [44]. In fact, lisuride slightly increased consumption in high ethanol drinking as well as previously ethanol-naïve rats, presumably as a result of autoregulatory decreases in DA activity produced by the D_2 agonist (see e.g., [93]), a condition that has been linked to increased ethanol preference and intake [48–51].

Table 1.

Model/procedure	Compound/dose	Abstinence	Effect	Species/strain	Reference
Conditioned reinstatement	D_1 antagonist — SCH 23390 (5, 10, 50 µg/kg)	16 days (Extinction)	Decrease	Wistar rats	[54]
	D_2 antagonist — Eticlopride (5, 10, 50 µg/kg)	16 days (Extinction)	Decrease	Wistar rats	[54]
Appetitive ethanol-seeking	D_2 antagonist — Raclopride (1, 3, 10 µg/rat; intra-NAcc)	24 h	Decrease	Long-Evans rats	[25]
	D_2 antagonist — Remoxipride (5, 10, 15 mg/kg)	24 h	Decrease	Long-Evans rats	[27]
Conditioned reinforcement	D_2 antagonist — Haloperidol (0.25 mg/kg)	24 h	Decrease	Sprague-Dawley rats	[113]
Conditioned place preference	D_1 antagonist — SCH 23390 (0.015, 0.03 mg/kg)		No Effect	DBA/2J mice	[30]
	D_2 antagonist — Raclopride (0.3, 0.6 mg/kg)		No Effect	DBA/2J mice	[30]
	D_2 antagonist — Haloperidol (0.05, 0.1 mg/kg)		No Effect	DBA/2J mice	[22]
	D_3 antagonist — U99194A (10, 20 mg/kg)		No Effect	DBA/2J mice	[30]
	D_3 antagonist — U99194A (10, 20 mg/kg)		Increase	Swiss-Webster mice	[10, 11]
Alcohol deprivation effect	D_2 antagonist — Haloperidol (1 mg/kg/day for 2 weeks)	14 days	Reduction	CBA x C57BL mice	[87]
	D_2 agonist — Lisuride (90 µg/kg/day for 8 weeks)	36 weeks	No effect	Wistar rats	[59]

Is dopamine neurotransmission a promising treatment target?

As elaborated above, the neurobiological literature reveals a significant role for DA in relapse risk associated with alcohol cue exposure as well as neuroadaptive changes in DA function. Preclinical data from studies targeting DA receptors to attenuate ethanol-seeking behavior are largely consistent with this literature and implicate DA neurotransmission as a treatment target. Controlled laboratory studies in humans further support this possibility with findings that blockade of D_2 receptors reduces craving induced by an ethanol priming dose, blocks the stimulant and euphorigenic effects of ethanol, and reduces ethanol intake. While the latter findings provide excellent support for a role of DA in acute ethanol reinforcement in humans it remains questionable whether direct pharmacological manipulation of DA transmission represents a promising pharmacotherapeutic approach for relapse prevention.

First, although preclinical evidence strongly suggests a potential for DA antagonists in preventing or ameliorating craving and relapse associated with alcohol cue exposure, the expectation of sustained therapeutic benefits with DA antagonist treatments is fraught with the complication that such treatments would exacerbate the DA hypoactivity that accompanies acute and protracted withdrawal, and that has been linked to increased relapse risk. Second, DA agonist treatments that, based on the preclinical literature, would be predicted to reduce susceptibility to relapse by ameliorating DA hypofunction are problematic as well. The D_2 agonist bromocriptine administered during the acute ethanol withdrawal phase reduced craving, anxiety, restlessness and depression in hospitalized alcoholics [41]. Longer bromocriptine treatment (six weeks) reduced craving and anxiety as well, particularly in alcoholics with the D_2 receptor A1 allele, but this treatment regimen failed to significantly reduce attrition rates [94]. Similarly, a long-acting bromocriptine preparation (Parlodel-LAR®) proved ineffective in reducing relapse rates in detoxified alcoholics [95], and chronic treatment with another D_2 agonist, lisuride, actually enhanced proclivity to relapse by decreasing the latency to resume drinking [93]. This effect, and the lack of anti-relapse efficacy of D_2 agonists in general, appears to be a consequence of DA autoregulatory changes associated with chronic DA agonist administration. More specifically, neuroendocrine measures of dopaminergic responsivity to D_2 stimulation in alcoholics suggest that chronic lisuride produces an autoinhibitory reduction in dopaminergic tone that likely exacerbates DA hypoactivity, thereby increasing the likelihood of relapse [93].

As more extensively discussed in the Chapter by Carai et al., these clinical findings provide further support for the hypothesis that DA transmission is an important player in the regulation of ethanol-seeking, but illustrate also that direct manipulation of this system, particularly in the context of chronic treatments likely to be required for relapse prevention, is associated with complications that limit therapeutic promise. It remains to be determined whether low-dose D_1 antagonist treatments or agents that act as partial agonists at the D_2

receptor may prove beneficial. The latter class of agents that exert functional agonist effects under conditions of low dopaminergic activity but exert antagonist actions when dopaminergic activity is high may offer some promise by ameliorating chronic ethanol-induced DA deficits while at the same time attenuating DA "surges" associated with ethanol cue exposure. Partial agonists including terguride and SDZ 208-911 reduce ethanol intake in rats with both acute and chronic administration [96], but a possible "anti-relapse" potential of these agents has not yet been established in appropriate animal models.

To properly evaluate the significance and potential of DA neurotransmission to serve as a target for anti-relapse medications, it is important to bear in mind that indirect modification of DA activity through agents that act on other neurochemical systems may perhaps provide an effective approach, eliminating problematic side- and secondary effects associated with chronic DA agonist or antagonist treatments. In fact, as briefly discussed below, regulation or modification of DA activity by other neurochemical systems is likely to contribute to the therapeutic actions of agents with established anti-craving or anti-relapse efficacy.

Effects of indirect manipulation of dopamine transmission on ethanol-seeking

Opiate antagonists

The nonselective opiate receptor antagonist naltrexone as well as mu-opioid and delta-opioid-selective antagonists effectively reverse the behavioral effects of ethanol-related stimuli in conditioned reinstatement [68, 70, 71, 97] and CPP [98, 99] tests. Paralleling these findings, naltrexone attenuates cue-induced ethanol craving [100, 101] and lowers relapse rates in alcoholic populations ([102–107]; see also Chapters by Cowen and O'Brien et al.).

Opiate antagonists blunt the stimulatory effects of systemically and self-administered ethanol on DA release in the nucleus accumbens [10, 108–110]. Interference with ethanol-induced activation of mesolimbic DA transmission and, consequently, ethanol reward has therefore been implicated as a mechanism by which opiate antagonists reduce ethanol intake. This naltrexone-induced decrease in the rewarding efficacy of ethanol also may limit ethanol intake during a "lapse" in abstinent alcoholics and thereby lower the incidence of "full-blown" relapse.

A second mechanism for the therapeutic effects of naltrexone involving interactions with DA transmission may be interference with the neurochemical effects of alcohol cues hypothesized to underlie conditioned cue reactivity and craving. Anticipation of ethanol and exposure to alcohol-related contextual stimuli can increase extracellular DA levels in the nucleus accumbens [5, 10–12], and alcohol cue exposure activates DA-rich brain regions in abstinent alcoholics [27]. In view of the ability of opiate antagonists to block ethanol-induced stimulation of DA release, it is possible that these agents also interfere with the DA-activating actions of ethanol cues. Direct evidence supporting this hypothesis is still lacking. However, it is not unlikely that such effects may contribute to the anti-craving effects of opiate antagonists in humans and their inhibitory effects on conditioned ethanol-seeking behavior in animals.

Acamprosate

Acamprosate reduces relapse during ethanol withdrawal in humans and lowers ethanol-seeking and intake in animal models (e.g., [111, 112]; see also Chapters by de Witte et al. and Mann). The mechanism(s) underlying these effects are not well understood, but are thought to involve reduction of neural hyperexcitability associated with ethanol withdrawal via restoration of normal *N*-methyl-D-aspartate receptor function [112–114]. Recent evidence indicates, however, that acute acamprosate treatment also dose-dependently delays and suppresses ethanol's stimulatory effects on DA release in the nucleus accumbens [115]. This suggests that acamprosate may reduce relapse during withdrawal by attenuating the ability of ethanol to activate the mesolimbic DA reward pathway. Acamprosate, however, did not attenuate appetitive-phase responding (i.e., ethanol-seeking elicited and maintained by the incentive-motivational effects of a drug-related environment) in rats [116]. Thus, the DA inhibitory effect of acamprosate and its consequences on ethanol intake may be limited to the direct pharmacological actions of ethanol, without interfering with the presumably DA-mediated effects of alcohol-related environmental stimuli on ethanol-seeking behavior and craving.

Acknowledgements
The author would like to thank Mike Arends for his editorial assistance.

References

1 Brodie MS, Shefner SA, Dunwiddie TV (1990) Ethanol increases the firing rate of dopamine neurons of the rat ventral tegmental area *in vitro*. *Brain Res* 508: 65–69
2 Brodie MS, Pesold C, Appel SB (1999) Ethanol directly excites dopaminergic ventral tegmental area reward neurons. *Alcohol Clin Exp Res* 23: 1848–1852
3 Gessa GL, Muntoni F, Collu M, Vargiu L, Mereu G (1985) Low doses of ethanol activate dopaminergic neurons in the ventral tegmental area. *Brain Res* 348: 201–203
4 Imperato A, DiChiara G (1986) Preferential stimulation of dopamine release in the nucleus accumbens of freely moving rats by ethanol. *J Pharmacol Exp Ther* 239: 219–239
5 Weiss F, Lorang MT, Bloom FE, Koob GF (1993) Oral alcohol self-administration stimulates dopamine release in the rat nucleus accumbens: Genetic and motivational determinants. *J Pharmacol Exp Ther* 267: 250–258
6 Koob GF, Vaccarino F, Amalric M, Bloom FE (1987) Positive reinforcement properties of drugs: Search for neural substrates. In: J Engel, L Oreland, DH Ingvar, B Pernow, S Rossner, LA Pellborn (eds): *Brain Reward Systems and Abuse*. Raven Press, New York, 35–50
7 Samson HH (1992) The function of brain dopamine in ethanol reinforcement. In: RR Watson (ed.): *Alcohol and Neurobiology: Receptors, Membranes, and Channels*. CRC Press, Boca Raton, FL, 91–107

8 Spanagel R, Weiss F (1999) The dopamine hypothesis of reward: past and current status. *Trends Neurosci* 22: 521–527
9 Wise RA, Rompre PP (1989) Brain dopamine and reward. *Annu Rev Psychol* 40: 191–225
10 Gonzales RA, Weiss F (1998) Suppression of ethanol-reinforced behavior by naltrexone is associated with attenuation of the ethanol-induced increase in dialysate dopamine levels in the nucleus accumbens. *J Neurosci* 18: 10663–10671
11 Melendez RI, Rodd-Henricks ZA, Engleman EA, Li TK, McBride WJ, Murphy JM (2002) Microdialysis of dopamine in the nucleus accumbens of alcohol-preferring (P) rats during anticipation and operant self-administration of ethanol. *Alcohol Clin Exp Res* 26: 318–325
12 Vavrousek-Jakuba E, Cohen CA, Shoemaker WJ (1992) Ethanol effects on dopamine receptors: *In vivo* binding following voluntary ethanol (ETOH) intake in rats. In: CA Naranjo, EM Sellers (eds: *Novel Pharmacological Interventions for Alcoholism*. Springer, New York, 372–374
13 O'Brien CP, Childress AR, Ehrman R, Robbins SJ (1998) Conditioning factors in drug abuse: Can they explain compulsion? *J Psychopharmacol* 12: 15–22
14 Cooney NL, Litt MD, Morse PA, Bauer LO, Gaupp L (1997) Alcohol cue reactivity, negative-mood reactivity, and relapse in treated alcoholic men. *J Abnorm Psychol* 106: 243–250
15 Kaplan RF, Cooney NL, Baker LH, Gillespie RA, Meyer RE, Pomerleau OF (1985) Reactivity to alcohol-related cues: Physiological and subjective responses in alcoholics and nonproblem drinkers. *J Stud Alcohol* 46: 267–272
16 Laberg JC (1986) Alcohol and expectancy: Subjective, psychophysiological and behavioral responses to alcohol stimuli in severely, moderately and non-dependent drinkers. *Br J Addict* 81: 797–808
17 McCusker CG, Brown K (1990) Alcohol-predictive cues enhance tolerance to and precipitate "craving" for alcohol in social drinkers. *J Stud Alcohol* 51: 494–499
18 McCusker CG, Brown K (1995) Cue-exposure to alcohol-associated stimuli reduces autonomic reactivity, but not craving and anxiety, in dependent drinkers. *Alcohol Alcoholism* 30: 319–327
19 Monti PM, Binkoff JA, Abrams DB, Zwick WR, Nirenberg TD, Liepman MR (1987) Reactivity of alcoholics and nonalcoholics to drinking cues. *J Abnorm Psychol* 96: 122–126
20 Monti PM, Rohsenow DJ, Rubonis AV, Niaura RS, Sirota AD, Colby SM, Abrams DB (1993) Alcohol cue reactivity: effects of detoxification and extended exposure. *J Stud Alcohol* 54: 235–245
21 Staiger PK, White JM (1991) Cue reactivity in alcohol abusers: Stimulus specificity and extinction of the responses. *Addict Behav* 16: 211–221
22 Miller NS, Gold MS (1994) Dissociation of "conscious desire" (craving) from and relapse in alcohol and cocaine dependence. *Ann Clin Psychiat* 6: 99–106
23 Tiffany ST, Carter BL (1998) Is craving the source of compulsive drug use? *J Psychopharmacol* 12: 23–30
24 Everitt BJ, Wolf ME (2002) Psychomotor stimulant addiction: a neural systems perspective. *J Neurosci* 22: 3312–3320
25 Ikemoto S, Panksepp J (1999) The role of nucleus accumbens dopamine in motivated behavior: a unifying interpretation with special reference to reward-seeking. *Brain Res Brain Res Rev* 31: 6–41
26 Robbins TW, Everitt BJ (1996) Neurobehavioural mechanisms of reward and motivation. *Curr Opin Neurobiol* 6: 228–236
27 Braus DF, Wrase J, Grusser S, Hermann D, Ruf M, Flor H, Mann K, Heinz A (2001) Alcohol-associated stimuli activate the ventral striatum in abstinent alcoholics. *J Neural Transm* 108: 887–894
28 Slawecki CJ, Hodge CW, Samson HH (1997) Dopaminergic and opiate agonists and antagonists differentially decrease multiple schedule responding maintained by sucrose/ethanol and sucrose. *Alcohol* 14: 281–294
29 Jaffe JH, Cascella NG, Kumor KM, Sherer MA (1989) Cocaine-induced cocaine craving. *Psychopharmacology* 97: 59–64
30 Ludwig AM, Wikler A, Stark LH (1974) The first drink: psychobiological aspects of craving. *Arch Gen Psychiatry* 30: 539–547
31 Modell JG, Mountz JM, Glaser FB, Lee JY (1993) Effect of haloperidol on measures of craving and impaired control in alcoholic subjects. *Alcohol Clin Exp Res* 17: 234–240
32 Enggasser JL, de Wit H (2001) Haloperidol reduces stimulant and reinforcing effects of ethanol in social drinkers. *Alcohol Clin Exp Res* 25: 1448–1456
33 Fadda F, Argiolas A, Melis MR, Serra G, Gessa GL (1980) Differential effect of acute and chron-

ic ethanol on dopamine metabolism in frontal cortex, caudate nucleus and substantia nigra. *Life Sci* 27: 979–986
34 Rothblat DS, Rubin E, Schneider JS (2001) Effects of chronic alcohol ingestion on the mesostriatal dopamine system in the rat. *Neurosci Lett* 300: 63–66
35 Tabakoff B, Hoffman PL (1978) Alterations in receptors controlling dopamine synthesis after chronic ethanol ingestion. *J Neurochem* 31: 1223–1229
36 Diana M, Pistis M, Carboni S, Gessa GL, Rossetti ZL (1993) Profound decrement of mesolimbic neuronal activity during ethanol withdrawal syndrome in rats: Electrophysiological and biochemical evidence. *Proc Natl Acad Sci USA* 90: 7966–7969
37 Shen RY, Chiodo LA (1993) Acute withdrawal after repeated ethanol treatment reduces the number of spontaneously active dopaminergic neurons in the ventral tegmental area. *Brain Res* 622: 289–293
38 Rossetti ZL, Hmaidan Y, Gessa GL (1992) Marked inhibition of mesolimbic dopamine release: A common feature of ethanol, morphine, cocaine and amphetamine abstinence in rats. *Eur J Pharmacol* 221: 227–234
39 Weiss F, Parsons LH, Schulteis G, Hyytiä P, Lorang MT, Bloom FE, Koob GF (1996) Ethanol self-administration restores withdrawal-associated deficiencies in accumbal dopamine and 5-hydroxytryptamine release in dependent rats. *J Neurosci* 16: 3474–3485
40 Rossetti ZL, Melis F, Carboni S, Diana M, Gessa GL (1992) Alcohol withdrawal in rats is associated with a marked fall in extraneuronal dopamine. *Alcohol Clin Exp Res* 16: 529–532
41 Borg V, Weinholdt T (1982) Bromocriptine in the treatment of the alcohol-withdrawal syndrome. *Acta Psychiatr Scand* 65: 101–111
42 Diana M, Pistis M, Muntoni A, Gessa G (1996) Mesolimbic dopaminergic reduction outlasts ethanol withdrawal syndrome: evidence of protracted abstinence. *Neuroscience* 71: 411–415
43 Bailey CP, Andrews N, McKnight AT, Hughes J, Little HJ (2000) Prolonged changes in neurochemistry of dopamine neurones after chronic ethanol consumption. *Pharmacol Biochem Behav* 66: 153–161
44 May T, Wolf U, Wolffgramm J (1995) Striatal dopamine receptors and adenylyl cyclase activity in a rat model of alcohol addiction: effects of ethanol and lisuride treatment. *J Pharmacol Exp Ther* 275: 1195–1203
45 Heinz A, Dettling M, Kuhn S, Dufeu P, Graf KJ, Kurten I, Rommelspacher H, Schmidt IG (1995) Blunted growth hormone response is associated with early relapse in alcohol-dependent patients. *Alcohol Clin Exp Res* 19: 62–65
46 Heinz A, Lichtenberg-Kraag B, Baum SS, Graf K, Kruger F, Dettling M, Rommelspacher H (1995) Evidence for prolonged recovery of dopaminergic transmission after detoxification in alcoholics with poor treatment outcome. *J Neural Transm Gen Sect* 102: 149–157
47 Wiesbeck GA, Weijers HG, Gross JP (2000) Craving for alcohol and dopamine receptor sensitivity in alcohol-dependent men and control subjects. *J Neural Transm* 107: 691–699
48 George SR, Fan T, Ng GY, Jung SY, O'Dowd BF, Naranjo CA (1995) Low endogenous dopamine function in brain predisposes to high alcohol preference and consumption: Reversal by increasing synaptic dopamine. *J Pharmacol Exp Ther* 273: 373–379
49 McBride WJ, Murphy JM, Gatto GJ, Levy AD, Lumeng L, Li TK (1991) Serotonin and dopamine systems regulating alcohol intake. *Alcohol Alcoholism Suppl* 1: 411–416
50 McBride WJ, Li TK (1998) Animal models of alcoholism: neurobiology of high alcohol-drinking behavior in rodents. *Crit Rev Neurobiol* 12: 339–369
51 Thanos PK, Volkow ND, Freimuth P, Umegaki H, Ikari H, Roth G, Ingram DK, Hitzemann R (2001) Overexpression of dopamine D2 receptors reduces alcohol self-administration. *J Neurochem* 78: 1094–1103
52 Sinclair JD, Senter RJ (1967) Increased preference for ethanol in rats following alcohol deprivation. *Psychonomic Sci* 8: 11–12
53 Sinclair JD, Li TK (1989) Long and short alcohol deprivation: Effects on AA and P alcohol-preferring rats. *Alcohol* 6: 505–509
54 Rankin HR, Hodgson R, Stockwell T (1979) The concept of craving and its measurement. *Behav Res Ther* 17: 389–396
55 Wolffgramm J, Heyne A (1995) From controlled drug intake to loss of control: the irreversible development of drug addiction in the rat. *Behav Brain Res* 70: 77–94
56 Kornet M, Goosen C, Van Ree JM (1991) Effects of naltrexone on alcohol consumption during chronic alcohol drinking and after a period of imposed abstinence in free-choice drinking rhesus

monkeys. *Psychopharmacology* 104: 367–376
57 Burish TG, Maisto SA, Cooper AM, Sobell MB (1981) Effects of voluntary short-term abstinence from alcohol on subsequent drinking patterns of college students. *J Stud Alcohol* 42: 1013–1020
58 Ludwig AM, Wikler A (1974) "Craving" and relapse to drink. *Q J Stud Alcohol* 35: 108–130
59 O'Donnell PJ (1984) The abstinence violation effect and circumstances surrounding relapse as predictors of outcome status in male alcoholic outpatients. *J Psychol* 117: 257–262
60 Spanagel R, Hölter SM, Allingham K, Landgraf R, Zieglgänsberger W (1996) Acamprosate and alcohol: I. Effects on alcohol intake following alcohol deprivation in the rat. *Eur J Pharmacol* 305: 39–44
61 Wolffgramm J (1991) An ethopharmacological approach to the development of drug addiction. *Neurosci Biobehav Rev* 15: 515–519
62 Thielen RJ, Engleman EA, Rodd ZA, Murphy JM, Lumeng L, Li TK, McBride WJ (2004) Ethanol drinking and deprivation alter dopaminergic and serotonergic function in the nucleus accumbens of alcohol-preferring rats. *J Pharmacol Exp Ther* 309: 216–225
63 Nestby P, Vanderschuren LJ, De Vries TJ, Mulder AH, Wardeh G, Hogenboom F, Schoffelmeer AN (1999) Unrestricted free-choice ethanol self-administration in rats causes long-term neuroadaptations in the nucleus accumbens and caudate putamen. *Psychopharmacology (Berl)* 141: 307–314
64 Lê A, Shaham Y (2002) Neurobiology of relapse to alcohol in rats. *Pharmacol Ther* 94: 137–156
65 Shaham Y, Shalev U, Lu L, De Wit H, Stewart J (2003) The reinstatement model of drug relapse: history, methodology and major findings. *Psychopharmacology (Berl)* 168: 3–20
66 Bienkowski P, Kostowski W, Koros E (1999) Ethanol-reinforced behaviour in the rat: effects of naltrexone. *Eur J Pharmacol* 374: 321–327
67 Bienkowski P, Kostowski W, Koros E (1999) The role of drug-paired stimuli in extinction and reinstatement of ethanol-seeking behaviour in the rat. *Eur J Pharmacol* 374: 315–319
68 Ciccocioppo R, Martin-Fardon R, Weiss F (2002) Effect of selective blockade of mu(1) or delta opioid receptors on reinstatement of alcohol-seeking behavior by drug-associated stimuli in rats. *Neuropsychopharmacology* 27: 391–399
69 Ciccocioppo R, Economidou D, Fedeli A, Angeletti S, Weiss F, Heilig M, Massi M (2003) Attenuation of ethanol self-administration and of conditioned reinstatement of alcohol-seeking behaviour by the antiopioid peptide nociceptin/orphanin FQ in alcohol-preferring rats. *Psychopharmacology (Berl)* 172: 170–178
70 Ciccocioppo R, Lin D, Martin-Fardon R, Weiss F (2003) Reinstatement of ethanol-seeking behavior by drug cues following single *versus* multiple ethanol intoxication in the rat: effects of naltrexone. *Psychopharmacology (Berl)* 168: 208–215
71 Katner SN, Magalong JG, Weiss F (1999) Reinstatement of alcohol-seeking behavior by drug-associated discriminative stimuli after prolonged extinction in the rat. *Neuropsychopharmacology* 20: 471–479
72 Katner SN, Weiss F (1999) Ethanol-associated olfactory stimuli reinstate ethanol-seeking behavior after extinction and modify extracellular dopamine levels in the nucleus accumbens. *Alcohol Clin Exp Res* 23: 1751–1760
73 Ciccocioppo R, Angeletti S, Weiss F (2001) Long-lasting resistance to extinction of response reinstatement induced by ethanol-related stimuli: role of genetic ethanol preference. *Alcohol Clin Exp Res* 25: 1414–1419
74 Liu X, Weiss F (2002) Reversal of ethanol-seeking behavior by D1 and D2 antagonists in an animal model of relapse: differences in antagonist potency in previously ethanol-dependent *versus* nondependent rats. *J Pharmacol Exp Ther* 300: 882–889
75 Samson HH, Slawecki CJ, Sharpe AL, Chappell A (1998) Appetitive and consummatory behaviors in the control of ethanol consumption: a measure of ethanol seeking behavior. *Alcohol Clin Exp Res* 22: 1783–1787
76 Samson HH, Sharpe AL, Denning C (1999) Initiation of ethanol self-administration in the rat using sucrose substitution in a sipper-tube procedure. *Psychopharmacology (Berl)* 147: 274–279
77 Samson HH, Czachowski CL, Slawecki CJ (2000) A new assessment of the ability of oral ethanol to function as a reinforcing stimulus. *Alcohol Clin Exp Res* 24: 766–773
78 Czachowski CL, Chappell AM, Samson HH (2001) Effects of raclopride in the nucleus accumbens on ethanol seeking and consumption. *Alcohol Clin Exp Res* 25: 1431–1440
79 Czachowski CL, Santini LA, Legg BH, Samson HH (2002) Separate measures of ethanol seeking and drinking in the rat: effects of remoxipride. *Alcohol* 28: 39–46

80 Wilson AW, Costall B, Neill JC (2000) Manipulation of operant responding for an ethanol-paired conditioned stimulus in the rat by pharmacological alteration of the serotonergic system. *J Psychopharmacol* 14: 340–346
81 Cunningham CL, Howard MA, Gill SJ, Rubinstein M, Low MJ, Grandy DK (2000) Ethanol-conditioned place preference is reduced in dopamine D2 receptor-deficient mice. *Pharmacol Biochem Behav* 67: 693–699
82 El-Ghundi M, George SR, Drago J, Fletcher PJ, Fan T, Nguyen T, Liu C, Sibley DR, Westphal H, O'Dowd BF (1998) Disruption of dopamine D1 receptor gene expression attenuates alcohol-seeking behavior. *Eur J Pharmacol* 353: 149–158
83 Risinger FO, Freeman PA, Rubinstein M, Low MJ, Grandy DK (2000) Lack of operant ethanol self-administration in dopamine D2 receptor knockout mice. *Psychopharmacology (Berl)* 152: 343–350
84 Risinger FO, Freeman PA, Greengard P, Fienberg AA (2001) Motivational effects of ethanol in DARPP-32 knock-out mice. *J Neurosci* 21: 340–348
85 Cunningham CL, Malott DH, Dickinson SD, Risinger FO (1992) Haloperidol does not alter expression of ethanol-induced conditioned place preference. *Behav Brain Res* 50: 1–5
86 Dickinson SD, Lee EL, Rindal K, Cunningham CL (2003) Lack of effect of dopamine receptor blockade on expression of ethanol-induced conditioned place preference in mice. *Psychopharmacology (Berl)* 165: 238–244
87 Boyce JM, Risinger FO (2000) Enhancement of ethanol reward by dopamine D3 receptor blockade. *Brain Res* 880: 202–206
88 Boyce JM, Risinger FO (2002) Dopamine D3 receptor antagonist effects on the motivational effects of ethanol. *Alcohol* 28: 47–55
89 Kling-Petersen T, Ljung E, Wollter L, Svensson K (1995) Effects of dopamine D3 preferring compounds on conditioned place preference and intracranial self-stimulation in the rat. *J Neural Transm Gen Sect* 101: 27–39
90 Boyce-Rustay JM, Risinger FO (2003) Dopamine D3 receptor knockout mice and the motivational effects of ethanol. *Pharmacol Biochem Behav* 75: 373–379
91 Roberts AJ, Heyser CJ, Cole M, Griffin P, Koob GF (2000) Excessive ethanol drinking following a history of dependence: animal model of allostasis. *Neuropsychopharmacology* 22: 581–594
92 Salimov RM, Salimova NB, Shvets LN, Maisky AI (2000) Haloperidol administered subchronically reduces the alcohol-deprivation effect in mice. *Alcohol* 20: 61–68
93 Schmidt LG, Kuhn S, Smolka M, Schmidt K, Rommelspacher H (2002) Lisuride, a dopamine D2 receptor agonist, and anticraving drug expectancy as modifiers of relapse in alcohol dependence. *Prog Neuropsychopharmacol Biol Psychiatry* 26: 209–217
94 Lawford BR, Young RM, Rowell JA, Qualichefski J, Fletcher BH, Syndulko K, Ritchie T, Noble EP (1995) Bromocriptine in the treatment of alcoholics with the D2 dopamine receptor A1 allele. *Nat Med* 1: 337–341
95 Naranjo CA, Dongier M, Bremner KE (1997) Long-acting injectable bromocriptine does not reduce relapse in alcoholics. *Addiction* 92: 969–978
96 Bono G, Balducci C, Richelmi P, Koob GF, Pulvirenti L (1996) Dopamine partial receptor agonists reduce ethanol intake in the rat. *Eur J Pharmacol* 296: 233–238
97 Liu X, Weiss F (2002) Additive effect of stress and drug cues on reinstatement of ethanol seeking: exacerbation by history of dependence and role of concurrent activation of corticotropin-releasing factor and opioid mechanisms. *J Neurosci* 22: 7856–7861
98 Cunningham CL, Henderson CM, Bormann NM (1998) Extinction of ethanol-induced conditioned place preference and conditioned place aversion: effects of naloxone. *Psychopharmacology (Berl)* 139: 62–70
99 Gerrits MAFM, Patkina N, Zvartau EE, van Ree JM (1995) Opioid blockade attenuates acquisition and expression of cocaine-induced place preference conditioning in rats. *Psychopharmacology (Berl)* 119: 92–98
100 Monti PM, Rohsenow DJ, Hutchison KE, Swift RM, Mueller TI, Colby SM, Brown RA, Gulliver SB, Gordon A, Abrams DB (1999) Naltrexone's effect on cue-elicited craving among alcoholics in treatment. *Alcohol Clin Exp Res* 23: 1386–1394
101 Rohsenow DJ, Monti PM, Hutchison KE, Swift RM, Colby SM, Kaplan GB (2000) Naltrexone's effects on reactivity to alcohol cues among alcoholic men. *J Abnorm Psychol* 109: 738–742
102 Anton RF, Moak DH, Waid LR, Latham PK, Malcolm RJ, Dias JK (1999) Naltrexone and cognitive behavioral therapy for the treatment of outpatient alcoholics: results of a placebo-con-

trolled trial. *Am J Psychiatry* 156: 1758–1764
103 Anton RF, Moak DH, Latham PK, Waid LR, Malcolm RJ, Dias JK, Roberts JS (2001) Posttreatment results of combining naltrexone with cognitive-behavior therapy for the treatment of alcoholism. *J Clin Psychopharmacol* 21: 72–77
104 Monterosso JR, Flannery BA, Pettinati HM, Oslin DW, Rukstalis M, O'Brien CP, Volpicelli JR (2001) Predicting treatment response to naltrexone: the influence of craving and family history. *Am J Addict* 10: 258–268
105 O'Brien CP, Volpicelli LA, Volpicelli JR (1996) Naltrexone in the treatment of alcoholism: a clinical review. *Alcohol* 13: 35–39
106 O'Malley SS, Jaffe AJ, Chang G, Schottenfeld RS, Meyer RE, Rounsaville B (1992) Naltrexone and coping skills therapy for alcohol dependence: a controlled study. *Arch Gen Psychiatry* 49: 881–887
107 Volpicelli JR, Alterman AI, Hayashida M, O'Brien CP (1992) Naltrexone in the treatment of alcohol dependence. *Arch Gen Psychiatry* 49: 876–880
108 Acquas E, Meloni M, Di Chiara G (1993) Blockade of δ-opioid receptors in the nucleus accumbens prevents ethanol-induced stimulation of dopamine release. *Eur J Pharmacol* 230: 239–241
109 Benjamin D, Grant ER, Pohorecky LA (1993) Naltrexone reverses ethanol-induced dopamine release in the nucleus accumbens in awake, freely moving rats. *Brain Res* 621: 137–140
110 Widdowson PS, Holman RB (1992) Ethanol-induced increase in endogenous dopamine release may involve endogenous opiates. *J Neurochem* 59: 157–163
111 Mason BJ, Ownby RL (2000) Acamprosate for the treatment of alcohol dependence: a review of double-blind, placebo-controlled trials. *CNS Spectrums* 5: 58–69
112 Spanagel R, Zieglgansberger W (1997) Anti-craving compounds for ethanol: new pharmacological tools to study addictive processes. *Trends Pharmacol Sci* 18: 54–59
113 Littleton J (1995) Acamprosate in alcohol dependence: How does it work? *Addiction* 90: 1179–1188
114 Wilde MI, Wagstaff AJ (1997) Acamprosate. A review of its pharmacology and clinical potential in the management of alcohol dependence after detoxification. *Drugs* 53: 1038–1053
115 Olive MF, Nannini MA, Ou CJ, Koenig HN, Hodge CW (2002) Effects of acute acamprosate and homotaurine on ethanol intake and ethanol-stimulated mesolimbic dopamine release. *Eur J Pharmacol* 437: 55–61
116 Czachowski CL, Legg BH, Samson HH (2001) Effects of acamprosate on ethanol-seeking and self-administration in the rat. *Alcohol Clin Exp Res* 25: 344–350

Drugs for Relapse Prevention of Alcoholism
Edited by R. Spanagel and K. Mann

Dopaminergic compounds: clinical data

Gerhard A. Wiesbeck

Department of Substance Use Disorders, Psychiatric University Hospital, Wilhelm-Klein-Str. 27, 4025 Basel, Switzerland

Introduction

There is reliable evidence from animal research that alcohol, like other drugs of abuse, exerts its reinforcing effects by acting on the brain dopaminergic systems [1–5]. Neuroadaptive transformations (sensitization) of these systems are considered to cause the transition from controlled to uncontrolled use of alcohol [6, 7].

It was assumed that the primary role of dopamine systems in reward is to mediate the so-called subjective pleasurable and hedonic effects of addictive drugs. This opinion was expressed most explicitly in the anhedonia hypothesis of dopamine function [8]. Others offered contradicting evidence in that dopamine systems do not mediate the pleasurable or euphoric effects. Rather sensitization-like persisting neuroadaptations of dopamine systems may result in a hyperactive reaction to the effects of drugs, even after long-term abstinence (incentive-sensitization hypothesis) [9]. Dopamine neurons seem to respond to salient stimuli and are thus involved in the anticipation of reward [1, 10].

There is increasing evidence from human studies implicating dopaminergic neurotransmission in alcohol dependence. For example, after the administration of apomorphine, which is a dopamine receptor agonist, maximal growth hormone (GH) response was significantly reduced in alcoholics who were two months abstinent, when compared to controls [11]. Additionally, only alcoholics with a positive family history for alcoholism (those with a biological loading) revealed an impaired dopamine receptor function [13]. Finally, relapsed alcoholics were characterized by a more blunted GH response to apomorphine than abstinent alcoholics, suggesting that a reduced sensitivity of dopamine receptors is associated with a higher risk for relapse [14].

There is additional evidence regarding the involvement of the brain dopaminergic systems in alcohol dependence from studies employing brain imaging techniques such as photon emission computed tomography (PECT) or positron emission tomography (PET) in humans [15]. For example, striatal dopamine D2 receptor density and affinity to dopamine were found to be reduced in alcoholics compared to controls [16]. This finding is supported by another PET study showing decreased dopamine D2 receptor availability in alcoholics [17]. PECT studies reported decreased striatal dopamine transporter densities in late-onset alcoholics [18] and interestingly, suggested a relationship between low striatal dopamine D2 receptor levels and the risk for early relapse in detoxified alcoholics [19].

Finally, there are genetic studies giving support to the implication of dopamine systems in alcoholism [20, 21]. For example, bromocriptine, a dopamine agonist, was reported to reduce craving selectively in alcoholics carrying the A1 allele of the dopamine D2 receptor gene [22] and recently it has been hypothesized that the D2 dopamine receptor gene is a reinforcement or reward gene for the effects of alcohol and other drugs of abuse [23]. Given this importance of dopamine in alcoholism, there is a legitimate interest in dopaminergic agents as treatments for alcohol dependence. Reports from animal studies suggest that dopamine agonists and antagonists both decrease alcohol consumption (see Chapter 15). The present chapter selectively reviews dopaminergic compounds experienced in human studies (Tab. 1). It discusses their putative mechanisms of action and their efficacy.

Table 1. Dopaminergic compounds tested in randomised, double-blind, placebo-controlled trials with an intention-to-treat analysis and outcome parameters related to abstinence

Dopaminergic compound	Result	Reference
Lisuride	A sig. higher relapse rate under lisuride compared to placebo	Schmidt et al 2002
Tiapride	No difference between tiapride and placebo treatment	Bender et al (personal communication)
Bromocriptine	No difference between bromocriptine and placebo treatment	Naranjo et al 1997
Flupenthixol	A sig. higher relapse rate under flupenthixol compared to placebo	Wiesbeck et al 2001

Lisuride

The ergot derivative lisuride is a dopamine agonist which exerts its activity primarily at postsynaptic dopamine D2 receptor sites. It also has weak dopamine D1 antagonistic as well as serotonin 5-HT1a agonistic effects. The substance is approved for the treatment of Parkinson's disease and hyperprolactinemia. Animal research and human studies on acute withdrawal suggested favorable effects in psychostimulant addiction as well [24–26].

In a double-blind, placebo-controlled trial, 120 detoxified alcohol-dependent subjects were randomly assigned a low dose of lisuride (1.0 mg/day), a high dose of lisuride (1.8 mg/day) or placebo. Time-to-first-drink within six months of treatment was defined as the primary parameter of efficacy. In contrast to the original hypothesis, lisuride treatment significantly reduced the period of abstinence as compared to placebo (effect size: 0.51). Additionally, the authors found a significant effect of drug expectancy on outcome, with the best outcome in patients who were expecting lisuride but had received placebo [27].

In summary, there is only one randomized controlled trial on lisuride. It reports a significantly better outcome under placebo. Therefore, lisuride cannot be recommended for relapse prevention in alcoholics.

Tiapride

Tiapride is a benzamide derivative which is often categorized as an atypical neuroleptic. It selectively antagonizes dopamine D2 receptors while lacking affinity for dopamine D1 receptors. There are regional differences in the binding of tiapride within the central nervous system (CNS). The substance preferentially binds to extrastriatal receptors, particularly in the hippocampus. Tiapride is characterized by antidyskinetic properties as well as by anxiolytic effects that have been shown in several animal models, including those involving ethanol withdrawal [27].

Tiapride has demonstrated clinical efficacy in ameliorating symptoms of acute alcohol withdrawal [28, 29] and earlier reports suggested promising effects in alcohol relapse prevention as well [30–33]. However, due to their less rigorous methodology, the level of evidence of those early reports is limited.

In a double-blind, placebo-controlled trial, 100 detoxified alcohol-dependent subjects were randomly assigned to either tiapride (300 mg/day) or placebo. Tiapride proved significantly better than placebo in maintaining abstinence and in reducing alcohol intake on heavy drinking days, after both three and six months of treatment [34]. The authors did not apply an intention-to-treat analysis but restricted their statistical evaluation to solely studying completers. Regarding relapse prevention, their study's level of evidence is limited.

So far, there is only one study on tiapride using a rigorous methodology in combination with an intention-to-treat analysis. Here, 300 detoxified alcohol-dependent subjects were randomly assigned to either tiapride (300 mg/day) or placebo. Primary parameters of efficacy were time-to-first-drink and relapse rate after six months of double-blind treatment. The final result revealed no difference between tiapride and placebo treatment (Bender et al., personal communication).

In summary, earlier reports on tiapride are of poor quality and their positive results have not been confirmed by a study with a more rigorous methodology. So far, there is no convincing evidence for a superiority of tiapride to placebo in alcohol relapse prevention.

Bromocriptine

Bromocriptine, an ergot alkaloid derivative, is structurally related to dopamine and it activates postsynaptic dopaminergic receptors. Since bromocriptine has been used to treat Parkinson's disease for more than two decades, there exists broad clinical experience of this drug. Bromocriptine is classified as a potent dopamine D2 receptor agonist with partial dopamine D1 antagonist activity [35].

First experiences with respect to alcoholism were related to the treatment of withdrawal. Some authors suggested that the drug ameliorates withdrawal symptoms in chronic alcoholics, whilst others did not confirm these findings [36, 37]. Using bromocriptine as a challenge drug, [38] there was reported evidence for a reduced dopamine D2 receptor sensitivity in alcoholics.

The first study on relapse prevention was conducted in 50 chronic alcoholics using a double-blind, placebo-controlled design [39]. Oral bromocriptine (3×2.5 mg/day for 3 months, increased to 3×5 mg/day during months 4–6) reduced both craving and the number of patients drinking. However, alcohol consumption data were not reliably collected and the evaluation was solely restricted to completers.

Another study was performed so as to confirm and further extend these findings. In a double-blind, placebo-controlled trial, the dose of oral bromocriptine was gradually increased to 3×2.5 mg/day, and was given over a period of seven weeks. A total of 84 subjects were enrolled; however, the analysis of efficacy was restricted to only 38 treatment completers. Though there was a marked improvement in almost all parameters between baseline and the end of treatment, bromocriptine was not superior to placebo with respect to alcohol consumption, drinking days per week, craving or GGT activity [40].

Powell et al. [41] conducted a double-blind, placebo-controlled trial of nortriptyline and bromocriptine in 216 male alcoholics, who were subtyped into three groups according to the diagnosis (alcoholism only; alcoholism and affective/anxiety disorder; alcoholism and antisocial personality disorder). The only significant effect found, after six months of treatment, was with the antisocial personality disorder patients, who were receiving nortriptyline. In this group 64% of the trial completers remained sober (bromocriptine: 34%; placebo: 11%).

In a pharmacogenetic approach, bromocriptine (3×2.5 mg/day) or placebo were double-blindly administered to 83 alcoholics with either the A1 allele (A1/A1 and A1/A2 genotypes) or only the A2 allele (A2/A2 genotype) of the dopamine D2 receptor gene [21]. The greatest improvement in craving was reported in bromocriptine-treated alcoholics carrying the A1 allele, while attrition was highest in the placebo-treated A1 alcoholics. The study indicates that pharmacogenetic strategies for subtyping alcoholics may be useful for treatment trials. However, since abstinence was not the goal of this study, it hence does not shed light on the question of whether bromocriptine might, in fact, be useful for relapse prevention.

So far, there is only one study [42] attempting to answer that question, by using a randomized, double-blind, placebo-controlled design in combination with an intention-to-treat analysis: 366 alcohol-dependent subjects were assigned to one of three treatment groups (long-acting injectable preparations of bromocriptine 25 mg/month or 50 mg/month or placebo). The primary variables of efficacy were time-to-first-drink and time-to-relapse (defined as drinking on ≥ five days per month and ≥ three drinks per day) assessed after six months of treatment. The intention-to-treat analysis revealed no differences between placebo and any dose of bromocriptine. Relapse rates to any drinking were 73%, 71% and 76% (bromocriptine 25 mg, bromocriptine 50 mg, placebo), while relapse rates to heavy drinking were 35%, 42% and 38% (bromocriptine 25 mg, bromocriptine 50 mg, placebo).

In summary, earlier reports on bromocriptine suffer from methodological shortcomings and their data have not been conclusive. So far, there is no study demonstrating a superiority of bromocriptine to placebo in maintaining abstinence or in reducing alcohol consumption. Therefore, bromocriptine cannot be recommended for relapse prevention in alcoholism.

Flupenthixol

The thioxanthene neuroleptic flupenthixol is an antipsychotic drug that antagonizes dopamine binding at a number of receptor subtypes, primarily at D1, D2, D3 and with less affinity at D4 receptors. It also affects serotonin and noradrenaline binding [43]. Besides anecdotal reports and small-sample-size studies in cocaine addicts, there are only a few studies investigating flupenthixol treatment in alcoholism [44]. For example, a six-month treatment of 21 schizophrenic patients with comorbid alcoholism resulted in a significant reduction of alcohol consumption in the intra-individual pre-/post-treatment comparison [45].

So far, there is only one randomized, double-blind, placebo-controlled study in alcoholism. In this trial 281 alcohol-dependent women and men, without comorbid psychiatric disorder, received either 10 mg flupenthixol decanoate or placebo as intramuscular injections. Primary efficacy parameters, which were based on absolute abstinence, were rated after six months of active treatment as well as after another six-month follow-up period. Subjects treated with flupenthixol revealed a significantly higher relapse rate (85.2%) after six months than those treated with placebo (65.5%). Flupenthixol was also inferior to placebo regarding the cumulative abstinence duration and the relapse rate after 12 months [46].

Two re-analyses were then calculated using these data. The first was done according to the Lesch typology [47]. It revealed that the negative outcome under flupenthixol treatment was restricted to patients belonging to Type I and Type III, indicating the importance of specific patients' characteristics and their relevance for relapse [48, 49]. The second re-analysis, which was done according to sex, discovered a significant gender-related effect. While men had an almost four-fold higher risk to relapse under flupenthixol (odds ratio: 3.95), this risk was barely elevated in women (odds ratio: 1.51). That is, the unfavorable outcome under flupenthixol was restricted to male alcoholics [50].

In summary, there is only one double-blind, placebo-controlled trial with an intention-to-treat analysis and outcome variables based on abstinence.

According to this study, flupenthixol treatment was associated with a significantly higher relapse rate as compared to placebo. Therefore, this substance cannot be recommended for relapse prevention in alcoholism.

References

1 Spanagel R, Weiss F (1999) The dopamine hypothesis of reward: past and current status. *Trends Neurosci* 22: 521–527
2 Olds ME (1990) Enhanced dopamine receptor activation in accumbens and frontal cortex has opposite effects on medial forebrain bundle self-stimulation. *Neuroscience* 35: 313–325
3 Samson HH, Hodge CW, Tolliver G et al. (1993) Effect of D1 and D2 dopamine agonists and antagonists on ethanol-reinforced behaviour: the involvement of the nucleus accumbens. *Brain Res Bull* 30: 133–141
4 Grace AA (2000) The tonic/phasic model of dopamine system regulation and its implications for understanding alcohol and psychostimulant craving. *Addiction* 95: S119–S128
5 Heinz A (2002) Dopaminergic dysfunction in alcoholism and schizophrenia – psychopathological and behavioural correlates. *Eur Psychiatry*17: 9–16
6 Hunt WA, Lands WEM (1992) A role for behavioural sensitisation in uncontrolled ethanol intake. *Alcohol* 9: 327–328
7 Robinson TE, Berridge KC (1993) the neural basis of drug craving: an incentive-sensitisation theory of addiction. *Brain Res Rev* 18: 247–291
8 Wise RA (1996) Neurobiology of addiction. *Curr Opin Neurobiol* 6: 243–251
9 Berridge KC, Robinson TE (1998) What is the role of dopamine in reward hedonic impact, reward learning, or incentive salience? *Brain Res Brain Res Rev* 28: 309–369
10 Schultz W, Dayan P, Montague PR (1997) A neural substrate of prediction and reward. *Science 275*: 1593–1599
11 Balldin J, Berggren UC, Lindstedt G (1992) Neuroendocrine evidence for reduced dopamine receptor sensitivity in alcoholism. *Alcohol Clin Exp Res* 16: 71–74
12 Balldin J, Berggren UC, Lindstedt et al. (1993) Further neuroendocrine evidence for reduced D2 receptor function in alcoholism. *Drug Alcohol Depend* 32: 159–162
13 Wiesbeck GA, Mauerer C, Thome J et al. (1995) Alcohol dependence, family history and D2 dopamine receptor function as neuroendocrinologically assessed with apomorphine. *Drug Alcohol Depend* 40: 49–53
14 Heinz A, Dettling M, Kuhn S et al. (1995) Blunted growth hormone response is associated with early relapse in alcohol-dependent patients. *Alcohol Clin Exp Res* 19: 62–65
15 Volkow ND, Fowler JS, Wang GJ (2003) Positron emission tomography and single-photon emission computed tomography in substance use research. Seminars in *Nucl Med* 33: 114–128
16 Hietela J, West C, Syvälahti E et al. (1994) Striatal D2 dopamine receptor binding characteristics *in vivo* in patients with alcohol dependence. *Psychopharmacology* 116: 285–290
17 Volkow ND, Wang GJ, Fowler JS et al. (1996) Decreases in dopamine receptors but not in dopamine transporters in alcoholics. *Alcohol Clin Exp Res* 20: 1594–1598
18 Repo E, Kuikka JT, Bergstrom KA et al. (1999) Dopamine transporter and D2-receptor density in late-onset alcoholism. *Psychopharmacology* 147: 314–318
19 Guardia J, Catafau AM; Batle F et al. (2000) Stratal dopaminergic D2 receptor density measured by [(123)I]iodobenzamide SPECT in the prediction of treatment outcome of alcohol-dependent patients. *Am J Psychiatry* 157: 127–129
20 Sander T, Harms H, Podschus J et al. (1995) Dopamine D1-, D2-, and D3 receptor genes in alcohol dependence. *Psych Genetics* 5: 177–181
21 Thome J, Gerwitz JC, Weijers HG et al. (2000) Genome polymorphism and alcoholism. *Pharmacogenomics* 1: 63–71
22 Lawford BR, Young RM, Rowell JA et al. (1995) Bromocriptine in the treatment of alcoholics with the D2 dopamine receptor A1 allele. *Nat Med* 1: 337–341
23 Noble EP (2003) D2 Dopamine receptor gene in psychiatric and neurologic disorders and its phenotypes. *Am J Med Genet* 116B: 103–125
24 Pulvirenti L, Koob GF (1993) Lisuride reduces psychomotor retardation during withdrawal from

chronic intravenous amphetamine self-administration in rats. *Neuropsychopharmacology* 8: 213–218
25 Pulvirenti L, Koob GF (1994) Dopamine receptor agonists, partial agonists and psychostimulant addiction. *Trends Pharmacol Sci* 15: 374–379
26 Gillin JC, Pulvirenti L, Withers N et al. (1994) The effects of lisuride on mood and sleep during acute withdrawal in stimulant abusers: a preliminary report. *Biol Psychiatry* 35: 843–849
27 Schmidt LG, Kuhn S, Smolka M et al. (2002) Lisuride, a dopamine D2 receptor agonist, and anticraving drug expectancy as modifiers of relapse in alcohol dependence. *Prog Neuropsychopharmacol Biol Psychiatry* 26: 209–217
28 Peters DH, Faulds D (1994) Tiapride – A review of its pharmacology and therapeutic potential in the management of alcohol dependence syndrome. *Drugs* 47: 1010–1032
29 Agricola R, Mazzarino M, Urani R (1982) Treatment of acute alcohol withdrawal syndrome with carbamazepine: a double blind comparison with tiapride. *J Int Med Res* 10: 160–165
30 Lepola U, Kokko S, Nuutila J et al. (1984) Tiapride and chlordiazepoxide in acute alcohol withdrawal. A controlled clinical trial. *Int J Clin Pharmacol Res* 5: 321–326
31 Bonnaffoux D, Jeanparis F, Bourgeat P et al. (1982) Follow-up treatment after alcohol withdrawal: results after two years in 200 cases. *La Semaine des hopitaux de Paris* 58: 2079–2081
32 Delamaire D, Carpentier MC, Eudier F et al. (1986) Essai ambulatoire du tiapridal après sevrage alcoholique hospitalier. Etude randomisée de cinquante et une hépatopathies alcooliques en double insu. *La Semaine des hopitaux de Paris* 62: 3109–3114
33 Sabourin A (1985) Effects of various therapeutic procedures on maintenance of abstinence in alcoholic patients: a statistical analysis. *La Semaine des hopitaux de Paris* 61: 669–676
34 Shaw GK, Majumdar SK, Waller S et al. (1987) Tiapride in the long-term management of alcoholics of anxious or depressive temperament. *Br J Psychiat* 150: 164–168
35 Shaw GK, Waller S, Majumdar SK et al. (1994) Tiapride in the prevention of relapse in recently detoxified alcoholics. *Br J Psychiat* 165: 515–523
36 Borg V, Weinholdt T (1982) Bromocriptine in the treatment of alcohol withdrawal syndrome. *Acta Psychiatr Scand* 65: 101–111
37 Burroughs AK, Morgan MY, Sherlock S (1985) Double-blind controlled trial of bromocriptine, chlordiazepoxide and chlormethiazole for alcohol withdrawal symptoms. *Alcohol Alcoholism* 20: 263–271
38 Farren CK, Clare AW, Ziedonis D et al. (1995) Evidence for reduced dopamine D2 receptor sensitivity in postwithdrawal alcoholics. *Alcohol Clin Exp Res* 19: 1520–1524
39 Borg V (1983) Bromocriptine in the prevention of alcohol abuse. *Acta Psychiatr Scand* 68: 100–110
40 Dongier M, Vachon L, Schwartz G (1991) Bromocriptine in the treatment of alcohol dependence. Alcoholism: *Clin Exp Res* 15: 970–977
41 Powell BJ, Campbell JL, Landon JF et al. (1995) A double-blind, placebo-controlled study of nortriptyline and bromocriptine in male alcoholics subtyped by comorbid psychiatric disorders. *Alcohol Clin Exp Res* 19: 462–468
42 Naranjo CA, Dongier M, Bremner KE (1997) Long-acting injectable bromocriptine does not reduce relapse in alcoholics. *Addiction 92*: 969–978
43 Glaser T, Fassbender M, Mauler F (1998) Receptor binding of flupenthixol and other neuroleptics. PT06004, *XXIst CINP Congress Glasgow*
44 Schilkrut R, Cabrera J, Morales E et al. (1988) Neuroleptics in the treatment of drug dependence in schizophrenics: A study with flupenthixol decanoate. *Psychopharmacology* 96: 342
45 Soyka M, Aichmuller C, v Bardeleben U et al. (2003) Flupenthixol in relapse prevention in schizophrenics with comorbid alcoholism: results from an open clinical study. *Eur Addict Res* 9: 65–72
46 Wiesbeck GA, Weijers H-G, Lesch OM et al. (2001) Flupenthixol decanoate and relapse prevention in alcoholics: results from a placebo-controlled study. *Alcohol Alcoholism* 36: 329–334
47 Walter H, Ramskogler K, Semler et al. (2001) Dopamine and alcohol relapse: D1 and D2 antagonists increase relapse rates in animal studies and inclinical trials. *J Biomed Sci* 8: 83–88
48 Lesch OM, Benda N, Böning J et al. (1999) Alcohol dependent patients characteristics and their relevance for relapse prevention. *Alcohol Alcoholism* 34: 433
49 Lesch OM, Böning J, Wiesbeck GA et al. (2000) Patients' characteristics and their implications for specific treatment. *Sucht* 46: 343
50 Wiesbeck GA, Weijers H-G, Wodarz N et al. (2003) Gender related differences in pharmacological relapse prevention with flupenthixol decanoate. *Arch Women's Mental Health* 6: 259–262

Drugs for Relapse Prevention of Alcoholism
Edited by R. Spanagel and K. Mann

Baclofen: preclinical data

Mauro A.M. Carai[1], Roberta Agabio[1], Giovanni Addolorato[2], Gian L. Gessa[1, 3] and Giancarlo Colombo[3]

[1] *Department of Neuroscience, University of Cagliari, Cagliari, Italy*
[2] *Institute of Internal Medicine, Catholic University, Rome, Italy*
[3] *C.N.R. Institute of Neuroscience, Section of Cagliari, Viale Diaz 182, I-09126 Cagliari, Italy*

Introduction

Alcohol interacts, as a receptor modulator capable of selectively altering specific neurochemical processes, with multiple brain receptor systems. The behavioral consequences of alcohol ingestion are thought to be the summation of its effects at these receptor systems. The contribution of each receptor system to the behavioral effects of alcohol varies as the alcohol dose/concentration is increased, providing a neurochemical basis for the dose-dependent nature of alcohol effects. Accordingly, it may be conceived that at a given alcohol dose/concentration, a specific receptor system is more sensitive to alcohol than others, thus resulting in a more prominent contribution to a particular behavioral effect of alcohol.

Recent experimental results (reviewed in the present chapter) as well as preliminary clinical data (reviewed in another chapter included in this book [1]) suggest that the $GABA_B$ receptor may be considered a novel player among the receptor systems controlling different aspects of alcohol drinking behavior, including alcohol consumption, alcohol relapse and alcohol appetitive properties. These preclinical and clinical findings feature $GABA_B$ receptor agonists as promising agents in the pharmacotherapy of alcoholism.

Effect of $GABA_B$ receptor agonists on alcohol intake and alcohol motivational properties in rats

Previous studies on the ability of $GABA_B$ receptor agonists to alter alcohol intake (under the non-operant, 2-bottle choice regimen) and alcohol self-administration (under operant procedures) have yielded mixed results. Indeed, it has been reported that the prototype $GABA_B$ receptor agonist, baclofen, reduces [2, 3], produces no change on [4], or even stimulates [5, 6] voluntary alcohol intake in rats, when given a choice between two bottles containing an alcohol solution and water, respectively. In the self-administration studies (where rats were trained to press a lever to gain access to alcohol), and depending on the drug dose or the experimental procedure used, baclofen has been found to stimulate [7] or decrease [7–9] operant responding for alcohol in rats. Furthermore, when observed, the inhibition of alcohol-motivated lever pressing was not selective, being accompanied by a proportional decrease in lever pressing for an alternative reinforcer such as sucrose [8]. In a study using the "sipper tube" model of alcohol access, which permits some separation between the appetitive and consummatory aspects of alcohol self-administration, baclofen decreased alcohol-seeking behavior (measured as the time spent in achieving the response requirement to gain access to the alcohol solution), while it increased alcohol-consummatory behavior (the amount of alcohol actually consumed) in rats [10].

More consistent results have been obtained in Sardinian alcohol-preferring (sP) rats, one of the few rat lines selectively bred worldwide for high alcohol preference and consumption. Indeed, recent work (as reviewed below) has demonstrated that the pharmacological activation of the $GABA_B$ receptor in sP rats resulted in the suppression of a) acquisition and maintenance of alcohol drinking behavior, b) alcohol motivational properties, and c) alcohol relapse-like drinking.

In the "acquisition" study [11], the $GABA_B$ receptor agonists, baclofen and CGP 44532, were administered to alcohol-naive sP rats, i.e., rats which had never consumed alcohol before the start of the experiment. Adult, male sP rats were singly housed and injected intraperitoneally with either baclofen (0, 1 and 3 mg/kg) or CGP 44532 (0, 0.1, 0.3 and 1 mg/kg) once a day, for a total of 10 consecutive days. Alcohol (10%, v/v) and water were offered under the standard, homecage 2-bottle "alcohol *versus* water" choice with unlimited access for 24 h/day, immediately following the first injection of baclofen or CGP 44532. Food was available *ad libitum*.

In vehicle-treated rats, acquisition of alcohol drinking behavior had a rapid onset, reaching an average daily intake of 5–6 g/kg (i.e., the amount of alcohol ordinarily consumed on a daily basis by sP rats) within 4–7 days. In contrast, daily alcohol intake was dose-dependently suppressed in baclofen- and CGP 44532-dosed rats throughout the 10-day treatment period. Interestingly, in the rat groups treated with the highest dose of each drug (3 mg/kg baclofen and 1 mg/kg CGP 44532), daily alcohol intake was on average lower than 1.5 g/kg. Reduction in alcohol intake, induced by both $GABA_B$ receptor agonists, was associated with a full compensatory increase in daily water intake, so that the total daily fluid intake (i.e., the sum of alcohol solution and water consumed) remained unchanged. Daily food intake (a variable usually recorded in pharmacological studies on alcohol intake as a signal of animal malaise or nonselectivity of drug action) tended to be higher in the rat groups treated with baclofen or CGP 44532, as compared to the vehicle-treated groups. This is likely to be due to the fact that in the control group, part of the total calorie intake was provided by alcohol. On completion of the treatment, daily alcohol intake progressively increased in the 3 mg/kg baclofen-treated group as well as in the 1 mg/kg CGP 44532-treated group, reaching control values after 10–14 days. Water intake diminished consistently.

A separate experiment [12] investigated the effects of baclofen on voluntary alcohol intake in alcohol-experienced sP rats, i.e., rats in which the consumption of pharmacologically relevant doses of alcohol was already established before baclofen administration. Alcohol-experienced rats are thought to represent a model of the "maintenance" or "active drinking" phase of human alcoholism. In this study, adult male sP rats were singly housed and exposed to alcohol for approximately 2 months before starting with drug treatment. Alcohol (10%, v/v) was offered under the 2-bottle "alcohol *versus* water" regimen with unlimited access for 24 h/day. Food was readily available. Alcohol intake averaged 6 g/kg/day across the 2-month period of alcohol exposure, which preceded the start of the experiment. Baclofen was injected intraperitoneally at doses of 0, 2.5, 5 and 10 mg/kg, once a day for a total of 14 consecutive days.

Alcohol intake in vehicle-treated rats averaged between 5.5 and 7 g/kg throughout the 14-day treatment period, whereas those animals treated with baclofen experienced a dose-dependent reduction of up to 40–50% in daily alcohol consumption. However, tolerance to the reducing effect of baclofen on alcohol intake progressively developed on continuing treatment. An increase in daily water intake fully compensated the reduction in alcohol consumption, with the total daily fluid intake remaining virtually unchanged. Food intake was significantly altered only by treatment with 10 mg/kg baclofen. This effect, however, was present only during the first half of the treatment period.

The results of the "acquisition" and "maintenance" experiments suggest that stimulation of the $GABA_B$ receptor by baclofen and/or CGP 44532 results in a virtually complete blockade of the disclosure and experience of those effects of alcohol that sustain alcohol drinking behavior, which is otherwise a phenomenon with a rapid onset and stable maintenance in sP rats, as indicated in control rats by the constant daily intake of pharmacologically relevant amounts of alcohol from the very beginning of alcohol exposure.

The above-mentioned experiments, performed with the 2-bottle choice paradigm, focused on the effect of baclofen and CGP 44532 on some consummatory aspects of alcohol ingestive behavior in sP rats. Subsequently, we extended to the appetitive, or motivational, properties of alcohol the investigation on the anti-alcohol effect of $GABA_B$ receptor agonists. Specifically, we evaluated the effect of baclofen on the extinction responding for alcohol, defined as the maximal amount of "work" that a rat trained to lever-press for alcohol is willing to perform to obtain alcohol [13]. Extinction responding has been proposed to represent an index of the appetitive strength of alcohol [14, 15]: the more the rat "works" on the lever, the stronger its motivation to gain access to alcohol. Recent work [16] has shown that sP rats trained to lever-press for alcohol displayed a) high values of extinction responding for alcohol, and b) a high and positive correlation between extinction responding and alcohol self-administration of the preceding session. Thus it is confirmed that alcohol possesses strong motivational capacities in sP rats, highlighting the suitability of this rat line for the planned study.

In the baclofen study [13], adult male sP rats were initially trained to lever-press for oral alcohol (15%, v/v) in daily 30-min sessions under a fixed ratio 4 (FR4) schedule (i.e., every 4 consecutive presses of the lever resulted in the presentation of a 0.1 ml drop of alcohol solution). After approximately 20 sessions, all rats displayed a robust and stable lever-pressing behavior, which resulted in a mean alcohol intake of 0.6 g/kg/session, with blood alcohol levels in the range of 40–50 mg%. A separate group of sP rats, trained to lever-press for 3% (w/v) sucrose under identical conditions, was included in the study to assess the specificity of the baclofen action on extinction responding for alcohol. Extinction responding for alcohol or sucrose was defined as the maximal number of lever presses attained by each rat in the absence of alcohol or sucrose reinforcement. More specifically, during extinction sessions, rats were exposed to the operant chamber for 30 min with lever-pressing not resulting in any alcohol or sucrose presentation. On test sessions, baclofen was injected intraperitoneally at doses of 0, 1, 2 and 3 mg/kg. Each dose of baclofen was tested in every single rat of both groups under a latin-square design.

Extinction responding for alcohol and sucrose in saline-treated rats averaged 54.8 ± 8.4 and 55.6 ± 13.2 (mean ± SEM), respectively, suggesting that 15% alcohol and 3% sucrose had comparable motivational properties in sP rats. Pretreatment with baclofen resulted in a dose-dependent suppression of extinction responding for alcohol: lever pressing of the rat groups treated with 1, 2 and 3 mg/kg baclofen was 64, 88 and 98% lower, respectively, than that observed in saline-treated rats. Seven out of eight rats in the 3 mg/kg baclofen-group, in fact, completely avoided pressing the lever. In the "sucrose" experiment, only the highest dose of baclofen tested significantly affected extinction responding for sucrose; extinction values in the rat groups treated with 1, 2 and 3 mg/kg baclofen were 23, 46 and 99% lower, respectively, than those observed in saline-treated rats. Taken together, these results suggest that baclofen was more potent in reducing the motivation for alcohol than for sucrose.

A separate experiment found that the above doses of baclofen did not affect spontaneous motor activity in sP rats, when tested in an open-field arena, suggesting that the suppressing effects of baclofen on extinction responding were indeed secondary to its ability in reducing the appetitive strength of alcohol and, furthermore, not due to muscle-relaxant or sedative properties of the drug.

These results suggest the involvement of the $GABA_B$ receptor in the neural system mediating alcohol reinforcement in sP rats, and that the reducing effects of baclofen on alcohol intake in sP rats may be secondary to its ability in suppressing the motivational properties of alcohol.

In conclusion, there appear to be some discrepancies of present-day data in the literature (indicating mixed effects of baclofen on alcohol intake and self-administration [2–10]) and those collected in sP rats (which consistently depict a reducing effect of baclofen and CGP 44532 on alcohol preference, consumption and motivational properties). Differences in the baclofen dose-range, route of baclofen administration, and the procedure of alcohol exposure may account, at least in part, for these discrepancies. However, the strain of rats used in these studies may be – to our understanding – an important factor in explaining these differences. Rats of the sP lines apparently possess a genetically determined sensitivity to the reducing effect of $GABA_B$ receptor agonists on alcohol intake and reinforcement, which may not be present in other rat strains. This peculiar sensitivity makes sP rats a proper animal model for investigations concerning the role of the $GABA_B$ receptor in the control of alcohol-ingestive behavior. Furthermore, the inconsistencies among the data generated with sP and other rat strains are not completely surprising, as a wide heterogeneity has historically been found among the different lines of alcohol-preferring rats with regard to the efficacy of certain drugs in reducing alcohol intake and self-administration. Thus, this replicates, to some extent, the different efficacies of some pharmacotherapies among the different types of alcoholics (e.g. [17–19]).

Effect of $GABA_B$ receptor agonists on relapse-like behaviors in rats

With a deeper focus on the scope of the present book, we report here the results of a series of experiments designed to investigate the effects of baclofen and CGP 44532 on the so-called alcohol deprivation effect (ADE), i.e., the transient increase in alcohol intake which occurs in several animal species following a period of abstinence from alcohol. This phenomenon has been proposed to model the loss of control over alcohol and the episodes of alcohol relapse seen in human alcoholics (see [20, 21]). Rats of the sP line appear to constitute a proper animal model for pharmacological investigations of ADE, as they have been found to display a pronounced ADE during the first hour of re-access to alcohol following its deprivation [22, 23]. Furthermore, ADE in sP rats has been found to be reduced with naltrexone (this laboratory, unpublished observations), a drug reported to possess some efficacy in reducing the likelihood of relapse in alcoholics (see [24]), providing evidence for the predictive value of this animal model with regard to human pathology.

In experiments carried out to test the effect of baclofen [25] and CGP 44532 on ADE, individually housed adult male sP rats were initially offered alcohol (10%, v/v) and water under the standard 2-bottle choice with unlimited access for 8 consecutive weeks. Subsequently, rats were divided into 2 groups (matched for alcohol intake over the last 7 days): one group was deprived of alcohol for 14 consecutive days, during which water was the sole fluid available (alcohol-deprived rats); the second group continued to have unlimited access to alcohol and water (alcohol-nondeprived rats), consuming an average of approximately 6 g/kg/day alcohol. At the end of the deprivation phase, 30 min before lights off, rats of both groups (alcohol-deprived and -nondeprived) were further divided into 4 subgroups (n = 7–8 in both experiments) and acutely injected with 0, 1, 1.7 and 3 mg/kg baclofen, or with 0, 0.03, 0.1 and 0.17 mg/kg CGP 44532. Both drugs were injected intraperitoneally. Alcohol was re-presented at lights off and its consumption was recorded 60 min later (previous studies indicated that 60 min after alcohol representa-

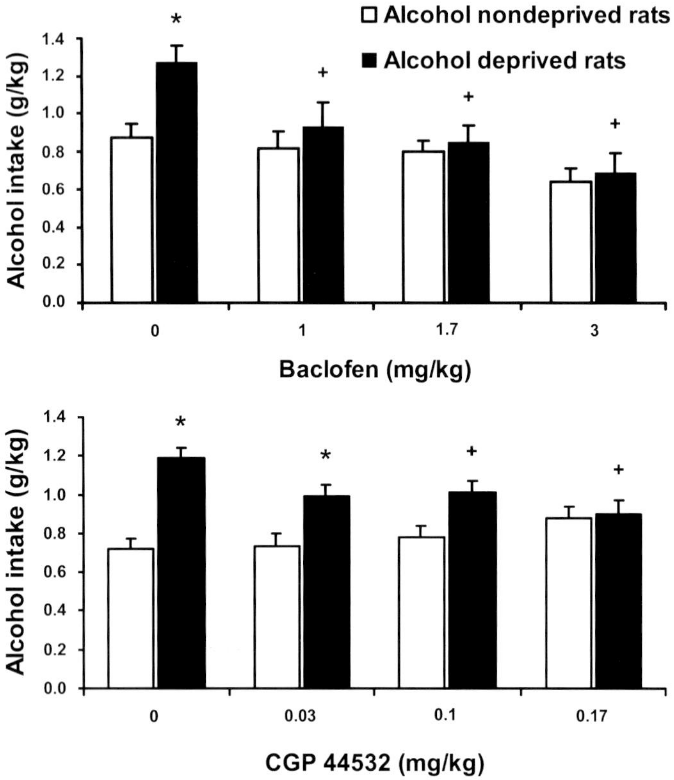

Figure 1. Suppressing effect of the $GABA_B$ receptor agonists, baclofen (top) and CGP 44532 (bottom), on the alcohol deprivation effect (ADE) in sP rats given alcohol under the 2-bottle choice regimen. Each bar is the mean ± SEM of n = 7–8. *: $P < 0.05$ with respect to saline-treated alcohol-nondeprived rats; +: $P < 0.05$ with respect to saline-treated alcohol-deprived rats (Newman-Keuls test). Top panel reprinted from Drug Alcohol Dep 70: 105–108, 2003, with permission from Elsevier.

tion is the time interval during which ADE is maximal in sP rats [22, 23]). Standard rat chow was available throughout the studies.

In both experiments, alcohol intake was 50–80% higher in saline-treated alcohol-deprived rats than in saline-treated alcohol-nondeprived rats (Fig. 1); this increase in alcohol intake was indicative of the development of a marked ADE. This increase in alcohol intake was eliminated by baclofen and CGP 44532. Indeed, all doses of baclofen (Fig. 1, top panel) and the two highest doses of CGP 44532 (Fig. 1, bottom panel) resulted in a virtually complete suppression of the extra intake of alcohol produced by alcohol deprivation. Importantly, no dose of baclofen or CGP 44532 affected water and food intake, tending to exclude that the action of baclofen and CGP 44532 on ADE was due to their muscle-relaxant and sedative effects. Accordingly, complementary experiments on motor activity found that the doses of baclofen and CGP 44532 that suppressed ADE neither affected the time spent moving, nor the distance traveled, nor the number of rearings (measures of horizontal and vertical motor activities in rodents) in alcohol-consuming sP rats tested in an open-field arena.

As demonstrated for several other alcohol-related behavioral responses, multiple receptor systems are likely to be involved in the expression of ADE. Indeed, drugs acting at the glutamate [26–31], opioid [32–34], dopamine [35], $GABA_A$/benzodiazepine [36], and cannabinoid [37] receptors have also been reported to modulate ADE in rats and mice. The results of the present study include the $GABA_B$ receptor in the neural substrate mediating ADE.

Finally, because of the predictive validity of ADE as an experimental model of alcohol relapse, the results of the present study suggest that baclofen may possess some efficacy in preventing relapse in human alcoholics. The results of preliminary, clinical surveys, reviewed by Addolorato and colleagues [1], apparently support this hypothesis, suggesting that baclofen may constitute a novel medication for alcoholism.

References

1 Addolorato G, Abenavoli L, Leggio L, De Lorenzi G, Ferrulli A, Caputo F, Agabio R, Gessa GL, Colombo G, Gasbarrini G (2004) Baclofen: clinical data. In: R Spanagel, K Mann (eds): *Drugs for Relapse Prevention of Alcoholism*. Birkhäuser, Basel, 171–180

2 Daoust M, Saligaut C, Lhuintre JP, Moore N, Flipo JL, Boismare F (1987) GABA transmission, but not benzodiazepine receptor stimulation, modulates ethanol intake by rats. *Alcohol* 4: 469–472

3 Perfumi M, Santoni M, Ciccocioppo R, Massi M (2002) Blockade of γ-aminobutyric acid receptors does not modify the inhibition of ethanol intake induced by *Hypericum perforatum* in rats. *Alcohol Alcoholism* 37: 540–546

4 Tomkins DM, Fletcher PJ (1996) Evidence that $GABA_A$ but not $GABA_B$ receptor activation in the dorsal raphe nucleus modulates ethanol intake in Wistar rats. *Behav Pharmacol* 7: 85–93

5 Smith BR, Robidoux J, Amit Z (1992) GABAergic involvement in the acquisition of voluntary ethanol intake in laboratory rats. *Alcohol Alcoholism* 27: 227–231

6 Smith BR, Boyle AEL, Amit Z (1999) The effects of $GABA_B$ agonist baclofen on the temporal and structural characteristics of ethanol intake. *Alcohol* 17: 231–240

7 Petry NM (1997) Benzodiazepine-GABA modulation of concurrent ethanol and sucrose reinforcement in the rat. *Exp Clin Psychopharmacol* 5: 183–194

8 Anstrom KK, Cromwell HC, Markowski T, Woodward DJ (2003) Effect of baclofen on alcohol and sucrose self-administration in rats. *Alcohol Clin Exp Res* 27: 900–908

9 Janak PH, Gill TM (2003) Comparison of the effects of allopregnanolone with direct GABAergic agonists on ethanol self-administration with and without concurrently available sucrose. *Alcohol* 30: 1–7

10 Czachowski CL, Legg BH, Stansfield KH (2002) Effects of the GABA(B) agonist, baclofen, on ethanol- and sucrose-seeking and self-administration. *Alcohol Clin Exp Res* 26: 115A

11 Colombo G, Serra S, Brunetti G, Atzori G, Pani M, Vacca G, Addolorato G, Froestl W, Carai MAM, Gessa GL (2002) The $GABA_B$ receptor agonists baclofen and CGP 44532 prevent acquisition of alcohol drinking behaviour in alcohol-preferring rats. *Alcohol Alcoholism* 37: 499–503

12 Colombo G, Agabio R, Carai MAM, Lobina C, Pani M, Reali R, Addolorato G, Gessa GL (2000) Ability of baclofen in reducing alcohol intake and withdrawal severity: I – Preclinical evidence. *Alcohol Clin Exp Res* 24: 58–66

13 Colombo G, Vacca G, Serra S, Brunetti G, Carai MAM, Gessa GL (2003) Baclofen suppresses motivation to consume alcohol in rats. *Psychopharmacology* 167: 221–224

14 Samson HH, Chappell A, Czachowski C, Sharpe A (2001) Measuring ethanol-seeking behavior: the effect of using repeated extinction trials. *Alcohol* 24: 205–209

15 Samson HH, Czachowski C, Chappell A, Legg B (2003) Measuring the appetitive strength of ethanol: use of an extinction trial procedure. *Alcohol* 31: 77–86

16 Vacca G, Serra S, Brunetti G, Carai MAM, Samson HH, Gessa GL, Colombo G (2002) Operant self-administration of ethanol in Sardinian alcohol-preferring rats. *Alcohol Clin Exp Res* 26: 1678–1685
17 Kranzler HR, Burleson JA, Brown J, Babor TF (1996) Fluoxetine treatment seems to reduce the beneficial effects of cognitive-behavioral therapy in type B alcoholics. *Alcohol Clin Exp Res* 20: 1534–1541
18 Johnson BA, Ait-Daoud N, Prihoda TJ (2000) Combining ondansetron and naltrexone effectively treats biologically predisposed alcoholics: from hypotheses to preliminary clinical evidence. *Alcohol Clin Exp Res* 24: 737–742
19 Pettinati HM, Volpicelli JR, Kranzler HR, Luck G, Rukstalis MR, Cnann A (2000) Sertraline treatment for alcohol dependence: interactive effects of medication and alcoholic subtype. *Alcohol Clin Exp Res* 24: 1041–1049
20 Boening JA-L, Lesch OM, Spanagel R, Wolffgramm J, Narita M, Sinclair D, Mason BJ, Wiesbeck GA (2001) Pharmacological relapse prevention in alcohol dependence: from animal models to clinical trials. *Alcohol Clin Exp Res* 25: 127S–131S
21 McBride WJ, Le A-D, Noronha A (2002) Central nervous system mechanisms in alcohol relapse. *Alcohol Clin Exp Res* 26: 280–286
22 Agabio R, Carai MAM, Lobina C, Pani M, Reali R, Vacca G, Gessa GL, Colombo G (2000) Development of short-lasting alcohol deprivation effect (ADE) in Sardinian alcohol-preferring rats. *Alcohol* 21: 59–62
23 Serra S, Brunetti G, Vacca G, Lobina C, Carai MAM, Gessa GL, Colombo G (2003) Stable preference for high ethanol concentrations after alcohol deprivation in Sardinian alcohol-preferring (sP) rats. *Alcohol* 29: 101–108
24 Streeton C, Whelan G (2001) Naltrexone, a relapse prevention maintenance treatment of alcohol dependence: a meta-analysis of randomized controlled trials. *Alcohol Alcoholism* 36: 544–552
25 Colombo G, Serra S, Brunetti G, Vacca G, Carai MAM, Gessa GL (2003) Suppression by baclofen of alcohol deprivation effect in Sardinian alcohol-preferring (sP) rats. *Drug Alcohol Dep* 70: 105–108
26 Salimov RM, Salimova NB (1993) L-Glutamate abolishes differential responses to alcohol deprivation in mice. *Alcohol* 10: 251–257
27 Hölter SM, Danysz W, Spanagel R (1996) Evidence for alcohol anti-craving properties of memantine. *Eur J Pharmacol* 314: R1–R2
28 Spanagel R, Hölter SM, Allingham K, Landgraf R, Zieglgänsberger W (1996) Acamprosate and alcohol: I. Effects on alcohol intake following alcohol deprivation in the rat. *Eur J Pharmacol* 305: 39–44
29 Hölter SM, Landgraf R, Zieglgänsberger W, Spanagel R (1997) Time course of acamprosate action on operant ethanol self-administration after ethanol deprivation. *Alcohol Clin Exp Res* 21: 862–868
30 Heyser CJ, Schulteis G, Durbin P, Koob GF (1998) Chronic acamprosate eliminates the alcohol deprivation effect while having limited effects on baseline responding for ethanol in rats. *Neuropsychopharmacology* 18: 125–133
31 Hölter SM, Danysz W, Spanagel R (2000) Novel uncompetitive *N*-methyl-D-aspartate (NMDA)-receptor antagonist MRZ 2/579 suppresses ethanol intake in long-term ethanol-experienced rats and generalizes to ethanol cue in drug discrimination procedure. *J Pharmacol Exp Ther* 292: 545–552
32 Cowen MS, Rezvani AH, Jarrott B, Lawrence AJ (1999) Ethanol consumption by Fawn-Hooded rats following abstinence: effect of naltrexone and changes in μ-opioid receptor density. *Alcohol Clin Exp Res* 23: 1008–1014
33 Hölter SM, Spanagel R (1999) Effects of opiate antagonist treatment on the alcohol deprivation effect in long-term ethanol-experienced rats. *Psychopharmacology* 145: 360–369
34 Hölter SM, Henniger MSH, Lipkowski AW, Spanagel R (2000) Kappa-opioid receptors and relapse-like drinking in long-term ethanol-experienced rats. *Psychopharmacology* 153: 93–102
35 Salimov RM, Salimova NB, Shvets LN, Maisky AI (2000) Hapoleridol administered subchronically reduces the alcohol-deprivation effect in mice. *Alcohol* 20: 61–68
36 Schmitt U, Waldhofer S, Weigelt T, Hiemke C (2002) Free-choice ethanol consumption under the influence of GABAergic drugs in rats. *Alcohol Clin Exp Res* 26: 457–462
37 Carai MAM, Lobina C, Gessa GL, Colombo G (2004) Cannabinoid receptor antagonists: a perspective. In: R Spanagel, K Mann (eds): *Drugs for Relapse Prevention of Alcoholism*. Birkhäuser, Basel, 181–187

Drugs for Relapse Prevention of Alcoholism
Edited by R. Spanagel and K. Mann

Baclofen: clinical data

Giovanni Addolorato[1], Ludovico Abenavoli[1], Lorenzo Leggio[1], Giosuè DeLorenzi[1], Anna Ferrulli[1], Fabio Caputo[2], Roberta Agabio[3], Gian Luigi Gessa[3, 4], Giancarlo Colombo[4] and Giovanni Gasbarrini[1]

[1] *Institute of Internal Medicine, Catholic University, Gemelli Hospital, L.go A. Gemelli 8, 00168 – Rome, Italy*
[2] *"G. Fontana", Center for the Study and Treatment of Alcohol Addiction, University of Bologna, Bologna, Italy*
[3] *Department of Neuroscience, University of Cagliari, Cagliari, Italy*
[4] *C.N.R. Institute of Neuroscience, c/o "Bernard B. Brodie" Department of Neuroscience, University of Cagliari, Viale Diaz 182, I-09126 Cagliari, Italy*

Introduction

Alcohol abuse and alcoholism represent a world wide problem, both from a medical and a social point of view. Alcohol dependence affects nearly 10% of the general population both in the United States [1] and in Europe [2], with the high prevalence rate of alcohol-related problems only highlighting the public health importance of this disorder.

The application of pharmacological means in the treatment of patients suffering from alcohol abuse and alcoholism represents an ever-growing field with which to enhance alcohol abstinence and prevent relapse, while also complementing psychosocial interventions which have been used for many years.

In recent years, several drugs useful in the treatment of alcohol addiction have been tested both in pre-clinical and clinical studies. Among them baclofen (beta-(4-chlorophenyl)-γ-aminobutyric acid) has shown promising results.

Baclofen is a lipophilic derivative of GABA and a stereoselective $GABA_B$ receptor agonist [3]. At present, it is used clinically in order to control spasticity [4].

Recent pre-clinical studies (reviewed in another chapter included in this book [5]) and clinical data (reviewed in the present chapter) have shown that baclofen has been effective in the treatment of alcohol addiction, both in emergencies and in relapse prevention, as well as in the treatment of other substance abuse.

Baclofen in alcohol dependence: relapse prevention

Clinical management of the alcohol addiction disorder is aimed at attaining relief from withdrawal syndromes and ensuring a smooth transition into a treatment program so as to achieve alcohol abstinence. Thus, relapse prevention represents the main objective of the treatment. This program includes both pharmacological as well as psychosocial approaches [6–9].

As far as the pharmacological approach is concerned, clinical studies point to baclofen as a new useful drug in the treatment of patients with alcohol problems. Krupitsky and co-workers [10] showed that baclofen is effective in reducing affective disorders in alcoholic patients. Affective disorders alongside alcohol dependence can increase withdrawal severity and vulnerability to relapse. The sample included in their study had 3 to 4 weeks of alcohol abstinence, with all subjects suffering from secondary affective disorders, in particular anxiety, and depression, or both. Patients were randomly divided into 4 groups and treated for a period of 3 weeks with 37.5 mg/day baclofen, 15 mg/day diazepam, 75 mg/day of amitriptyline, or placebo. The Zung Scale, the Minnesota Multiphasic Personality Inventory, Spielberger's State-Trait Anxiety, blood platelet MAO-B activity, plasma levels of dopamine, serotonin and GABA, as well as an electroencephalogram (EEG), were evaluated at the start and at the end of the study. The results of the post-treatment rating scale scores showed a significant decrease in anxiety and depression in all drug-treated patients as compared to the placebo group. However, while subjects treated with diazepam and amitriptyline experienced sedation, no side-effects were found in the baclofen-treated patients. The results indicated that selective ligands of $GABA_B$ receptors can be as effective in treating affective disorders in alcoholics as $GABA_A$ receptor ligands, and were associated with lower side-effects.

After promising data were obtained from an open pilot study, performed in a small sample of selected patients [11], the efficacy of baclofen was recently evaluated in patients affected by alcohol addiction in a controlled double-blind randomized study [12].

After 12–24 h of abstinence from alcohol, a total of 39 patients were randomly divided into two groups. The patients were treated with oral administration of baclofen or placebo for a total of 30 days, starting at a dose of 15 mg/day for the first three days and 30 mg/day for the subsequent 27 days. Each subject was checked as an outpatient every week and at each visit routine psychological support and counselling were provided, attended to by the same professional staff. Craving level was evaluated using the Italian version [13] of the Obsessive Compulsive Drinking Scale (OCDS) [14] at the start of the study (T0) and at each weekly outpatient visit. Abstinence from alcohol was measured on the basis of a patient's self-evaluation, a family member interview, as well as main biological markers of alcohol abuse. A self-reported alcohol intake was recorded as the average number of standard drinks consumed per day. Variation of state anxiety by the State and Trait Inventory test, Y1 axes, and of current depression by the Zung Self-Rating Depression Scale were recorded. The percentage of drop-outs was lower in the baclofen as compared to the placebo group. Furthermore, a significantly higher number of patients who achieved and maintained abstinence throughout the experimental period, were found among the group of patients treated with baclofen. The study showed a significant effect of the drug in reducing alcohol intake. In the baclofen group, the average number of daily drinks was virtually completely suppressed within the first week of treatment (Fig. 1). The OCDS craving score in the baclofen group was constantly lower than that monitored in the placebo group (Fig. 2; top panel). A significant effect of drug treatment on both the compulsive (Fig. 2; central panel) and obsessive (Fig. 2; bottom panel) drinking sub-scale of OCDS was found, with scores in the baclofen group consistently lower than those of the placebo one throughout the study. While lower scores of state anxiety were found in the baclofen group relative to the placebo one, no significant difference was observed regarding the depression score. Tolerability was fair in all patients and no systemic or single-organ event leading to drug cessation was reported. No patient reported euphoria or any other pleasant effects caused by the drug. No subject showed craving for the drug either. Furthermore, at drug discontinuation, no withdrawal syndromes or side-effects due to drug suspension were observed.

In conclusion, the study showed that the administration of low doses of baclofen in alcohol-dependent patients has a significantly higher efficacy, when compared to placebo, in inducing alcohol abstinence, reducing alcohol intake and alcohol craving in both components (obsessive and compulsive) as well as reducing state anxiety. The higher efficacy of baclofen with respect to placebo could be related to its anti-reward and anti-craving action. In fact, the drug was effective in quickly decreasing both compulsive and obsessive components of craving. The anti-craving effect of baclofen could be due to the abil-

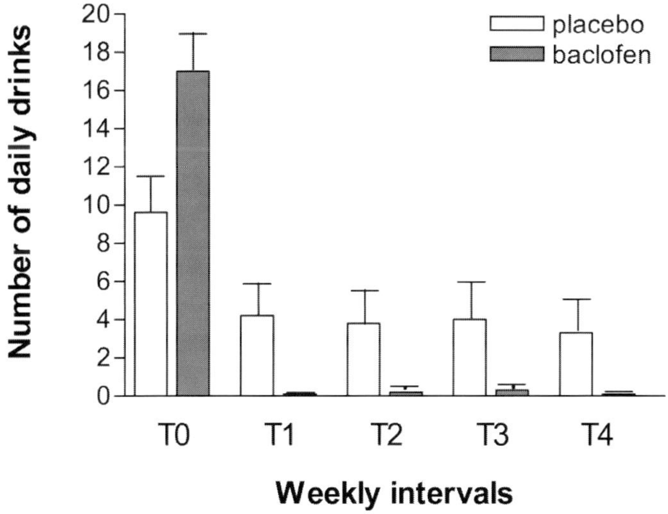

Figure 1. Number of daily drinks in baclofen and placebo groups at T0 (baseline) and over the four weekly visits (T1–T4).

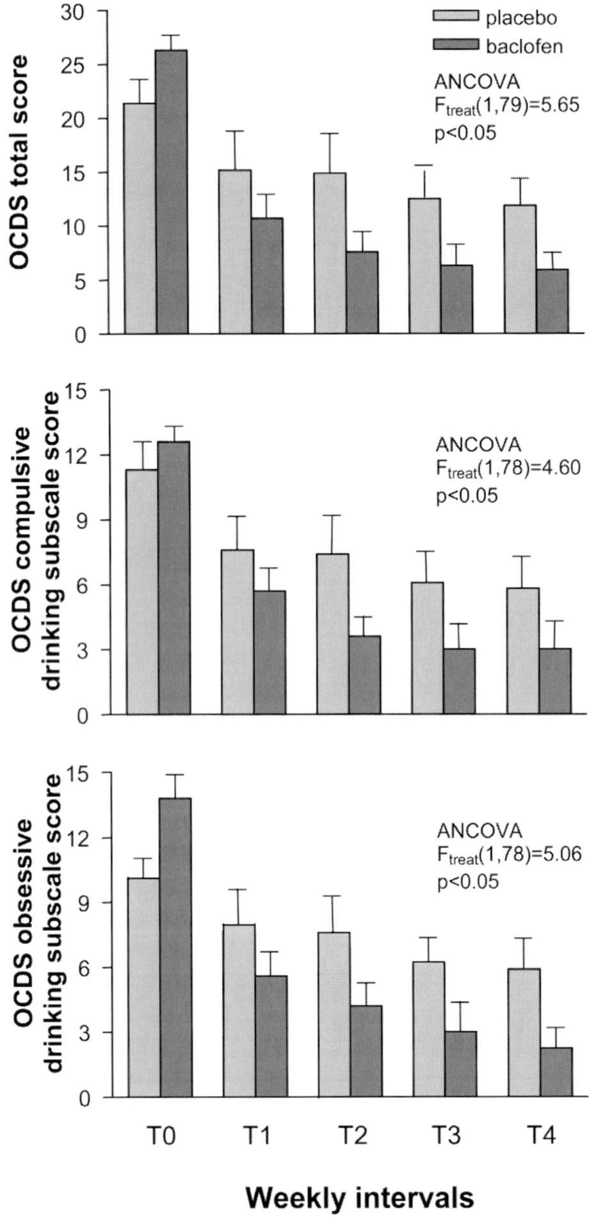

Figure 2. Obsessive-Compulsive Drinking Scale (OCDS) total (top panel), OCDS Compulsive Drinking subscale (centre panel) and OCDS Obsessive Drinking subscale (bottom panel) scores in baclofen and placebo groups at T0 (baseline) and over the four weekly visits (T1- T4).

ity of the drug to interfere with the neuronal substrates mediating the reinforcing properties of ethanol [5].

These data indicate that baclofen could have an important role in the treatment of patients affected by alcoholism, especially taking into account the safety of the drug. In particular, the ability of the drug in reducing the main components of craving, suppressing alcohol intake, and reducing anxiety as well as in suppressing alcohol withdrawal symptoms (see below) further emphasize its efficacy as a drug useful in relapse prevention.

Alcohol withdrawal syndrome

Recent preliminary data showed that baclofen could be effective in the treatment of alcohol-dependent patients affected by severe alcohol withdrawal syndrome (AWS) [15] also complicated by delirium tremens (DT) [16]. In particular, patients with a Clinical Institute of Withdrawal Assessment revised (CIWA-Ar) scale score higher than 20 points, showing the presence of severe AWS that required pharmacological treatment, were evaluated. Baclofen was orally administered at 10 mg, every 8 h. The CIWA-Ar was applied every hour for 4–8 h. A rapid decrease of the CIWA-Ar score and a marked improvement of AWS symptoms were observed in the first few hours after baclofen administration in all treated patients (Fig. 3). In particular, a rapid decrease of some withdrawal symptoms such as anxiety, agitation and depression was observed. Furthermore, these data could be of interest, as it has already been shown that a rapid decrease of these factors may facilitate the patient's transition into a long-term rehabilitation program [17].

A case of AWS complicated by DT, and successfully treated with baclofen (25 mg orally administered every 8 h for the first three days, then a tapering off of the dose to 10 mg every 8 h), has recently been reported [16]. In this case as well, a rapid decrease of the CIWA-Ar score and a marked improvement of AWS and DT symptoms were observed starting from the first hour on after baclofen administration, due to dissipation of nausea, vomiting, tactile disturbances and hallucinations, and a reduction in other symptoms. After drug discontinuation, AWS and DT symptoms did not recur. Furthermore, in AWS and DT treatment, baclofen was manageable, without resulting in any significant side-effects. In particular, no sedation or respiratory disorders were present in the treated patients.

Moreover, as it has been shown that baclofen is effective in reducing alcohol craving and intake and in inducing and maintaining abstinence from alcohol as well as in suppressing AWS (if the present data are confirmed), then the use of a single drug like baclofen could be more appropriate, as opposed to various drugs, for the management of AWS and DT, and then followed by a program to maintain alcoholic patients in long-term alcohol abstinence [16].

Future studies are needed to confirm the present data. At present, the use of baclofen for the treatment of AWS and DT in clinical practice is inappropriate, until its safety and effectiveness can be examined in controlled clinical trials.

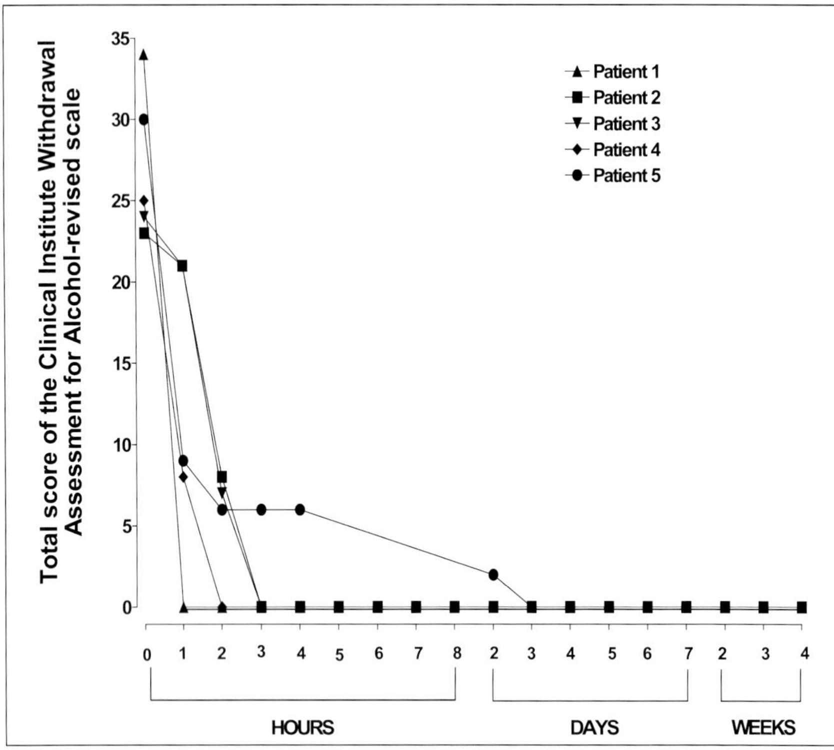

Figure 3. Scores on the revised Clinical Institute of Withdrawal Assessment for Alcohol scale of the patients included in the study during the 30-day observation period. Baclofen (10 mg, orally) was administered. CIWA-Ar was administered immediately after the observation at time 0; subsequently, baclofen (10 mg, orally) was administered every 8 h for 30 consecutive days. The withdrawal scale was administered every hour for the first 4–8 h, once a day from day 3 to 7 and once a week from week 2 to week 4.

Baclofen in other substance dependence

Heroin

The first studies examining the use of baclofen in opioid-dependent patients were carried out by Krystal and co-workers [18]. In an open label pilot study, five opiate-dependent patients underwent a baclofen-assisted opiate detoxification after an abrupt discontinuation of methadone. Baclofen was administered in 80 mg/day doses, and all patients reported some reduction in discomfort. However, 3 out of the 5 patients did not manage to complete the detoxification process with baclofen, primarily because of an insufficient suppression of vomiting, myalgias and headache. These patients completed their detoxification with clonidine. Therefore, the findings suggest a limited use of baclofen as a primary treatment choice for opiate dependence.

Akhondzadeh and coworkers [19] compared the efficacy of baclofen with clonidine regarding acute detoxification of opioid-dependent subjects. The authors found that baclofen proved as effective as clonidine in the management of physical symptoms of opiate withdrawal syndromes. Further, baclofen showed a significantly higher efficacy with respect to clonidine in the management of mental symptoms. In a subsequent randomised, double-blind controlled study, the same authors showed that there was no significant difference between the baclofen and clonidine treatment in terms of dropout and overall side-effects among the opioid-dependent subjects treated [20]. However, the low incidence of hypotension in patients treated with baclofen could suggest that this drug may be suitable for outpatient ambulatory treatment of opiate withdrawal syndrome [20].

Recently, Assadi and co-workers evaluated the efficacy of baclofen in maintenance treatment of opioid-addicted patients [21]. In this study, 40 opioid-dependent patients were detoxified and randomly assigned to receive baclofen (60 mg/day) or placebo. Treatment retention was significantly higher in the baclofen group. Baclofen showed a significantly higher efficacy with respect to placebo in terms of opiate withdrawal syndrome and depressive symptoms. A generally favorable response was also found in the baclofen group as regards some outcome parameters, including opioid craving and self-reported opioid and alcohol use. However, no difference was found in the rates of opioid-positive urine tests.

Cocaine

Preclinical studies have shown that baclofen a) prevents the development of cocaine-induced behavioral sensitization, b) abolishes the motor stimulant actions of cocaine, and c) suppresses the intravenous self-administration of cocaine [22, 23]. Shoptaw and co-workers carried out a randomized placebo-controlled trial of baclofen with respect to cocaine dependence [24]. This screening trial evaluated whether baclofen demonstrated sufficient clinical efficacy so as to recommend an adequately powered trial of the medication as a pharmacotherapy for cocaine dependence. Project findings showed an initial clinical efficacy of baclofen over placebo in reducing cocaine use, when administered concurrently with thrice-weekly drug abuse counselling sessions. The effects of baclofen were particularly apparent for those participants with chronic levels of cocaine use at baseline and provide support for a full-scale efficacy trial for baclofen, especially among this subgroup of patients.

Very interesting data have been reported in recent imaging studies performed in cocaine-dependent patients [23]. These subjects were given baclofen (10–20 mg twice daily) for 7–10 days prior to a Positron Emission Tomography (PET) session, with results showing a substantial blunting of cue-induced craving and no limbic anterior cingulate and amygdalar activation to cocaine (*versus* non-drugs) videos [23]. These data indicate that craving is reduced and limbic activation eliminated in cocaine-dependent patients treated with baclofen, suggesting that baclofen may help protect against cue-induced craving and that its benefits can potentially be sustained for years.

Finally, Lile and co-workers [25] showed that pre-treatment with baclofen does not influence acute behavioral effects (e.g., the reinforcing, subject-rated or cardiovascular effects) of intranasal cocaine administration in humans.

Nicotine

The mesolimbic dopamine system has been implicated in the reinforcing effects of nicotine, a drug which appears to act, at least in part, through the ventral tegmental area (VTA). Nicotine activates a circuitry in which μ-opioid receptors are situated, especially GABAergic elements [26]. In 2001, Cousins and coworkers [27] studied the acute effect of a single dose of baclofen on cigarette smoking, craving for nicotine, cigarette taste, and smoking satisfaction. Baclofen did not change the number of cigarettes smoked by the subjects, nor did it change ratings of nicotine craving. However, baclofen did alter the sensory properties of smoked cigarettes. It also produced mild sedative-like subjective effects, such as increases in feeling 'relaxed'. Thus, although baclofen did not reduce cigarette craving or smoking, it did produce some mood-altering effects and changes in sensory aspects of smoking that may facilitate smoking cessation.

Conclusions

A growing number of both pre-clinical and clinical studies support the idea that $GABA_B$ compounds may attenuate the craving and the reinforcing effects of alcohol, cocaine, heroin and nicotine [28]. The findings reviewed in the present chapter suggest that the $GABA_B$ agonist baclofen may offer a powerful method for controlling drug abuse in humans.

Acknowledgement
Supported by grants from "Associazione Ricerca in Medicina", Bologna-Roma, Italy.

References

1 McGinnis JM, Foege WH (1993) Actual causes of death in the United States. *JAMA* 270: 2207–2212
2 Hupkens C, Knibbe R, Drop M (1993) Alcohol consumption in the European Community: uniformity and diversity in drinking patterns. *Addiction* 88: 1391–1404
3 Allan AM, Harris A (1989) *A new alcohol antagonist:* Phaclofen. *Life Sci* 45: 1771–1779
4 Davidoff RA (1985) Antispasticity drugs: mechanisms of action. *Ann Neurol* 17: 107–116
5 Carai MAM, Agabio R, Addolorato G, Gessa GL, Colombo G (2004) Baclofen: preclinical data. In: R Spanagel, K Mann (eds): *Drug for Relapse Prevention of Alcoholism*. Birkhäuser, Basel,

163–170
6 Addolorato G, Viaggi M, Gentilini L, Castelli E, Nicastro P, Stefanini GF, Gasbarrini G (1993) Alcohol addiction: evaluation of the therapeutic effectiveness of self-help group in the maintenance of abstinence from alcohol. *Alcologia Eur J Alcohol Stud* 5: 261–263
7 Caputo F, Addolorato G, Dall'Aglio C, Baudanza P, Bernardi M, Gasbarrini G, Stefanini GF (1998) Are brief and long-term alcohol intervention really defined? *Addiction* 93: 1104
8 Garbutt JC, West SL, Carey T, Lohr, Crews FT (1999) Pharmacological treatment of alcohol dependence: a review of the evidence. *JAMA* 14: 1318–1325
9 Addolorato G, Armuzzi A, Gasbarrini G and Alcoholism Treatment Study Group (2002) Pharmacological approaches to the management of alcohol addiction. *Eur Rev Med Pharmacol Sci* 6: 89–97
10 Krupitsky EM, Burakov AM, Ivanov VB, Krandashova GF, Lapin IP, Grinenko A, Borodkin Yu S (1993) Baclofen administration for the treatment of effective disorders in alcoholic patients. *Drug Alcohol Depend* 33(2): 157–163
11 Addolorato G, Caputo F, Capristo E, Colombo G, Gessa GL, Gasbarrini G (2000) Ability of baclofen in reducing alcohol craving and intake: II-preliminary clinical evidence. *Alcohol Clin Exp Res* 24: 67–71
12 Addolorato G, Caputo F, Capristo E, Domenicali M, Bernardi M, Janiri L, Agabio R, Colombo G, Gessa GL, Gasbarrini G (2002) Baclofen efficacy in reducing alcohol craving and intake: a preliminary double-blind randomized controlled study. *Alcohol Alcoholism* 37: 504–508
13 Janiri L, Calvosa F, Dario T, Ruggeri A, Pozzi G, Addolorato G, De Risio S (2004) The Italian version of the obsessive-compulsive drinking scale: validation, comparison with the other versions and difference between type 1- and type 2-like alcoholics. *Drug Alcohol Dep* 74: 187–195
14 Anton RF, Moak DH, Latham P (1995) The Obsessive Compulsive Drinking Scale: a self-rated instrument for the quantification of thoughts about alcohol and drinking behavior. *Alcohol Clin Exp Res* 19: 92–99
15 Addolorato G, Caputo F, Capristo E, Janiri L, Bernardi M, Agabio R, Colombo G, Gessa GL, Gasbarrini G (2002) Rapid suppression of alcohol withdrawal syndrome by baclofen. *Am J Med* 112: 226–229
16 Addolorato G, Leggio L, Abenavoli L, DeLorenzi G, Parente A, Caputo F, Janiri L, Capristo E, Rapaccini GL, Gasbarrini G (2003) Suppression of alcohol delirium tremens by baclofen administration: a case report. *Clin Neuropharmacol* 26: 258–262
17 O'Connor PG, Schotternfeld RS (1998) Patients with alcohol problems. *N Engl J Med* 338: 592–602
18 Krystal JH, Mc Dougle CJ, Kosten TR, Price LH, Aghajanian GK, Charney DS (1992) Baclofen-assisted detoxification from opiates. A pilot study. *J Subst Abuse Treat* 9: 139–142
19 Ahmadi-Abhari SA, Akhondzadeh S, Assadi SM, Shabestari OL, Farzanehgan ZM, Kamlipour A (2001) Baclofen *versus* clonidine in the treatment of opiates withdrawal, side-effects aspects: a double-blind randomized controlled trial. *J Clin Pharm Ther* 26: 67–71
20 Akhondzadeh S, Ahmadi-Abhari SA, Assadi SM, Shabestari OL, Kashani AR, Farzanehgan ZM (2000) Double-bind randomized controlled trial of baclofen *versus* clonidine in the treatment of opiate withdrawal. *J Clin Pharm Ther* 25: 347–353
21 Assadi SM, Radgoodarzi R, Ahmadi-Abhari SA (2003) Baclofen for maintenance tretment of opioid dependence: a randomized double-blind placebo-controlled clinical trial. *BMC Psychiatry* 3: 16
22 Kalivas PW and Stewart J (1994) Dopamine transmission in the initiation and expression of drug and stress-induced sensitization of motor activity. *Brain Res Rev* 16: 223–244
23 Brebner K, Childress AR, Roberts DC (2002) A potential role for GABA (B) agonists in the treatment of psychostimulant addiction. *Alcohol Alcoholism* 37: 478–484
24 Shoptaw S, Yang X, Rotheram-Fuller EJ, Hsieh YC, Charuvastra VC, Ling W (2003) Randomized placebo-controlled trial of baclofen for cocaine dependence: preliminary effects for individual with chronic patterns of cocaine use. *J Clin Psy* 64: 1440–1448
25 Lile JA, Stoops WW, Allen TS, Glaser PE, Hays LR, Rush CR (2004) Baclofen does not alter the reinforcing, subject-rated or cardiovascular effects of intranasal cocaine in humans. *Psychopharmacology* 171: 441–449
26 Carrigall WA, Coen KM, Adamson KL, Chow BL, Zhang J (2000) Response of nicotine self-administration in the rat to manipulations of mu-opioid and gamma-aminobutyric acid receptors in the ventral tegmental area. *Psychopharmacology* 149: 107–114

27 Cousins MS, Stamat HM, de Wit H (2001) Effects of a single dose of baclofen on self-reported subjective effects and tobacco smoking. *Nicotine Tob Res* 3: 409
28 Cousins MS, Roberts DCS, de Wit H (2002) $GABA_B$ receptor agonist for the treatment of drug addiction: a review of recent findings. *Dug Alcohol Dep* 65: 209–220

Drawings for Relapse Prevention of Alcoholism
Edited by R. Spanagel and K. Mann

Cannabinoid receptor antagonists: a perspective

Mauro A.M. Carai[1], Carla Lobina[1], Gian Luigi Gessa[1, 2] and Giancarlo Colombo[2]

[1] *Department of Neuroscience, University of Cagliari, Cagliari, Italy*
[2] *C.N.R. Institute of Neuroscience, Section of Cagliari, Viale Diaz 182, 09126 Cagliari, Italy*

Introduction

Cannabinoid CB_1 receptors are G-protein coupled receptors, located in different brain areas, including the cerebral cortex, hippocampus, basal ganglia, limbic structures and cerebellum (see [1]). Together with their endogenous ligands (anandamide and 2-arachidonyl-glycerol- identified to date), cannabinoid CB_1 receptors constitute the so-called "endocannabinoid system". When activated, cannabinoid CB_1 receptors suppress the neuronal release of different excitatory and inhibitory neurotransmitters, including acetylcholine, noradrenaline, dopamine, serotonin, GABA, glutamate and aspartate (see [1]). As a consequence, cannabinoid CB_1 receptors have been implicated in physiological functions such as cognition, control of movement, pain perception, and emotional responses as well as appetite (see [1]).

Furthermore, in recent years, accumulating lines of experimental evidence have suggested the possible involvement of the brain cannabinoid CB_1 receptor system in the neural circuitry controlling alcohol intake, including alcohol relapse-like drinking, and alcohol reinforcing properties. Specifically, the acute administration of cannabinoid receptor agonists CP 55,940 and WIN 55,212-2 has been found to stimulate alcohol intake in selectively bred Sardinian alcohol-preferring (sP) rats [2] and enhance the break-point for beer (i.e., an index of the appetitive or motivational properties of alcohol) in Wistar rats [3]. These effects were completely blocked via pre-treatment with the selective cannabinoid CB_1 receptor antagonist SR 141716, indicating that they were indeed mediated by the cannabinoid CB_1 receptor [2, 3].

Conversely, the administration of SR 141716 has repeatedly been reported to produce opposite effects. Specifically, SR 141716 has been found to:

a) Reduce voluntary alcohol intake and alcohol preference under the home cage, 2-bottle "alcohol *versus* water" choice paradigm in alcohol-experienced C57BL/6 mice [4], sP rats [5] and Wistar rats [6], which may model the "maintenance" or "active drinking" phase of human alcoholism;

b) Suppress the acquisition of alcohol drinking behavior in alcohol-naive sP rats, that is rats which had never consumed alcohol before the start of the experiment [7];
c) Suppress the alcohol deprivation effect (ADE; a model of alcohol relapses in human alcoholics) in sP rats (see below);
d) Decrease the oral self-administration of alcohol in unselected rats tested under operant procedures (i.e., experimental paradigms in which rats were trained to press a lever to gain access to alcohol) [8] (see also L. Parsons and G. Koob's contribution to [9]);
e) Attenuate the appetitive or motivational properties of alcohol, as revealed by a decrease in the probability of the completion of response requirements for alcohol in operant procedures using unselected rats [8, 10] and by the suppression of extinction responding for alcohol (i.e., the maximal amount of "work" which a rat trained to lever-press for alcohol is willing to perform to obtain alcohol) in sP rats [11].

In agreement with the above results on the effect of the pharmacological blockade of the cannabinoid CB_1 receptor by SR 141716, CB_1 receptor knockout mice tested under the 2-bottle choice paradigm displayed significantly lower levels of alcohol preference and consumption in comparison to the wild-type mice ([12–15]; see however [16]).

Taken together, these results suggest that the cannabinoid CB_1 receptor is one of multiple receptor systems implicated in the mediation of the behavioral responses to alcohol.

Effect of SR 141716 on relapse-like behavior in alcohol-preferring rats

The possible anti-relapse properties of SR 141716 have been tested by investigating its effect on ADE in alcohol-preferring sP rats. ADE is defined as the transient increase in alcohol intake which occurs in several animal species after a period of abstinence from alcohol, and ADE has been proposed to model the loss of control over alcohol and the episodes of alcohol relapse of human alcoholics (see [17, 18]). Notably, ADE is a relevant feature of alcohol drinking behavior of sP rats. Indeed, after a proper period of exposure to alcohol and a subsequent period of alcohol deprivation, sP rats display a pronounced ADE during the first hour of re-access to alcohol [19, 20]. Furthermore, ADE in sP rats is reduced by the anti-relapse agent naltrexone (this laboratory, unpublished observations). These results make sP rats a suitable animal model, with predictive validity for pharmacological investigations on ADE.

In the experiment with SR 141716 [21], adult male sP rats were individually housed and continuously (24 h/day) offered alcohol (10%, v/v) and water under the standard home cage 2-bottle choice paradigm with unlimited access for 4 consecutive weeks. Subsequently, rats were divided into two groups (matched for alcohol intake over the last seven days). One group was deprived of alcohol for 15 consecutive days, during which water was the sole fluid available (alcohol-deprived rats). The second group continued to have unlimited access to alcohol and water (alcohol-nondeprived rats), consuming an average of approximately 6 g/kg/day alcohol. At the end of the deprivation phase, rats of both groups (alcohol-deprived and -nondeprived) were further divided into four subgroups (n = 12–13) and acutely and intraperitoneally injected with 0, 0.3, 1 and 3 mg/kg SR 141716, 30 min before lights off. Alcohol was re-presented at lights off and its consumption was recorded 1 h later. Alcohol intake was recorded 60 min later (previous studies have indicated that the time interval of 60 min after alcohol re-presentation is the interval during which ADE is maximal in sP rats [19, 20]). Standard rat chow was available throughout the study.

The results of the study showed that alcohol intake was approximately two times higher in vehicle-treated alcohol-deprived rats than in vehicle-treated alcohol-nondeprived rats, indicative of the development of a robust ADE (Fig. 1). This extra intake of alcohol was, however, eliminated by all doses of SR 141716 (Fig. 1). Indeed, alcohol intake in the alcohol-deprived rat groups treated with all doses of SR 141716 was a) significantly lower than that recorded in alcohol-deprived vehicle-dosed rats, and b) not significantly different from that recorded in the corresponding alcohol-nondeprived SR 141716-treated groups.

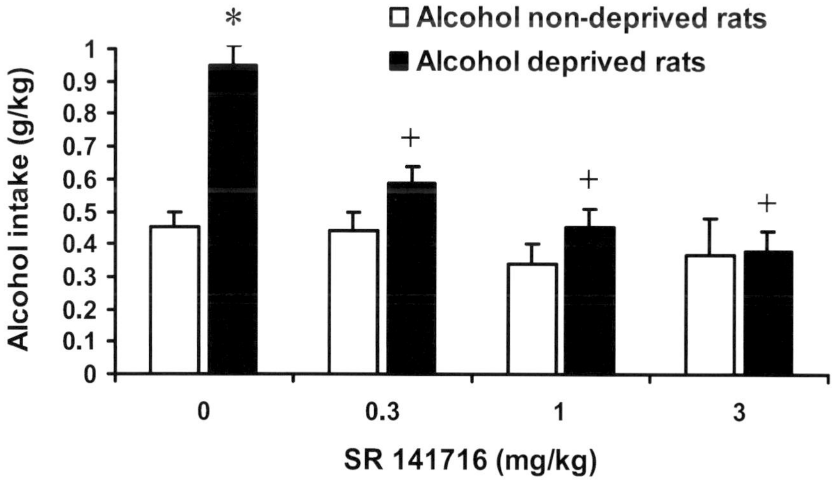

Figure 1. Suppressing effect of the cannabinoid CB_1 receptor antagonist SR 141716 on the alcohol deprivation effect (ADE) in sP rats with alcohol given under the 2-bottle choice regimen. Each bar is the mean ± SEM of n = 12–13. *: $P < 0.05$ with respect to vehicle-treated alcohol-nondeprived rats; +: $P < 0.05$ with respect to vehicle-treated alcohol-deprived rats (Newman-Keuls test). Reprinted from *Eur J Pharmacol* 443: 95–97, 2002, with permission from Elsevier.

These results suggest that the cannabinoid CB_1 receptor is likely part of a neural substrate mediating ADE. These results, together with the apparent safety and potential use in humans of SR 141716 (Rimonabant) [22], also suggest that SR 141716 should be tested for its ability in preventing alcohol relapse in human alcoholics.

Effect of the combination of SR 141716 plus naloxone on relapse-like behavior in alcohol-preferring rats

Recent clinical surveys suggest that positive therapeutic outcomes in the treatment of alcohol relapse may be achieved by using a combination of drugs. For example, a recent, double-blind placebo-controlled study demonstrated that the combination of naltrexone and acamprosate tended to be more effective than the single application of either drug in the prevention of alcohol relapse in alcoholics [23]. Following this line of thought, this laboratory recently tested the hypothesis that a combination of SR 141716 with the opioid receptor antagonist naloxone would result in a potentiation of the reducing effect of each single drug on ADE in alcohol-preferring sP rats.

The procedure employed in this experiment was similar to that used in the study testing SR 141716 alone (see above). Specifically, adult male sP rats were singly housed and continuously (24 h/day) offered alcohol (10%, v/v) and water under the home cage 2-bottle regimen for 8 consecutive weeks. After the initial period of access to alcohol and water, rats were divided into two groups matched for similar daily alcohol consumption and preference over the last 7 days. One group was deprived of alcohol for 14 consecutive days, during which water was the sole fluid available (alcohol-deprived rats). The second group continued to have unlimited access to alcohol and water (alcohol-nondeprived rats). At the end of the deprivation phase, rats of both groups were divided into four subgroups (n = 16), matched for body weight, and acutely treated with: a) SR 141716 vehicle plus naloxone vehicle; b) 0.05 mg/kg SR 141716 plus naloxone vehicle; c) SR 141716 vehicle plus 0.01 mg/kg naloxone; d) 0.05 mg/kg SR 141716 plus 0.01 mg/kg naloxone. The SR 141716 and naloxone doses were chosen so as to be ineffective when given alone. Drugs were administered intraperitoneally 30 min before alcohol presentation (which coincided with lights off). Alcohol was re-presented at lights off. Alcohol intake was recorded 60 min later. Standard rat chow was available throughout the study.

As expected, after alcohol re-presentation, alcohol intake was higher by approximately 50% in control alcohol-deprived rats as compared to control alcohol-nondeprived rats (Fig. 2), indicating the development of ADE. When given alone, neither 0.05 mg/kg SR 141716 nor 0.01 mg/kg naloxone affected the extra amount of alcohol intake, which constitutes ADE. In contrast, their combination reduced ADE by approximately 30% (Fig. 2).

These results are in close agreement with those of recent studies, demonstrating that SR 141716 and the opioid receptor antagonist naltrexone syner-

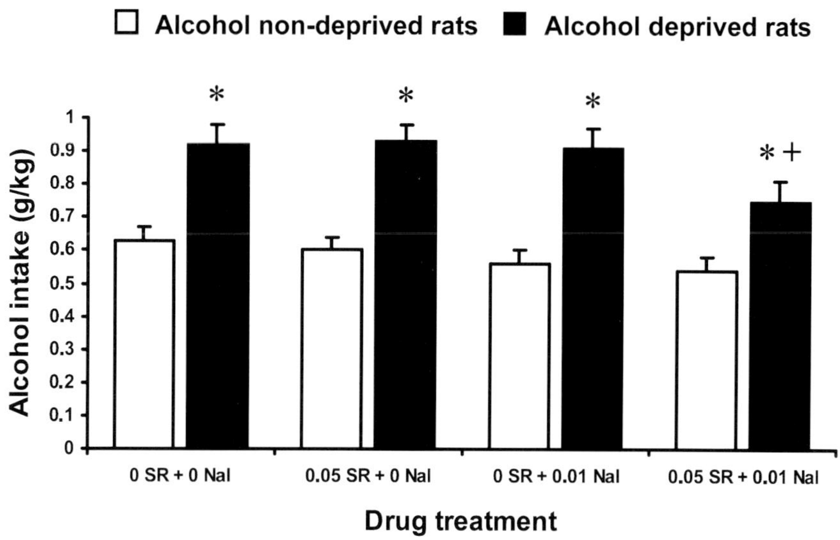

Figure 2. Reducing effect of the combination of the cannabinoid CB_1 receptor antagonist SR 141716, and the opioid receptor antagonist naloxone, on the alcohol deprivation effect (ADE) in sP rats with alcohol given under the 2-bottle choice regimen. Each bar is the mean ± SEM of n = 16. *: $P < 0.05$ with respect to alcohol-nondeprived rats treated with the corresponding doses of SR 141716 and naloxone; +: $P < 0.05$ with respect to alcohol-deprived rats treated with SR 141716 vehicle plus naloxone vehicle (Newman-Keuls test).

gistically reduce the break-point for alcohol in unselected rats [24] as well as acquisition of alcohol drinking behavior in alcohol-preferring sP rats (this laboratory, unpublished observations), providing further support to the hypothesized existence of functional interactions between the cannabinoid and opioid receptor systems in relation to alcohol drinking behavior [2, 25]. Accordingly, previous work has demonstrated that the stimulating effects of cannabinoids and morphine on alcohol intake were blocked by both SR 141716 and naloxone [2, 25]. Finally, the extension to food intake of the synergistic reducing effect of SR 141716 and naloxone in rats [26, 27] suggests that the functional relationship between cannabinoid and opioid receptor systems pertains not only to alcohol intake but to different ingestive behaviors as well.

The possible generalization to human alcoholics of the reducing effect of the combination of SR 141716 and naloxone on alcohol relapse-like drinking in sP rats would indicate a new therapeutic strategy.

References

1 Iversen L (2003) Cannabis and the brain. *Brain* 126: 1252–1270
2 Colombo G, Serra S, Brunetti G, Gomez R, Melis S, Vacca G, Carai MAM, Gessa GL (2002) Stimulation of voluntary ethanol intake by cannabinoid receptor agonists in ethanol-preferring sP rats. *Psychopharmacology* 159: 181–187

3 Gallate JE, Saharov T, Mallet PE, McGregor IS (1999) Increased motivation for beer in rats following administration of a cannabinoid CB_1 receptor agonist. *Eur J Pharmacol* 370: 233–240
4 Arnone M, Maruani J, Chaperon F, Thiébot M-E, Poncelet M, Soubrié P, Le Fur G (1997) Selective inhibition of sucrose and alcohol intake by SR 141716, an antagonist of central cannabinoid (CB1) receptors. *Psychopharmacology* 132: 104–106
5 Colombo G, Agabio R, Fà M, Guano L, Lobina C, Loche A, Reali R, Gessa GL (1998) Reduction of voluntary ethanol intake in ethanol-preferring sP rats by the cannabinoid antagonist SR 141716. *Alcohol Alcoholism* 33: 126–130
6 Lallemand F, Soubrié PH, De Witte PH (2001) Effects of CB_1 cannabinoid receptor blockade on ethanol preference after chronic ethanol administration. *Alcohol Clin Exp Res* 25: 1317–1323
7 Serra S, Carai MAM, Brunetti G, Gomez R, Melis S, Vacca G, Colombo G, Gessa GL (2001) The cannabinoid receptor antagonist SR 141716 prevents acquisition of drinking behaviour in alcohol-preferring rats. *Eur J Pharmacol* 430: 369–371
8 Freedland CS, Sharpe AL, Samson HH, Porrino LJ (2001) Effects of SR 1417161A on alcohol and sucrose self-administration. *Alcohol Clin Exp Res* 25: 277–282
9 Hungund BL, Basavarajappa BS, Vadasz C, Kunos G, Rodriguez de Fonseca F, Colombo G, Serra S, Parsons L, Koob GF (2002) Ethanol, endocannabinoids, and the cannabinoidergic signaling system. *Alcohol Clin Exp Res* 26: 565–574
10 Gallate JE, McGregor IS (1999) The motivation for beer in rats: effects of ritanserin, naloxone and SR 141716. *Psychopharmacology* 142: 302–308
11 Colombo G, Vacca G, Serra S, Carai MAM, Gessa GL (2004) Suppressing effect of the cannabinoid CB_1 receptor antagonist, SR 141716, on alcohol's motivational properties in alcohol-preferring rats. *Eur J Pharmacol* 498: 119–123
12 Hungund BL, Szakall I, Adam A, Basavarajappa BS, Vadasz C (2003) Cannabinoid CB1 receptor knockout exhibit markedly reduced voluntary alcohol consumption and lack alcohol-induced dopamine release in the nucleus accumbens. *J Neurochem* 84: 698–704
13 Poncelet M, Maruani J, Calassi R, Soubrié P (2003) Overeating, alcohol and sucrose consumption decrease in CB1 receptor deleted mice. *Neurosci Lett* 343: 216–218
14 Wang L, Liu J, Harvey-White J, Zimmer A, Kunos G (2003) Endocannabinoid signaling via cannabinoid receptor 1 is involved in ethanol preference and its age-dependent decline in mice. *Proc Natl Acad Sci USA* 100: 1393–1398
15 Naassila M, Pierrefiche O, Ledent C, Daoust M (2004) Decreased alcohol self-administration and increased alcohol sensitivity and withdrawal in CB1 receptor knockout mice. *Neuropharmacology* 46: 243–253
16 Racz I, Bilkei-Gorzo A, Toth ZE, Michel K, Palkovits M, Zimmer A (2003) A critical role for the cannabinoid CB_1 receptors in alcohol dependence and stress-stimulated ethanol drinking. *J Neurosci* 23: 2453–2458
17 Boening JA-L, Lesch OM, Spanagel R, Wolffgramm J, Narita M, Sinclair D, Mason BJ, Wiesbeck GA (2001) Pharmacological relapse prevention in alcohol dependence: from animal models to clinical trials. *Alcohol Clin Exp Res* 25: 127S–131S
18 McBride WJ, Le A-D, Noronha A (2002) Central nervous system mechanisms in alcohol relapse. *Alcohol Clin Exp Res* 26: 280–286
19 Agabio R, Carai MAM, Lobina C, Pani M, Reali R, Vacca G, Gessa GL, Colombo G (2000) Development of short-lasting alcohol deprivation effect (ADE) in Sardinian alcohol-preferring rats. *Alcohol* 21: 59–62
20 Serra S, Brunetti G, Vacca G, Lobina C, Carai MAM, Gessa GL, Colombo G (2003) Stable preference for high ethanol concentrations after alcohol deprivation in Sardinian alcohol-preferring (sP) rats. *Alcohol* 29: 101–108
21 Serra S, Brunetti G, Pani M, Vacca G, Carai MAM, Gessa GL, Colombo G (2002) Blockade by the cannabinoid CB_1 receptor antagonist, SR 141716, of alcohol deprivation effect in alcohol-preferring rats. *Eur J Pharmacol* 443: 95–97
22 Le Fur G, Arnone M, Rinaldi-Carmona M, Barth F, Heshmati H (2001) SR 141716, a selective antagonist of CB_1 receptors and obesity. Abstract book, 2001 Meeting of the International Cannabinoid Research Society, San Lorenzo de El Escorial, Spain, 101
23 Kiefer F, Jahn H, Tarnasker T, Helwig H, Briken P, Holbach R, Kampf P, Stracke R, Baehr M, Naber D et al. (2003) Comparing and combining naltrexone and acamprosate in relapse prevention of alcoholism. A double-blind, placebo-controlled study. *Arch Gen Psychiatry* 60: 92–99
24 Gallate JE, Mallet PE, McGregor IS (2004) Combined low dose treatment with opioid and

cannabinoid receptor antagonists synergistically reduces the motivation to consume alcohol in rats. *Psychopharmacology* 173: 210–216
25 Vacca G, Serra S, Brunetti G, Carai MAM, Gessa GL, Colombo G (2002) Boosting effect of morphine on alcohol drinking is suppressed not only by naloxone but also by the cannabinoid CB_1 receptor antagonist, SR 141716. *Eur J Pharmacol* 445: 55–59
26 Kirkham TC, Williams CM (2001) Synergistic effect of opioid and cannabinoid antagonists on food intake. *Psychopharmacology* 153: 267–270
27 Rowland NE, Mukherjee M, Robertson K (2001) Effects of the cannabinoid receptor antagonist SR 141716, alone and in combination with dexfenfluramine or naloxone, on food intake in rats. *Psychopharmacology* 159: 111–116

Drugs for Relapse Prevention of Alcoholism
Edited by R. Spanagel and K. Mann

Neuropeptide Y antagonists: a perspective

Markus Heilig[1] and Todd E. Thiele[2]

[1] *Department of NEUROTEC, Karolinska Institute, M57 Huddinge University Hospital, S-14186, Stockholm, Sweden*
[2] *Department of Psychology, University of North Carolina, Davie Hall, CB# 3270, Chapel Hill, NC 27599–3270, USA*

Introduction

Over the last 15 years, a role has been firmly established for NPY as an endogenous anti-stress system, potentially also implicating this system in the pathophysiology of anxiety and depressive disorders (recently reviewed in, e.g., [1, 2]). More recently, converging evidence from genetically modified animals, pharmacological studies and human observations has indicated that endogenous NPY signalling is involved in regulation of voluntary alcohol intake, in particular in states of abnormally high drinking, and that targeting this system may offer an attractive mechanism for relapse prevention in alcoholism. In this chapter, we introduce the brain NPY system and its biology, and review the findings supporting a role for NPY receptor ligands in future treatment of alcoholism.

Basic biology of the central NPY system

NPY, named so because of its exclusively neuronal expression, and its terminal tyrosine (Y in the 1-letter aa code), is a 36 amino acid (aa) peptide with a C-terminal amide-group [3, 4]. It belongs to a family of peptides related to pancreatic polypeptide (PP; [5]). NPY is one of the most highly conserved neuroendocrine peptides known [6], which implies an important functional role. NPY-like peptides all consist of an N-terminal polyproline helix (residues 1–8) and an amphiphilic α-helix (residues 15–30), connected with a β-turn, creating a hairpin-like loop [7]. The preproNPY gene encodes a simple 97 aa precursor [8], which contains a 28 aa signal peptide and a 69 aa prohormone. Mature NPY (36 aa) is here flanked at its C-terminus by 33 amino acids, three of which are a motif necessary for NPY amidation, critical for virtually all actions of NPY. The peptide formed by the remaining 30 amino acids of the precursor has been named CPON (C flanking peptide of NPY). Although it also shows a high degree of sequence conservation, its function remains unknown [6].

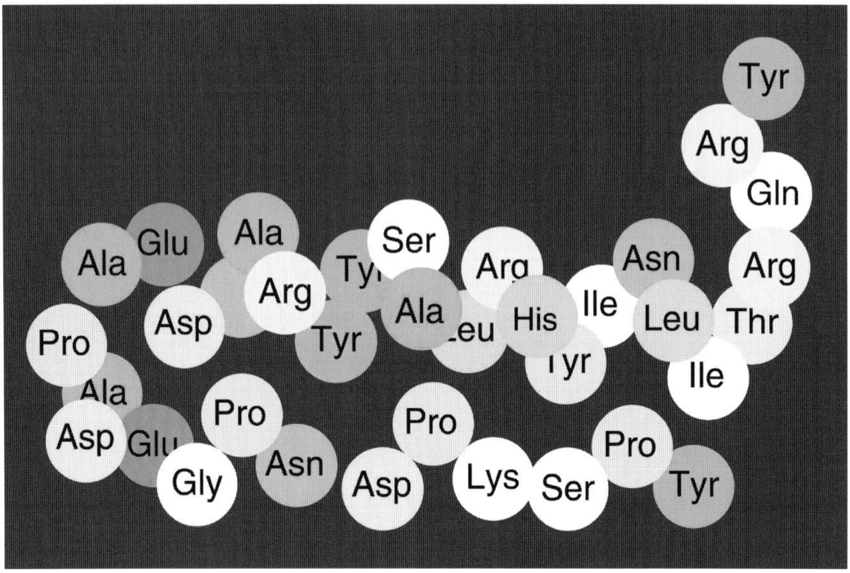

Figure 1. Structure of the NPY molecule. The hairpin loop structure is characteristic of the family of NPY related peptides. The full 1–36 a.a. sequence is required for activation of Y1 receptors, while shorter, truncated fragments are capable of activation of other receptor subtypes.

NPY-positive neurons are abundant in the central nervous system (CNS), including brain areas involved in emotionality [9–11]. Within the cerebral cortex and forebrain nuclei, NPY is mainly present in inhibitory interneurons. Coexistence of NPY with somatostatin and nitric oxide synthase is common in cortex and striatum. In addition, within the cortex and amygdala, NPY is extensively colocalized with GABA [11]. NPY-positive neurons in the hypothalamus, brainstem and spinal cord are more heterogeneous. An important NPY-ergic pathway from the arcuate nucleus to the ipsilateral paraventricular nucleus of the hypothalamus has been described [12] and is involved in feeding effects of NPY. More recently, a co-existence was also reported of NPY and agouti-gene-related protein (AGRP) in arcuate hypothalamic neurons. Among these, numerous cells project to telencephalic brain areas involved in emotionality, such as the amygdala complex [13].

Pharmacological experiments followed by molecular cloning have revealed a diversity of the NPY receptor family. NPY receptors cloned to date all belong to the superfamily of G-protein coupled receptors. The NPY-Y_1 receptor, one of the subtypes implicated in feeding effects of NPY [14–16], requires the intact NPY sequence for recognition and activation, and appears to be the subtype mediating anti-stress actions of NPY [17–21]. The Y_2 receptor subtype is also activated by C-terminal fragments of NPY such as NPY_{13-36} [22, 23]. A high number of Y_2 sites is found within the hippocampus. Activation of Y_2 receptors within this structure has been shown to suppress hippocampal

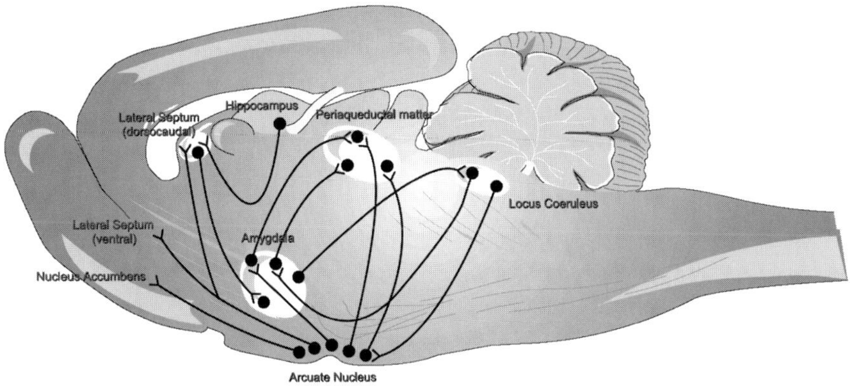

Figure 2. Brain NPY circuitry potentially related to regulation of alcohol intake.

glutamatergic transmission through presynaptic mechanisms [24, 25]. Behavioral consequences of Y_2 signalling in this area are unclear. Although the existence of a Y_3 receptor has been postulated on the basis of pharmacological experiments [26], this has not been confirmed by molecular studies [27]. Furthermore, a receptor termed Y_4 has been cloned, but appears to preferentially bind PP, and is therefore more appropriately referred to as a PP receptor [28, 29]. Finally, a Y_5 receptor with restricted hypothalamic expression has been cloned, and postulated to mediate the profound effects of NPY on feeding [30]. Subsequent work indicates that the Y_5 receptor probably shares this role with the Y_1 subtype.

NPY in EtOH-responses and EtOH seeking behavior

Central actions of NPY and alcohol show similarities

Studies carried out shortly following its isolation indicated shared properties between NPY and several classes of sedative compounds, including alcohol, benzodiazepines and barbiturates [31]. It was subsequently noted that icv infusion of NPY, as well as a Y_1 receptor agonist, produced electrophysiological and behavioral profiles similar to those induced by anxiolytic drugs such as EtOH and benzodiazepines [32]. Additionally, icv infusion of NPY and peripheral administration of EtOH to rats produced identical effects on event-related potential (ERP) profiles in response to auditory stimuli, both in cortex and amygdala. The effects of NPY and EtOH were additive [33]. Following 10–15 weeks of withdrawal from chronic exposure to EtOH vapor, icv infusion of NPY significantly decreased the amplitude of the N1 component of ERP in the amygdala of withdrawn Wistar rats when compared to controls, indicating that EtOH withdrawal augments brain sensitivity to NPY [34].

Together, these data suggest that electrophysiological responses to EtOH and EtOH withdrawal may be mediated, in part, by NPY signalling. A related study comparing the P and NP rats showed opposite electrophysiological activity in the amygdala following icv infusion of NPY [35]; this observation, together with those showing low NPY levels in P rats [36] strongly suggest that altered NPY signalling in the amygdala of P rats contributes to their high alcohol drinking.

Additional similarities between NPY and conventional sedatives, including alcohol are suggested by anti-convulsant actions of NPY [37, 38], mutual substitution of NPY and EtOH with regard to electrophysiological effects [33] and potentiation by NPY of barbiturate-induced sleep [39, 40]. The latter study mapped out NPY effects on sedation to the posterior hypothalamus, an area involved in the regulation of sleep–wake cycles. Y_1 mediation of NPY-induced sedation was demonstrated using Y_1-receptor knockouts. In contrast, anti-convulsant actions of NPY appear to be mediated through Y_5-receptors [41].

EtOH administration modulates central NPY signalling

EtOH administration influences NPY signalling. Relative to control animals that received an isocaloric diet as their sole source of calories, Long-Evans rats given access to a diet containing 6% EtOH for 12 weeks showed significant increases of NPY levels in the arcuate and ventromedial nuclei of the hypothalamus, the median eminence, and the suprachiasmatic nucleus [42]. Additionally, peripheral injection of 1.5 and 3.5 g/kg EtOH caused activation of NPY-containing neurons in the ventrolateral medulla of Long-Evans rats [43]. Wistar rats exposed to EtOH vapor for 14 h/day showed no differences in brain NPY expression after 7 weeks of exposure, but showed increased NPY expression in the hypothalamus 7 weeks after withdrawal from EtOH [44]. On the other hand, EtOH administration and withdrawal from EtOH have also been found to reduce NPY signalling. NPY mRNA levels in the arcuate nucleus of the hypothalamus were decreased when Sprague-Dawley rats were given a single peripheral injection of a 1.0 g/kg dose of EtOH [45]. More recently, Sprague-Dawley rats examined 24 h after withdrawal from a diet containing 9% EtOH (after 15 days of exposure) showed decreased NPY immunoreactivity in the cingulate gyrus, various regions of the cortex, the central and medial nuclei of the amygdala, and the paraventricular and arcuate nuclei of the hypothalamus [46]. Thus, withdrawal from EtOH is associated with reduced central NPY signalling. Consistent with this observation, a recent report found that icv infusion of NPY significantly attenuated EtOH withdrawal responses in Wistar rats [47]. The discrepancies between these studies (that is, either increases or decreases of NPY levels) may be related to rat strain differences, method of EtOH administration, technique for assessing NPY levels, or an interaction between these factors. An indication which might be helpful in relating these findings to the human clinical situation is a recent cDNA microarray study which examined the expression of approximately 10,000 genes in the frontal cortex and motor cortex of alcoholics and matched control samples. One of the most intriguing observations was that brain tissue from alcoholics had significantly lower NPY expression than brain tissue from controls [48]. Some caution in the interpretation is, however, warranted, since in this type of study it is unclear if low NPY levels were the result of chronic alcohol use, or reflect a pre-existing, genetically encoded susceptibility factors for triggering the disease.

Genetic models suggest involvement of NPY in alcohol drinking

The first genetic evidence linking NPY to alcoholism came from studies involving rats selectively bred for high alcohol drinking. Quantitative trait locus (QTL) analysis identified a region of chromosome 4 that significantly correlated with differences in alcohol drinking between the Indiana alcohol-preferring (P) and alcohol-nonpreferring (NP) rats. This chromosomal region includes the NPY precursor gene [49, 50]. Interestingly, another transcript differentially expressed between P and NP rats is also encoded by a gene in this region, suggesting the possibility of a commonly regulated haplotype block [51]. Subsequent research found that P rats had low levels of NPY in the amygdala, frontal cortex, and hippocampus relative to NP rats, but higher levels of NPY in the hypothalamus and cingulate cortex [36, 44]. High alcohol-drinking (HAD) rats, bred by a similar strategy as that used to generate the P rats, also had low levels of NPY in the amygdala compared with low alcohol-drinking (LAD) rats, and had lower levels of NPY in hypothalamic nuclei [36]. Hwang et al. concluded that the high alcohol drinking by the P and HAD rats are best explained by low levels of NPY in the amygdala. It should be noted, however, that QTL analyses with HAD and LAD rats failed to confirm a role for the NPY precursor gene [52]. More recently, alcohol-avoiding, ANA (Alko Non-Alcohol) line of rats was found to have high NPY mRNA in the hippocampal Cornu Ammonis (CA) region and the dentate gyrus when compared with the alcohol-preferring, AA (Alko Alcohol) line and nonselected Wistar rats. Additionally, NPY Y_2 receptor mRNA was reduced in the AA line, suggesting a role for the Y_2 receptor in modulating alcohol drinking [53].

The studies reviewed above provide suggestive, but only correlative evidence for a role of NPY transmission in regulation of alcohol drinking. Two complementary lines of intervention studies provide data supporting the notion that this relation is in fact causal. The first of these is based on genetic manipulations in rodents, leading to absent or excessive expression of NPY, or selective inactivation NPY receptor subtypes. Voluntary EtOH consumption and resistance to the sedative effects of EtOH were inversely related to NPY levels in knockout and transgenic mice [54]. These initial results were obtained in a mixed C57BL/6Jx129/SvEv genetic background. These mice also react with an exaggerated locomotor response to EtOH administration. It has subsequently been found that EtOH-associated phenotypes are partly dependent on the genetic background. Neither the resistance to sedative effects nor the potentiation of locomotor stimulation were found in an inbred 129/SvEv background, where the differences in voluntary EtOH consumption were also less marked. However, at the highest concentration of EtOH tested, 20%, increased voluntary consumption was found also in this background [55].

Regionality of central NPY overexpression appears to be crucial for modulation of EtOH drinking. This is highlighted by data from transgenic rats selectively overexpressing NPY in CA1 and CA2 regions of the hippocampus. Relative to control animals, these subjects were resistant to anxiety provoked by restraint-stress, and showed impairment of spatial memory acquisition. However, the NPY transgenic rats showed normal voluntary EtOH drinking [56].

Data from Y_1 receptor knockout mice ($Y_1^{-/-}$) provide further support for a role of the NPY system, and point to receptor mediation of NPY effects on EtOH intake. Except for slightly diminished food intake and the development of late-onset obesity due to low energy expenditure, these animals show normal gross phenotypic features [57]. However, $Y_1^{-/-}$ mice showed increased consumption of solutions containing 3%, 6%, and 10% (v/v) EtOH but displayed normal consumption of sucrose and quinine solutions. Furthermore, $Y_1^{-/-}$ mice were less sensitive to the sedative effects of 3.5 and 4.0 g EtOH/kg as measured by more rapid recovery from EtOH-induced sleep, even though plasma EtOH levels did not differ significantly between the genotypes following a 3.5 g/kg dose. Finally, male $Y_1^{-/-}$ mice showed normal EtOH-induced ataxia on a rotarod test following administration of a 2.5 g/kg dose [58].

The Y_2 receptor is a presynaptic autoreceptor and inhibits NPY release [59, 60]. Mutant mice lacking the Y_2 receptor ($Y_2^{-/-}$) have been shown to have increased food intake, body weight, and fat production but have a normal response to NPY-induced food intake [61]. It was hypothesized that if presynaptic Y_2 receptors are involved with modulating voluntary EtOH consumption and sensitivity, the $Y_2^{-/-}$ mice should exhibit EtOH-related phenotypes opposite to those found with the $Y_1^{-/-}$ mice. Thus, an absence of presynaptic inhibition of NPY release in $Y_2^{-/-}$ mice would augment NPY signalling, rendering mice with a similar phenotype as NPY overexpressing mice. Relative to wild-type ($Y_2^{+/+}$) mice, the $Y_2^{-/-}$ mice drank significantly less of solutions containing 3% and 6% EtOH, and had significantly lower EtOH preference at each concentration tested. On the other hand, $Y_2^{-/-}$ mice showed normal consumption of solutions containing either sucrose or quinine, normal time to recover from EtOH-induced sedation following 3.0 or 3.5 g/kg doses, and normal metabolism of EtOH following injection of a 3.0 g/kg dose [62].

Mutant mice lacking the NPY Y_5 receptor ($Y_5^{-/-}$) show late-onset obesity and increased food intake, have reduced sensitivity to NPY, and are seizure-prone [41]. When given access to solutions containing EtOH, $Y_5^{-/-}$ mice drank normal amounts of 3, 6, 10, and 20% (v/v) EtOH, but had increased sleep time following administration of 2.5 or 3.0 g EtOH/kg. However, the $Y_5^{-/-}$ mice also showed high plasma EtOH levels relative to wild-type mice following injection of a 3.0 g/kg dose [63].

Taken together, evidence from genetic animal models implies that low NPY signalling can promote high voluntary EtOH drinking while upregulation of NPY signalling can be protective against excessive consumption. Data from NPY receptor knockout mice suggest that voluntary consumption of EtOH is modulated by the Y_1 and Y_2 receptors, and that EtOH-induced sedation is modulated by Y_1, and perhaps Y_5, receptors. These results are consistent with several recent findings. First, like $Y_2^{-/-}$ mice which drink low amounts of EtOH, rats self-administer less EtOH following central infusion of a Y_2 receptor antagonist [64] (see below). Second, $Y_1^{-/-}$ mice are resistant to the sedative effects of EtOH, and recent studies found that $Y_1^{-/-}$ mice are resistant to sodium pentobarbital-induced sleep [39, 40].

NPY, alcoholism and human genetics

The human preproNPY gene is polymorphic. Most attention this far has been attracted by a thymidine (T) to cytosine (C) single nucleotide polymorphism (SNP) that is present at the 1128 position of the human *NPY* gene, resulting in a leucine-to-proline substitution (Leu7Pro) in the signal peptide of preproNPY [65]. Individuals with the *Leu7/Pro7* genotype have an average of 42% higher maximal increases of plasma NPY in response to physiological stress when compared with *Leu7/Leu7* individuals [66]. Interestingly, Finnish men with the Pro7 substitution reported 34% higher average alcohol consumption when compared to men not having this polymorphism [67]. It should be noted that consumption levels in this study were reported from non-dependent subjects, and the reported consumption levels were low, averaging app. 70 g/week. The relevance of these data for alcohol dependence is thus unclear. A subsequent study carried out in European–American men with carefully diagnosed alcoholism reported a 5–5.5% Pro7 allele frequency, while non-alcoholics had a Pro7 allele frequency of only 2.0%, leading to a statistically significant association between genotype and diagnosis [68]. However, results are mixed. A lower rather than higher frequency of the Pro7 allele has been reported in type 2 alcoholics compared to controls [69], while a more recent study found no difference of Pro7 allele frequency between diagnosed Caucasian alcoholics and ethnically matched controls from Finland and Sweden [70]. Furthermore, a meta-analysis performed in the latter study found that while the Pro7 allele frequencies in alcoholics were similar in each report, the allele frequencies in nonalcoholic control groups were very different between studies. This issue remains unresolved at present. Despite ambitious attempts to exclude this possibility by Lappalainen and colleagues, the discrepant results might be related to ethnic stratification. The Pro7 allele differs in frequency between ethnic groups, and is, e.g., entirely absent in Asians [71, 72].

The prepro NPY gene is polymorphic also at other positions. Among these, an SNP within a trk-B consensus sequence in the promoter region (–399C/T) is clearly functional, as C-containing alleles in this position confer higher transcriptional activity in neuronal cells. Both alleles have high frequencies in populations examined, and preliminary associations have been suggested both for schizophrenia [73] and treatment-refractory depression [94]. It is at present unknown whether this polymorphism plays a role in alcohol dependence of subtypes thereof.

Finally, a C to T substitution at position 5671, mapping to exon 3 of the *NPY* gene, has been described in a Japanese population. Although there was no association between genotype and a diagnosis of alcohol dependence, it was reported that the T-allele was found in a significantly higher frequency in alcoholic patients experiencing seizures [72]. Since this SNP is synonymous (i.e., it does not encode an amino acid substitution), its role is unclear. One possibility is that it is in linkage dysequilibrium with other, functional polymorphisms.

NPY and alcoholism: pharmacological mechanisms and strategies

Genetic modifications are powerful and highly selective tools, but have known limitations, in particular related to issues of genetic background, and compensatory mechanisms which can be activated in constitutive overexpressors or knockouts [74]. In the case of NPY and alcohol, however, genetic and neurochemical evidence is largely supported by emerging pharmacological studies. These have used icv infusion of NPY and other NPY receptor ligands to determine if NPY signalling regulates voluntary EtOH consumption, and thus directly point to possible future applications in the clinic. To correctly interpret the results of these studies, it is crucial to understand a basic fact of experimental alcohol research: laboratory rodents, and in particular genetically heterogeneous rats commonly used in pharmacological experiments, do not voluntarily consume sufficient amounts of EtOH to achieve pharmacological effects. Instead, they are likely to drink for other types of motivation, such as caloric content. Modifications of this baseline consumption are of little relevance for developing clinical treatments. On the other hand, states of excessive drinking can be induced either by genetic selection or behavioral manipulations [75]. This induced, excessive drinking component is selectively affected by clinically effective drugs (see, e.g. [76, 77]), and therefore the appropriate target for candidate anticraving/relapse-preventive drugs.

The general picture which has emerged against the background of this distinction is that exogenous NPY does not reliably regulate basal voluntary EtOH drinking and may even slightly increase it under normal conditions. In contrast, potentiation of NPY signalling potently suppresses EtOH drinking in states of excessive intake. Thus, in the first attempt with Golden Hamsters, icv infusion of NPY did not reliably alter drinking of a 5% EtOH solution [78].

More recently, Wistar rats were given icv infusion of various doses of NPY ranging from 2.5 to 15.0 μg in a within-subjects design. While 5.0 μg of NPY significantly increased consumption of a sucrose solution, none of the doses tested altered alcohol intake [79]. Similarly, neither third ventricle infusion of NPY nor direct infusion of NPY into the amygdala altered EtOH drinking in Wistar rats [80, 81]. In fact, direct infusion of femtomolar doses of NPY into the paraventricular nucleus of the (PVN) hypothalamus increased consumption of alcohol in Long-Evans rats, an effect which was blocked by pretreatment with the Y_1 receptor antagonist BIBP 3226 [82]. The PVN is known to mediate appetite effects of NPY, and this finding likely indicates that in "normal" rats, effects of NPY on EtOH consumption primarily reflect appetite modulation, since EtOH in addition to being an intoxicating agent is also a caloric nutrient. Recently, a report found that amygdalar infusion of BIBP 3226 decreased ethanol self-administration in Long-Evans rats [83], an effect that may be unrelated to the pharmacological effects of ethanol by this moderate alcohol-consuming strain.

On the other hand, icv infusion of both 5.0 and 10.0 μg doses of NPY significantly reduced voluntary consumption of an 8% EtOH solution in alcohol-preferring P rats; in these experiments, it again did not alter EtOH drinking of non-preferring NP or outbred Wistar rats [84]. The suppressing effect of NPY on EtOH intake in P rats is even more pronounced after a sequence of continuous access followed by a deprivation phase, a procedure known to increase the motivation to consume EtOH for its reinforcing properties [85]. More recently, the ability of NPY to selectively suppress excessive EtOH drinking was confirmed in an interesting manner in another genetically selected high-preferring line, HAD rats [86]. In this study, icv administration of NPY increased sucrose self-administration in both HAD and low-preferring LAD rats, but selectively suppressed EtOH-self-administration in the HAD line only. Interestingly, this was found despite the observation that the well-known anti-anxiety actions of NPY (1) were identical in the two lines. This indicates that, although altered emotionality may contribute to regulation of EtOH intake by NPY, the latter can also be modulated independently of the former. Recently, amygdalar infusion of a PKA inhibitor increased anxiety and ethanol drinking by Sprague Dawley rats, and caused local reductions of NPY levels. Elevated levels of anxiety and ethanol drinking were rescued by amygdalar co-administration of NPY. Consistent with the above observations, NPY did not affect ethanol consumption by rats not treated with the PKA inhibitor and which had normal (i.e., non-elevated) ethanol consumption [87].

Of particular relevance for development of NPY-based pharmacological treatments of alcohol dependence, EtOH self-administration has also been examined following central administration of the selective NPY-Y_2 antagonist BIIE0246. This compound is known to potentiate the release of endogenous NPY [59], an indirect approach which may circumvent the difficulties inherent in developing an agonist for NPY-Y_1 receptors. Initial experiments with BIIE0246 were carried out using regular rats, but under conditions of limited access operant self-administration which do produce significant blood alcohol concentrations. Icv administration of BIIE0246 dose-dependently suppressed self-administration. Interestingly, in follow-up experiments using rats with a history of dependence induced according to a recently published model [76], doses of BIIE0246 which were subthreshold in non-dependent animals were effective in suppressing self-administration in subjects with a history of dependence (Thorsell et al., in preparation). The same dissociation was observed using antisense mediated inhibition ("knockdown") of Y_2 receptor expression. Thus, Y_2 antagonism appears to offer an attractive strategy, which might selectively target states of excessive EtOH consumption.

Finally, it should be noted that peripheral administration of a Y_5 receptor antagonist delayed the onset of ethanol-reinforced responding but did not alter the amount of ethanol consumed by C57BL/6 mice in a 16-h session [88]. These findings, and the observation that Y_5 receptor knockout mice show normal ethanol consumption [63], do not provide a strong case for the Y_5 receptor in the modulation of ethanol consumption.

NPY and alcohol: conclusions and future directions

Research over more than 15 years has implicated NPY in mechanisms of emotionality and stress, identifying it as a potential therapeutic target for novel treatments in anxiety disorders and depression [1, 89–91]. More recent evidence identifies the NPY system as a highly interesting treatment target in alcoholism. In summary, EtOH and NPY have similar effects on brain electrophysiological activity, while CNS responses to EtOH involve central NPY signalling, as evidenced by altered central NPY levels and expression following administration of EtOH and EtOH withdrawal. Low NPY signalling in animal models predisposes to high EtOH drinking, while central administration of NPY selectively reduces excessive EtOH drinking but not drinking in 'normal' unselected animals. Activation of NPY Y_1 receptors appears to mediate NPY's suppression of excessive drinking; blockade or inactivation of Y_2 receptors leads to the same functional outcome, presumably through removal of Y_2-mediated presynapic inhibition of endogenous NPY release.

It can be hypothesized that central NPY activity is recruited in response to EtOH consumption, and that this NPY activation serves as a protective feedback mechanism to prevent high EtOH drinking. Animals with abnormally low NPY levels would not benefit from this feedback protection and drink excessive quantities of EtOH. Such a mechanism could also explain excessive drinking in alcoholics with low brain NPY expression. The regulatory role of NPY for regulation of excessive EtOH might in part also be related to effects of NPY on emotionality and stress responses. While adaptive in the short term, activation of these systems imposes an allostatic load on the organism if present over prolonged periods [92]. Negative emotionality and dysregulated stress responses are important factors in the development of dependence, and in one of its hallmarks, relapse [93]. NPY counteracts and buffers negative emotionality and stress responses [1, 89], and these effects of potentiating NPY transmission may be beneficial in the treatment of alcoholism.

Thus available data suggest that drugs targeting central NPY systems may become useful therapeutic agents in alcoholism. Agonists aimed at the Y_1 or, perhaps more realistically, antagonists of Y_2 receptors are particularly promising candidates. NPY-targeting drugs might turn out to be most useful in alcoholism with co-morbid anxiety and/or depression.

References

1 Heilig M, Thorsell A (2002) Brain neuropeptide Y (NPY) in stress and alcohol dependence. *Rev Neurosci* 13: 85–94
2 Thiele TE, Heilig M (2004) Behavioural effects of NPY. In: MC Michel (ed.): *NPY and related peptides*. Springer Verlag, Heidelberg, 252–282
3 Tatemoto K (1982) Neuropeptide Y complete amino acid sequence of the brain peptide. *Proc Natl Acad Sci USA* 79: 5485–5489
4 Tatemoto K, Carlquist M, Mutt V (1982) Neuropeptide Y—a novel brain peptide with structural similarities to peptide YY and pancreatic polypeptide. *Nature* 296: 659–660
5 Kimmel JR, Hayden LJ, Pollock HG (1975) Isolation and characterization of a new pancreatic polypeptide hormone. *J Biol Chem* 250: 9369–9376
6 Larhammar D, Söderberg C, Blomqvist AG (1993) Evolution of the neuropeptide Y family of peptides. In: WF Colmers, C Wahlestedt (eds) *The biology of neuropeptide Y and related peptides*. Humana Press, Totowa, NJ, 1–41
7 Schwartz TW, Fuhlendorff J, Kjems LL, Kristensen MS, Vervelde M, O'Hare M, Krstenansky JL, Bjornholm B (1990) Signal epitopes in the three-dimensional structure of neuropeptide Y. Interaction with Y1, Y2, and pancreatic polypeptide receptors. *Ann N Y Acad Sci* 611: 35–47
8 Minth CD, Bloom SR, Polak JM, Dixon JE (1984) Cloning, characterization, and DNA sequence of a human cDNA encoding neuropeptide tyrosine. *Proc Natl Acad Sci USA* 81: 4577–4581
9 De Quidt ME, Emson PC (1986) Distribution of neuropeptide Y-like immunoreactivity in the rat central nervous system—I. Radioimmunoassay and chromatographic characterisation. *Neuroscience* 18: 527–543
10 De Quidt ME, Emson PC (1986) Distribution of neuropeptide Y-like immunoreactivity in the rat central nervous system—II. Immunohistochemical analysis. *Neuroscience* 18: 545–618
11 Hendry SH (1993) Organization of Neuropeptide Y neurons in the mammalian central nervous system. In: Colmers WF, Wahlestedt C (eds): *The Biology of Neuropeptide Y and Related Peptides*. Humana Press, Totowa, 65–156
12 Bai FL, Yamano M, Shiotani Y, Emson PC, Smith AD, Powell JF, Tohyama M (1985) An arcuato-paraventricular and -dorsomedial hypothalamic neuropeptide Y-containing system which lacks noradrenaline in the rat. *Brain Res* 331: 172–175
13 Broberger C, Johansen J, Johansson C, Schalling M, Hokfelt T (1998) The neuropeptide Y/agouti gene-related protein (AGRP) brain circuitry in normal, anorectic, and monosodium glutamate-treated mice. *Proc Natl Acad Sci USA* 95: 15043–15048
14 Eva C, Keinanen K, Monyer H, Seeburg P, Sprengel R (1990) Molecular cloning of a novel G protein-coupled receptor that may belong to the neuropeptide receptor family. *FEBS Lett* 271: 81–84
15 Herzog H, Hort YJ, Ball HJ, Hayes G, Shine J, Selbie LA (1992) Cloned human neuropeptide Y receptor couples to two different second messenger systems. *Proc Natl Acad Sci USA* 89: 5794–5798
16 Larhammar D, Blomqvist AG, Yee F, Jazin E, Yoo H, Wahlestedt C (1992) Cloning and functional expression of a human neuropeptide Y/peptide YY receptor of the Y1 type. *J Biol Chem* 267: 10935–10938
17 Heilig M, McLeod S, Brot M, Heinrichs SC, Menzaghi F, Koob GF, Britton KT (1993) Anxiolytic-like action of neuropeptide Y: mediation by Y1 receptors in amygdala, and dissociation from food intake effects. *Neuropsychopharmacology* 8: 357–363

18 Wahlestedt C, Pich EM, Koob GF, Yee F, Heilig M (1993) Modulation of anxiety and neuropeptide Y-Y1 receptors by antisense oligodeoxynucleotides. *Science* 259: 528–531
19 Heilig M (1995) Antisense inhibition of neuropeptide Y (NPY)-Y1 receptor expression blocks the anxiolytic-like action of NPY in amygdala and paradoxically increases feeding. *Regul Peptides* 59: 201–205
20 Broqua P, Wettstein JG, Rocher MN, Gauthier-Martin B, Junien JL (1995) Behavioral effects of neuropeptide receptor agonists in the elevated plus-maze and fear-potentiated startle procedure. *Behav Pharmacol* 6: 215–222
21 Sajdyk TJ, Vandergriff MG, Gehlert DR (1999) Amygdalar neuropeptide Y Y-1 receptors mediate the anxiolytic-like actions of neuropeptide Y in the social interaction test. *Eur J Pharmacol* 368: 143–147
22 Sheikh SP, Håkansson R, Schwartz TW (1989) Y_1 and Y_2 receptors for neuropeptide Y. *FEBS Lett* 245: 209–214
23 Gerald C, Walker MW, Vaysse PJ, He C, Branchek TA, Weinshank RL (1995) Expression cloning and pharmacological characterization of a human hippocampal neuropeptide Y/peptide YY Y2 receptor subtype. *J Biol Chem* 270: 26758–26761
24 McQuiston AR, Colmers WF (1996) Neuropeptide Y2 receptors inhibit the frequency of spontaneous but not miniature EPSCs in CA3 pyramidal cells of rat hippocampus. *J Neurophysiol* 76: 3159–3168
25 Qian J, Colmers WF, Saggau P (1997) Inhibition of synaptic transmission by neuropeptide Y in rat hippocampal area CA1: modulation of presynaptic Ca^{2+} entry. *J Neurosci* 17: 8169–8177
26 Glaum SR, Miller RJ, Rhim H, Maclean D, Georgic LM, MacKenzie RG, Grundemar L (1997) Characterization of Y3 receptor-mediated synaptic inhibition by chimeric neuropeptide Y-peptide YY peptides in the rat brainstem. *Br J Pharmacol* 120: 481–487
27 Jazin EE, Yoo H, Blomqvist AG, Yee F, Weng G, Walker MW, Salon J, Larhammar D, Wahlestedt C (1993) A proposed bovine neuropeptide Y (NPY) receptor cDNA clone, or its human homologue, confers neither NPY binding sites nor NPY responsiveness on transfected cells. *Regul Peptides* 47: 247–258
28 Bard JA, Walker MW, Branchek TA, Weinshank RL (1995) Cloning and functional expression of a human Y4 subtype receptor for pancreatic polypeptide, neuropeptide Y, and peptide YY. *J Biol Chem* 270: 26762–26765
29 Lundell I, Blomqvist AG, Berglund MM, Schober DA, Johnson D, Statnick MA, Gadski RA, Gehlert DR, Larhammar D (1995) Cloning of a human receptor of the NPY receptor family with high affinity for pancreatic polypeptide and peptide YY. *J Biol Chem* 270: 29123–29128
30 Gerald C, Walker MW, Criscione L, Gustafson EL, Batzlhartmann C, Smith KE, Vaysse P, Durkin MM, Laz TM, Linemeyer DL et al. (1996) A receptor subtype involved in neuropeptide-Y-induced food intake. *Nature* 382: 168–171
31 Zini I, Merlo Pich E, Fuxe K, Lenzi PL, Agnati LF, Harfstrand A, Mutt V, Tatemoto K, Moscara M (1984) Actions of centrally administered neuropeptide Y on EEG activity in different rat strains and in different phases of their circadian cycle. *Acta Physiol Scand* 122: 71–77
32 Ehlers CL, Somes C, Lopez A, Kirby D, Rivier JE (1997) Electrophysiological actions of neuropeptide Y and its analogs: new measures for anxiolytic therapy? *Neuropsychopharmacology* 17: 34–43
33 Ehlers CL, Somes C, Cloutier D (1998) Are some of the effects of ethanol mediated through NPY? *Psychopharmacology* 139: 136–144
34 Slawecki CJ, Somes C, Ehlers CL (1999) Effects of chronic ethanol exposure on neurophysiological responses to corticotropin-releasing factor and neuropeptide Y. *Alcohol Alcoholism* 34: 289–299
35 Ehlers CL, Somes C, Lumeng L, Li TK (1999) Electrophysiological response to neuropeptide Y (NPY): in alcohol-naive preferring and non-preferring rats. *Pharmacol Biochem Behav* 63: 291–299
36 Hwang BH, Zhang JK, Ehlers CL, Lumeng L, Li TK (1999) Innate differences of neuropeptide Y (NPY) in hypothalamic nuclei and central nucleus of the amygdala between selectively bred rats with high and low alcohol preference. *Alcohol Clin Exp Res* 23: 1023–1030
37 Baraban SC, Hollopeter G, Erickson JC, Schwartzkroin PA, Palmiter RD (1997) Knock-out mice reveal a critical antiepileptic role for neuropeptide Y. *J Neurosci* 17: 8927–8936
38 Woldbye DP, Larsen PJ, Mikkelsen JD, Klemp K, Madsen TM, Bolwig TG (1997) Powerful inhibition of kainic acid seizures by neuropeptide Y via Y5-like receptors. *Nat Medicine* 3: 761–764

39 Naveilhan P, Canals JM, Valjakka A, Vartiainen J, Arenas E, Ernfors P (2001) Neuropeptide Y alters sedation through a hypothalamic Y1-mediated mechanism. *Eur J Neurosci* 13: 2241–2246
40 Naveilhan P, Canals JM, Arenas E, Ernfors P (2001) Distinct roles of the Y1 and Y2 receptors on neuropeptide Y-induced sensitization to sedation. *J Neurochem* 78: 1201–1207
41 Marsh DJ, Baraban SC, Hollopeter G, Palmiter RD (1999) Role of the Y5 neuropeptide Y receptor in limbic seizures. *Proc Natl Acad Sci USA* 96: 13518–13523
42 Clark JT, Keaton AK, Sahu A, Kalra SP, Mahajan SC, Gudger JN (1998) Neuropeptide Y (NPY) levels in alcoholic and food restricted male rats: implications for site-selective function. *Regul Peptides* 75–76: 335–345
43 Thiele TE, Cubero I, van Dijk G, Mediavilla C, Bernstein IL (2000) Ethanol-induced c-Fos expression in catecholamine- and neuropeptide Y-producing neurons in rat brainstem. *Alcohol Clin Exp Res* 24: 802–809
44 Ehlers CL, Li TK, Lumeng L, Hwang BH, Somes C, Jimenez P, Mathe AA (1998) Neuropeptide Y levels in ethanol-naive, alcohol-preferring, and nonpreferring rats and in Wistar rats after ethanol exposure. *Alcohol Clin Exp Res* 22: 1778–1782
45 Kinoshita H, Jessop DS, Finn DP, Coventry TL, Roberts DJ, Ameno K, Ijiri I, Harbuz MS (2000) Acute ethanol decreases NPY mRNA but not POMC mRNA in the arcuate nucleus. *Neuroreport* 11: 3517–3519
46 Roy A, Pandey SC (2002) The decreased cellular expression of neuropeptide Y protein in rat brain structures during ethanol withdrawal after chronic ethanol exposure. *Alcohol Clin Exp Res* 26: 796–803
47 Woldbye DPD, Ulrichsen J, Haugbol S, Bolwig TG (2002) Ethanol withdrawal in rats is attenuated by intracerebroventricular administration of neuropeptide Y. *Alcohol Alcoholism* 37: 318–321
48 Mayfield RD, Lewohl JM, Dodd PR, Herlihy A, Liu J, Harris RA (2002) Patterns of gene expression are altered in the frontal and motor cortices of human alcoholics. *J Neurochem* 81: 802–813
49 Bice P, Foroud T, Bo RH, Castelluccio P, Lumeng L, Li TK, Carr LG (1998) Genomic screen for QTLs underlying alcohol consumption in the P and NP rat lines. *Mamm Genome* 9: 949–955
50 Carr LG, Foroud T, Bice P, Gobbett T, Ivashina J, Edenberg H, Lumeng L, Li TK (1998) A quantitative trait locus for alcohol consumption in selectively bred rat lines. *Alcohol Clin Exp Res* 22: 884–887
51 Liang T, Spence J, Liu L, Strother WN, Chang HW, Ellison JA, Lumeng L, Li TK, Foroud T, Carr LG (2003) alpha-Synuclein maps to a quantitative trait locus for alcohol preference and is differentially expressed in alcohol-preferring and -nonpreferring rats. *Proc Natl Acad Sci USA* 100: 4690–4695
52 Foroud T, Bice P, Castelluccio P, Bo R, Miller L, Ritchotte A, Lumeng L, Li TK, Carr LG (2000) Identification of quantitative trait loci influencing alcohol consumption in the high alcohol drinking and low alcohol drinking rat lines. *Behav Genet* 30: 131–140
53 Caberlotto L, Thorsell A, Rimondini R, Sommer W, Hyytia P, Heilig M (2001) Differential expression of NPY and its receptors in alcohol-preferring AA and alcohol-avoiding ANA rats. *Alcohol Clin Exp Res* 25: 1564–1569
54 Thiele TE, Marsh DJ, Ste M, Bernstein IL, Palmiter RD (1998) Ethanol consumption and resistance are inversely related to neuropeptide Y levels. *Nature* 396: 366–369
55 Thiele TE, Miura GI, Marsh DJ, Bernstein IL, Palmiter RD (2000) Neurobiological responses to ethanol in mutant mice lacking neuropeptide Y or the Y5 receptor. *Pharmacol Biochem Behav* 67: 683–691
56 Thorsell A, Michalkiewicz M, Dumont Y, Quirion R, Caberlotto L, Rimondini R, Mathe AA, Heilig M (2000) Behavioral insensitivity to restraint stress, absent fear suppression of behavior and impaired spatial learning in transgenic rats with hippocampal neuropeptide Y overexpression. *Proc Natl Acad Sci USA* 97: 12852–12857
57 Pedrazzini T, Seydoux J, Kunstner P, Aubert JF, Grouzmann E, Beermann F, Brunner HR (1998) Cardiovascular response, feeding behavior and locomotor activity in mice lacking the NPY Y1 receptor [see comments]. *Nat Med* 4: 722–726
58 Thiele TE, Koh MT, Pedrazzini T (2002) Voluntary alcohol consumption is controlled via the neuropeptide Y Y1 receptor. *J Neurosci* 22: RC208
59 King PJ, Williams G, Doods H, Widdowson PS (2000) Effect of a selective neuropeptide Y Y(2) receptor antagonist, BIIE0246 on neuropeptide Y release. *Eur J Pharmacol* 396: R1–R3
60 King PJ, Widdowson PS, Doods HN, Williams G (1999) Regulation of neuropeptide Y release by neuropeptide Y receptor ligands and calcium channel antagonists in hypothalamic slices. *J*

Neurochem 73: 641–646
61 Naveilhan P, Hassani H, Canals JM, Ekstrand AJ, Larefalk A, Chhajlani V, Arenas E, Gedda K, Svensson L, Thoren P et al. (1999) Normal feeding behavior, body weight and leptin response require the neuropeptide Y Y2 receptor. *Nat Med* 5: 1188–1193
62 Thiele TE, Naveilhan P, Ernfors P (2000) Mutant mice lacking the Y2 neuropeptide Y (NPY) receptor consume less ethanol than wild-type mice. *Alcohol Clin Exp Res* 24: 97A
63 Thiele TE, Miura GI, Marsh DJ, Bernstein IL, Palmiter RD (2000) Neurobiological responses to ethanol in mutant mice lacking neuropeptide Y or the Y5 receptor. *Pharmacol Biochem Behav* 67: 683–691
64 Thorsell A, Rimondini R, Heilig M (2002) Blockade of central neuropeptide Y (NPY) Y2 receptors reduces ethanol self-administration in rats. *Neurosci Lett* 332: 1–4
65 Karvonen MK, Pesonen U, Koulu M, Niskanen L, Laakso M, Rissanen A, Dekker JM, Hart LM, Valve R, Uusitupa MIJ (1998) Association of a leucine(7)-to-proline(7) polymorphism in the signal peptide of neuropeptide Y with high serum cholesterol and LDL cholesterol levels. *Nat Med* 4: 1434–1437
66 Kallio J, Pesonen U, Kaipio K, Karvonen MK, Jaakkola U, Heinonen OJ, Uusitupa MI, Koulu M (2001) Altered intracellular processing and release of neuropeptide Y due to leucine 7 to proline 7 polymorphism in the signal peptide of preproneuropeptide Y in humans. *FASEB J* 15: 1242–1244
67 Kauhanen J, Karvonen MK, Pesonen U, Koulu M, Tuomainen TP, Uusitupa MI, Salonen JT (2000) Neuropeptide Y polymorphism and alcohol consumption in middle-aged men. *Am J Med Gen* 93: 117–121
68 Lappalainen J, Kranzler HR, Malison R, Price LH, Van Dyck C, Rosenheck RA, Cramer J, Southwick S, Charney D, Krystal J et al. (2002) A functional neuropeptide Y Leu7Pro polymorphism associated with alcohol dependence in a large population sample from the United States. *Arch Gen Psychiatry* 59: 825–831
69 Ilveskoski E, Kajander OA, Lehtimaki T, Kunnas T, Karhunen PJ, Heinala P, Virkkunen M, Alho H (2001) Association of neuropeptide Y polymorphism with the occurrence of type I and type 2 alcoholism. *Alcohol Clin Exp Res* 25: 1420–1422
70 Zhu G, Pollak L, Mottagui-Tabar S, Wahlestedt C, Taubman J, Virkkunen M, Goldman D, Heilig M (2003) NPY leu7pro and alcohol dependence in finnish and swedish populations. *Alcohol Clin Exp Res* 27: 19–24
71 Ding B, Bertilsson L, Wahlestedt C (2002) The single nucleotide polymorphism T1128C in the signal peptide of neuropeptide Y (NPY) was not identified in a Korean population. *J Clin Pharm Ther* 27: 211–212
72 Okubo T, Harada S (2001) Polymorphism of the neuropeptide Y gene: an association study with alcohol withdrawal. *Alcohol Clin Exp Res* 25: 59–62
73 Itokawa M, Arai M, Kato S, Ogata Y, Furukawa A, Haga S, Ujike H, Sora I, Ikeda K, Yoshikawa T (2003) Association between a novel polymorphism in the promoter region of the neuropeptide Y gene and schizophrenia in humans. *Neurosci Lett* 347: 202–204
74 Crawley JN (2000) *What's wrong with my mouse: behavioral phenotyping of transgenic and knockout mice*. Wiley and Sons, New York
75 McBride WJ, Li TK (1998) Animal models of alcoholism: neurobiology of high alcohol-drinking behavior in rodents. *Crit Rev Neurobiol* 12: 339–369
76 Rimondini R, Arlinde C, Sommer W, Heilig M (2002) Long-lasting increase in voluntary ethanol consumption and transcriptional regulation in the rat brain after intermittent exposure to alcohol. *FASEB J* 16: 27–35
77 Heyser CJ, Schulteis G, Durbin P, Koob GF (1998) Chronic acamprosate eliminates the alcohol deprivation effect while having limited effects on baseline responding for ethanol in rats. *Neuropsychopharmacology* 18: 125–133
78 Kulkosky PJ, Glazner GW, Moore HD, Low CA, Woods SC (1988) Neuropeptide Y: behavioral effects in the golden hamster. *Peptides* 9: 1389–1393
79 Slawecki CJ, Betancourt M, Walpole T, Ehlers CL (2000) Increases in sucrose consumption, but not ethanol consumption, following ICV NPY administration. *Pharmacol Biochem Behav* 66: 591–594
80 Katner SN, Slawecki CJ, Ehlers CL (2002) Neuropeptide Y administration into the amygdala does not effect ethanol consumption. *Alcohol* 28: 29–38
81 Katner SN, Slawecki CJ, Ehlers CL (2002) Neuropeptide Y administration into the third ventricle

does not increase sucrose or ethanol self-administration but does affect the cortical EEG and increases food intake. *Psychopharmacology* 160: 146–154
82 Kelley SP, Nannini MA, Bratt AM, Hodge CW (2001) Neuropeptide-Y in the paraventricular nucleus increases ethanol self-administration. *Peptides* 22: 515–522
83 Schroeder JP, Olive F, Koenig H, Hodge CW (2003) Intra-amygdala infusion of the NPY Y1 receptor antagonist BIBP 3226 attenuates operant ethanol self-administration. *Alcohol Clin Exp Res* 27: 1884–1891
84 Badia-Elder NE, Stewart RB, Powrozek TA, Roy KF, Murphy JM, Li TK (2001) Effect of neuropeptide Y (NPY) on oral ethanol intake in Wistar, alcohol-preferring (P), and -nonpreferring (NP) rats. *Alcohol Clin Exp Res* 25: 386–390
85 Gilpin NW, Stewart RB, Murphy JM, Li TK, Badia-Elder NE (2003) Neuropeptide Y reduces oral ethanol intake in alcohol-preferring (P) rats following a period of imposed ethanol abstinence. *Alcohol Clin Exp Res* 27: 787–794
86 Badia-Elder NE, Stewart RB, Powrozek TA, Murphy JM, Li TK (2003) Effects of neuropeptide Y on sucrose and ethanol intake and on anxiety-like behavior in high alcohol drinking (HAD) and low alcohol drinking (LAD) rats. *Alcohol Clin Exp Res* 27: 894–899
87 Pandey SC, Roy A, Zhang H (2003) The decreased phosphorylation of cyclic adenosine monophosphate (cAMP) response element binding (CREB) protein in the central amygdala acts as a molecular substrate for anxiety related to ethanol withdrawal in rats. *Alcohol Clin Exp Res* 27: 396–409
88 Schroeder JP, Iller KA, Hodge CW (2003) Neuropeptide-Y Y5 receptors modulate the onset and maintenance of operant ethanol self-administration. *Alcohol Clin Exp Res* 27: 1912–1920
89 Heilig M, Koob GF, Ekman R, Britton KT (1994) Corticotropin-releasing factor and neuropeptide Y: role in emotional integration. *Trends Neurosci* 17: 80–85
90 Kask A, Harro J, von Horsten S, Redrobe JP, Dumont Y, Quirion R (2002) The neurocircuitry and receptor subtypes mediating anxiolytic-like effects of neuropeptide Y. *Neurosci Biobehav Rev* 26: 259–283
91 Redrobe JP, Dumont Y, Quirion R (2002) Neuropeptide Y (NPY) and depression: from animal studies to the human condition. *Life Sci* 71: 2921–2937
92 McEwen BS (2000) Allostasis and allostatic load: implications for neuropsychopharmacology. *Neuropsychopharmacology* 22: 108–124
93 Koob GF, Le Moal M (1997) Drug abuse: hedonic homeostatic dysregulation. *Science* 278: 52–58
94 Heilig M, Zachrisson O, Thorsell A, Ehnvall A, Mottagui-Tabar S, Sjogren M, Asberg M, Ekman R, Wahlestedt C, Agren H (2004) Decreased cerebrospinal fluid neuropeptide Y (NPY) in patients with treatment refractory unipolar major depression: preliminary evidence for association with preproNPY gene polymorphism. *J Psychiatr Res* 38:113–121

Drugs for Relapse Prevention of Alcoholism
Edited by R. Spanagel and K. Mann

Glutamatergic compounds: a perspective

Daniel Bachteler and Rainer Spanagel

Central Institute of Mental Health, Department of Psychopharmacology, University of Heidelberg, (CIMH), J5, 68159 Mannheim, Germany

Introduction

It is well known that ethanol alters the synaptic action of glutamate – the major excitatory neurotransmitter in our brain. Thus, high intoxicating doses of ethanol decrease glutamate levels and reduce the activity of glutamate receptors, especially the NMDA receptors. As a consequence of chronic alcohol intake, adaptive changes in the glutamate system are observed, which result in a hyper-glutamatergic state. It is suggested that a hyper-glutamatergic state is a trigger for alcoholic patients to further drink alcohol during withdrawal or to relapse after a period of abstinence. This causal relationship of chronic alcohol intake and long-lasting adaptive changes within the glutamatergic system and its involvement in relapse mechanisms provides the basis for the glutamatergic theory of alcoholism [1, 2].

Glutamate´s postsynaptic actions are mediated either through the interaction with ionotropic glutamate receptor channels (iGluR), including NMDA, AMPA and Kainate receptors, or by G protein coupled metabotropic glutamate receptors (mGluR). Among these, the NMDAR complex represents one of the highest affinity targets for ethanol in the brain [3]. Its composition offers several potential sites to interfere with alcohol-related actions. These include, for example, blockade of the receptor channel, selective antagonism of subunits or special binding sites, or functional antagonism. Selective pharmacological manipulation of glutamate receptors has been shown to decrease alcohol intake and preference as well as relapse behavior, including the severity of withdrawal symptoms in various animal models. Moreover, current research demonstrates a functional interplay not only within, but also among receptor types, as for example NMDA and mGluR5. In addition to pharmacological targeting of glutamate receptors, interventions in glutamate release mechanisms as well as re-uptake processes could also represent a promising approach in interfering with relapse mechanisms.

In the following sections, we will review different glutamatergic compounds that have been tested preclinically in different animal models of alcoholism, including basic home cage drinking models such as the alcohol deprivation effect model and more sophisticated operant conditioning paradigms such as cue-induced reinstatement of ethanol-seeking behavior (for detailed description of these paradigms see Chapter by Spanagel).

Substances acting at ionotropic glutamate receptors (iGluR)

NMDA receptors

NMDAR is a heteromeric subunit complex, composed of NR1 and NR2 subunits (NR2A-D), each of which exists in several splice variants. At least six pharmacologically distinct binding sites have been recognized until the present, through which compounds may alter the receptor's activity. Functional receptors in the adult mammalian central nervous system (CNS) are formed by combinations of NR1 and NR2 subunits, which express the glycine and glutamate binding sites, respectively. Glutamate acts as an agonist and glycine as a co-agonist at the NMDAR. The endogenous polyamines spermine and spermidine also influence the activity of the NMDAR, in particular at the NR2B subunit-containing receptors [4]. Yet another important binding site lies within the NMDAR complex channel, where un-/non-competitive antagonists block the receptor in a use-dependent manner, i.e., the channel must first be opened by an agonist for the antagonist to bind [5].

Depending on the NMDA receptor subunit composition, acute alcohol inhibits NMDA receptor activation at behaviorally relevant concentrations [6]. For example, Lovinger and colleagues [7] showed in voltage-clamped hippocampal neurons that the ion current induced by the glutamate receptor agonist NMDA was concentration-dependent inhibited by ethanol. At an intoxicating concentration of 50 mM ethanol, the NMDA-activated current was inhibited by more than 60%. Despite intensive research, the site of action of ethanol on the NMDA receptor still remains unknown. However, recently Ren et al. [8] identified a residue in the fourth membrane-associated domain in the NR2A subunit that is suggested to interact with or forms part of a site of ethanol action.

Numerous glutamatergic compounds interfering with the different binding and modulatory sites of the NMDAR complex have been tested in animal models of alcoholism. Initially, some studies produced conflicting results: Rassnick et al. [9] described the attenuation of ethanol self-administration in a free-choice operant task by intra-nucleus accumbens microinjection of the competitive NMDAR antagonist AP-5 (2-amino-5-phosphopentanoic acid) without affecting water consumption. This selectivity was questioned by Shelton and Balster [10], who demonstrated that the competitive NMDA antagonist CPPene, as well as the non-competitive antagonist phencyclidine (PCP) decreased both ethanol and saccharin self-administration. No effects were reported by Danysz et al. [11] using a two-bottle free-choice ethanol-drinking paradigm with the non-competitive NMDAR antagonist MK-801 (dizocilpine). Only recently have competitive and uncompetitive antagonists been studied in either the alcohol deprivation effect model or the cue-induced reinstatement paradigm of ethanol-seeking behavior. In the following sections, two promising compounds, memantine and neramexane – both uncompetitive NMDAR antagonists – will be described in more detail.

Firstly, evidence of the alcohol anti-relapse properties of the low affinity, uncompetitive NMDAR antagonist memantine, originally developed for the treatment of moderate to severe Alzheimer´s dementia [12], was published by Hölter et al. [13]. In a long-term drinking paradigm, rats had voluntary access to different ethanol solutions and water in their home cages. After months of continuous access to ethanol, a deprivation phase was introduced with water as the only available liquid. Re-presentation of alcohol after 3 to 14 days led to a temporary increase in ethanol consumption and preference, known as the alcohol deprivation effect. Implanted with osmotic mini-pumps delivering 5.0 μl/h of memantine, rats no longer showed an alcohol deprivation effect after having regained access to alcohol, whereas saline control animals showed the characteristic increase in consumption (Fig. 1). A lack of sedation or behavioral impairment further demonstrated the potential usefulness of this compound. A similar effect of memantine on alcohol intake was described by Piasecki et al. [14] in a free-choice, limited access procedure. Administration of 4.5–24 mg/kg significantly, but not dose-dependently, affected alcohol intake in rats; however, only at 6 mg/kg did it selectively decrease consumption. Rats, operantly trained to lever press for ethanol, showed reduced responding at the 9 mg/kg dose in an extinction procedure. This effect, however, was not selective for ethanol, since water-reinforced responses diminished as well [14].

Initial characterization of the Alzheimer drug and amino-alkyl-cyclohexane neramexane (MRZ 2/579), which displays a pharmacokinetic and pharmacodynamic profile very similar to memantine, provided support for the assumption that neramexane could be useful in the treatment of alcohol abuse [15]. Thus, this compound produced selective effects on ethanol-seeking behavior [14], suppressed ethanol withdrawal seizures in a dose-dependent manner [16], and prevented the acquisition of ethanol-induced conditioned place preference (Kotlinska, pers. comm.). Further, neramexane´s effects on ethanol intake have been employed in operant and non-operant animal models. Using an extinction procedure, Bienkowski et al. [17] reported a selective and dose-dependent suppression of operant ethanol self-administration at 2.5–7.5 mg/kg without alterations in locomotor activity, given alone or in combination with ethanol. Voluntarily consuming ethanol experienced rats no longer showed an alcohol deprivation effect, when chronically administered with neramexane (9.6 mg/day) via osmotic minipumps [18]. The same rats were tested for a second alcohol deprivation effect three weeks later in the absence of the drug. The animals exhibited an alcohol deprivation effect to the same extent as control rats, showing that the effect of neramexane on relapse drinking behavior requires the presence of the drug and that a short-term treatment might only be of limited benefit. Long-term alcohol-experienced rats also underwent an oper-

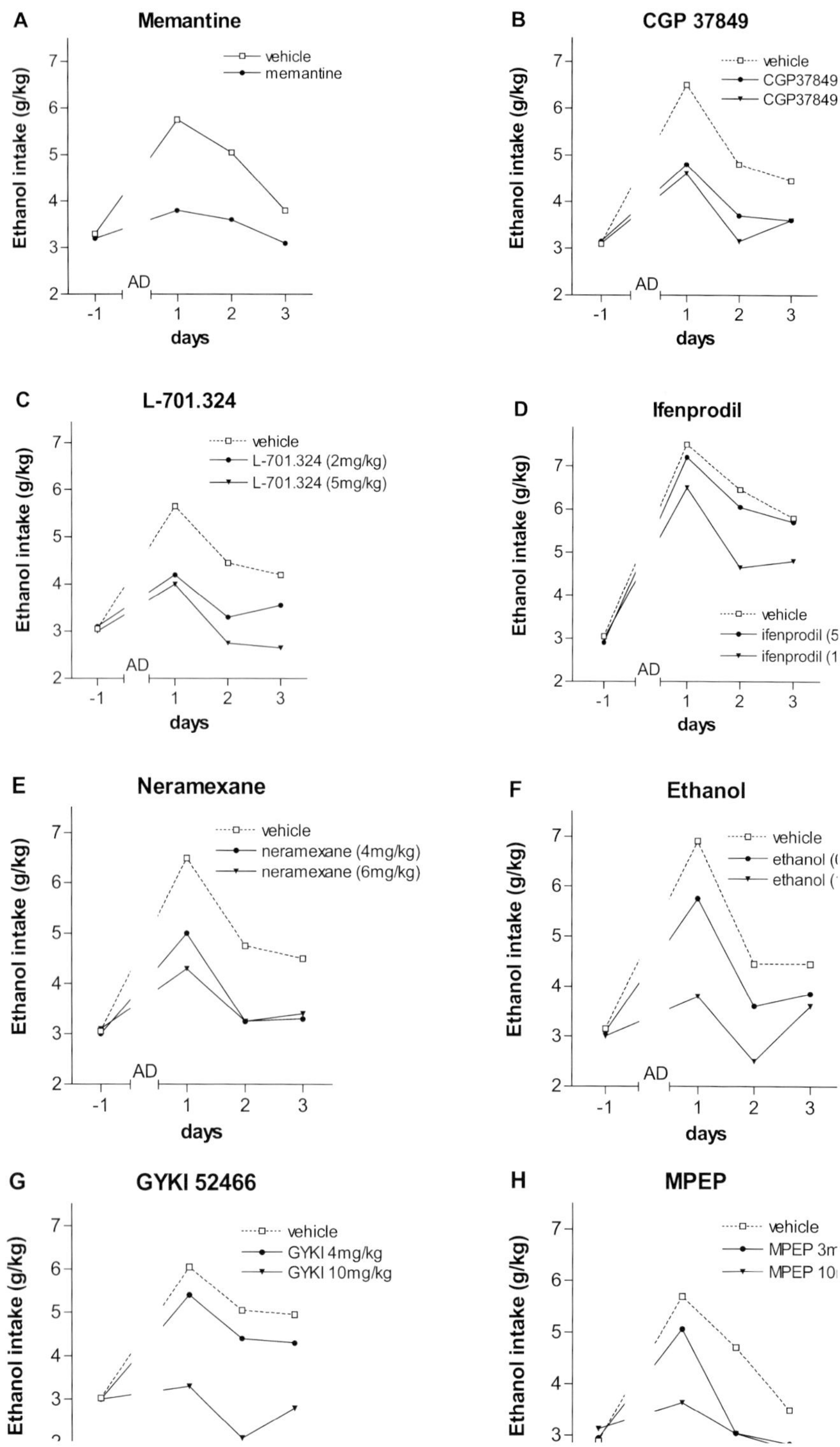
A
Memantine
vehicle
memantine
B
CGP 37849
vehicle
CGP37849
CGP37849
C
L-701.324
vehicle
L-701.324 (2mg/kg)
L-701.324 (5mg/kg)
D
Ifenprodil
vehicle
E
Neramexane
vehicle
neramexane (4mg/kg)
neramexane (6mg/kg)
F
Ethanol
vehicle
G
GYKI 52466
vehicle
GYKI 4mg/kg
GYKI 10mg/kg
H
MPEP
vehicle
Ethanol intake (g/kg)
days
AD

ant free-choice ethanol self-administration paradigm, in which they received acute treatment of neramexane (2 and 4 mg/kg). Herein, neramexane had a short-lasting reductive, selective effect on lever pressing for ethanol during the alcohol-deprivation effect, but not for water. More recently, Bienkowski et al. [16] tested the same compound in both an operant and non-operant two-bottle choice setup. Non-operant ethanol intake was not affected by a chronic 6-day infusion of neramexane via osmotic minipumps (9.6 mg/day). Only intermittent injections (5 mg/kg) resulted in a significant, progressive decrease in operant responding. This finding is in line with a recent study that showed a dose-dependent reduction of the alcohol deprivation effect following intermittent injections of neramexane (4 and 6 mg/kg; Fig. 1) [19]. Bachteler et al. [20] evaluated whether neramexane would also reduce craving for alcohol in the reinstatement paradigm, applying different doses (1, 2, 4 mg/kg) to rats. Animals were trained to lever press for ethanol in the presence of a distinct set of cues. After extinction, the animals were exposed to the respective cues that initiated reinstatement of responding. However, intraperitoneal administration of neramexane did not result in a significant reduction of lever presses on the ethanol lever at the used doses, but led to a suppression of lever presses on both levers at the highest dose, thereby displaying unspecific response behavior (Fig. 2). This finding is in line with a recent study using the high affinity non-competitive NMDA receptor antagonist MK-801 in the reinstatement paradigm. Thus doses up to 0.15 mg/kg of MK-801 did not reduce the number of lever presses on the ethanol lever following cue presentation. MK-801 rather had a tendency to increase responding, but this effect did not produce significance [21]. The latter results may suggest that neramexane and other un- or non-competitive NMDA receptor antagonists might be more useful in the treatment of relapse behavior than in craving (however, it is important to note that relapse often occurs in the context of craving) and further shows that the neurochemical substrates underlying the expression of an alcohol deprivation effect and reinstatement behavior are different.

In line with several drug discrimination experiments that show a substitution to the ethanol cue following the application of different uncompetitive NMDA receptor antagonists [22, 23], memantine and neramexane also exhibited a dose-dependent substitution to the ethanol cue [18, 22], suggesting that these compounds exert their effects on ethanol consumption, at least in part, by generalizing for some of the stimulus properties of ethanol. In other words

Figure 1. Alcohol deprivation effect. Total ethanol intake (g/kg/day) in rats before and after an alcohol deprivation period (AD) of two weeks. The average of three days measurements as a baseline drinking (–1) is shown. To test for an effect on the ADE, different compounds acting at the glutamatergic system were i.p.-administered. Animals were treated with A the uncompetitive NMDAR antagonist memantine (4.8 mg/day); B the competitive NMDAR antagonist CGP 37849 (2, 8 mg/kg); C the NMDA/glycine binding site antagonist L-701,324 (2, 5 mg/kg); D the NMDAR 2B subunit selective antagonists ifenprodil (5, 10 mg/kg); E the NMDA channel blocker neramexane (4, 6 mg/kg); F ethanol (0.8, 1.6 g/kg) as a functional NMDAR antagonist; G the AMPA receptor antagonist GYKI 52466 (4, 10 mg/kg); F the mGlu5 receptor antagonist MPEP (3, 10 mg/kg). Controls received saline. Data was adapted after [13; A], [19; B–F], [31, G], [42; H].

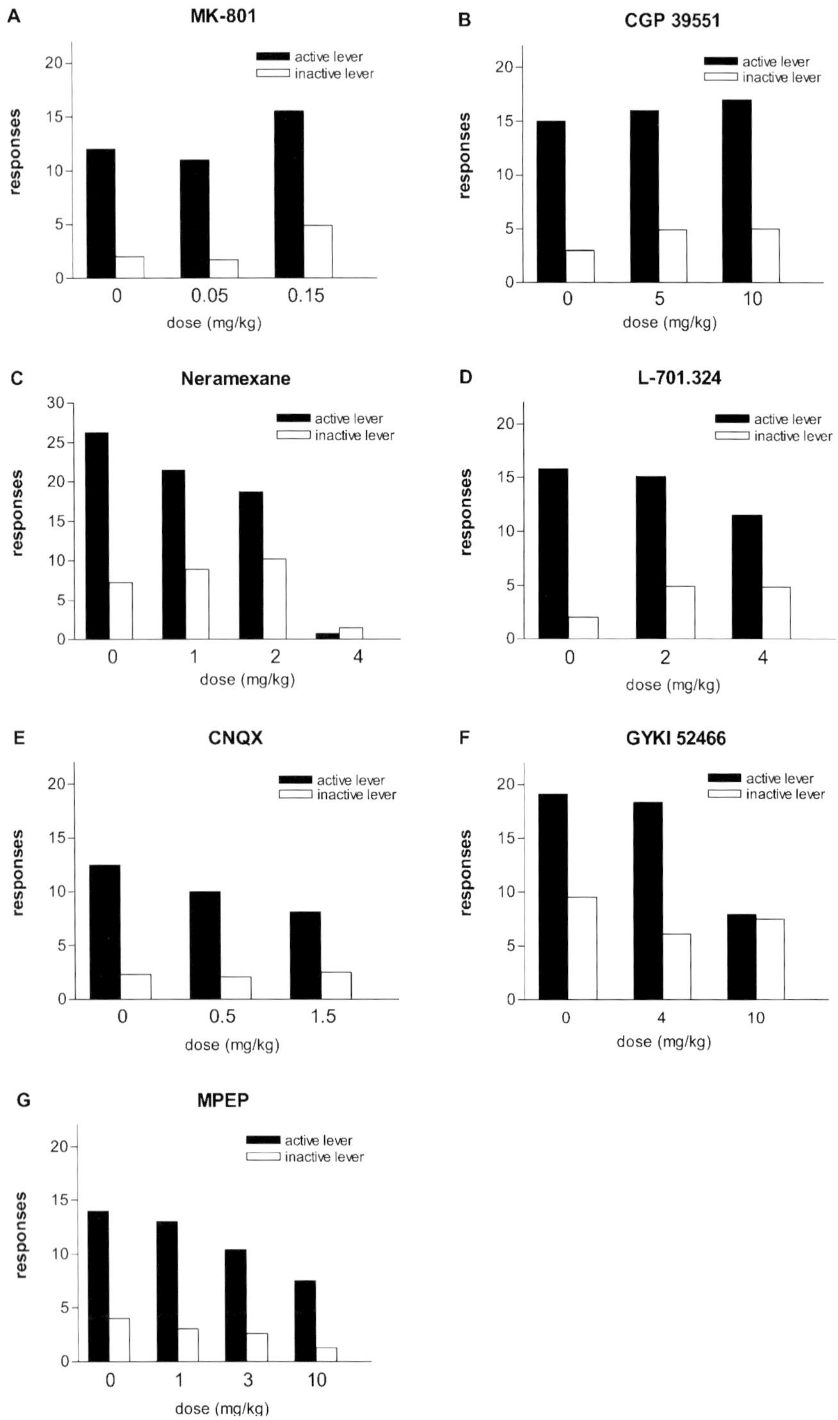
A
MK-801
active lever
inactive lever
responses
20
15
10
5
0
0
0.05
0.15
dose (mg/kg)
B
CGP 39551
active lever
inactive lever
responses
20
15
10
5
0
0
5
10
dose (mg/kg)
C
Neramexane
active lever
inactive lever
responses
30
25
20
15
10
5
0
0
1
2
4
dose (mg/kg)
D
L-701.324
active lever
inactive lever
responses
20
15
10
5
0
0
2
4
dose (mg/kg)
E
CNQX
active lever
inactive lever
responses
20
15
10
5
0
0
0.5
1.5
dose (mg/kg)
F
GYKI 52466
active lever
inactive lever
responses
20
15
10
5
0
0
4
10
dose (mg/kg)
G
MPEP
active lever
inactive lever
responses
20
15
10
5
0
0
1
3
10
dose (mg/kg)

uncompetitive NMDAR antagonists can be viewed as substitution drugs for the treatment of alcoholism. Based on this new concept, a multicenter, randomized, double-blind, placebo-controlled study with neramexane was launched in 2002. Surprisingly, neramexane treatment had no significant effect on either relapse rates or total number of abstinence days. However, compliance problems and the use of a very low dose of neramexane might have confounded the outcome of this study.

Very recently, several compounds acting at different sites of the NMDA receptor were comparatively tested in the alcohol deprivation effect model by Vengeliene et al. [19], including the competitive antagonist CGP 37849, the glycine binding site antagonist L-701.324, and ifenprodil. Repeated administration of these agents produced a significant dose-dependent reduction of alcohol intake in long-term drinking rats during the alcohol deprivation effect to a similar extent (Fig. 1). However, the effect of administration of ifenprodil on the alcohol deprivation effect was weaker but still significant. In all experiments, water intake was not affected.

Besides selective channel blockade of the NMDAR, the competitive and glycine binding sites are therefore interesting for drug development in alcohol dependence. However, it should be emphasized that in recent studies on cue-induced reinstatement of ethanol-seeking behavior, competitive and the already aforementioned uncompetitive NMDAR antagonists were not effective (Fig. 2) [20, 21], whereas L-701,324 dose-dependently reduced ethanol-seeking behavior (Fig. 2) [21]. Furthermore, since NMDARs are involved in almost all of the functions of the CNS, therapeutic intervention may potentially be associated with side-effects (e.g., sedation and cognitive disturbances).

AMPA/kaniate receptors

Synaptic plasticity, underlying associative (i.e., Pavlovian conditioning) and non-associative learning, depends on changes in the glutamatergic system, and more specifically to the AMPA glutamate receptor subtype [24, 25]. The ionotropic AMPA receptor is a heterodimer which may be formed by four different protein subunits, named A, B, C and D, and synaptic plasticity associated to this receptor is clearly dependent on its subunit composition [26]. Interestingly, neuroplastic changes associated to AMPA subunit composition have been proposed to underlie main features of addictive behavior, including drug seeking and relapse behavior. Thus, it has been shown that abstinence or

Figure 2. Cue-induced reinstatement in rats. Modulation of ethanol-responding under stimulus conditions associated with ethanol (active lever) and water (inactive lever) during a 30-min reinstatement session. Animals were pretreated with A the uncompetitive NMDAR antagonist MK-801 (0, 0.05, 0.15 mg/kg); B the competitive NMDAR antagonist CGP 37849 (0, 5, 10 mg/kg); C the NMDA channel blocker neramexane (0, 1, 2, 4 mg/kg); D the NMDA/glycine binding site antagonist L-701,324 (0, 2, 4 mg/kg); E the AMPA/Kainate receptor antagonist CQNX (0, 0.5, 1.5 mg/kg); F the AMPA receptor antagonist GYKI 52466 (0, 4, 10 mg/kg); G the mGlu5 receptor antagonist MPEP (0, 1, 3, 10 mg/kg). Data was adapted after [20; C], [21; A, B, D, E], [31; F], [42; G].

extinction after cocaine self-administration is accompanied with an up-regulation of the GluR-A and B/C sub-units within the nucleus accumbens [27, 28].

Interestingly, following chronic alcohol treatment, GluR-C, but not GluR-A or GluR-B, seems to be upregulated in several brain regions [29]. This observation is in accordance with studies in GluR-A and GluR-C knockout mice. GluR-A knockout mice did not differ from wild-type animals in ethanol intake and preference, nor in the alcohol deprivation effect [30]. However, in GluR-C knockout mice a blunted cue-induced reinstatement response and a lack of the alcohol deprivation effect were observed, when compared to wild-type controls [31]. In contrast, under operant and home cage drinking conditions, no difference in genotype could be observed, showing that the GluR-C subunit is not involved in acquisition and maintenance of alcohol reinforcement. For further pharmacological validation of this genetic model, the AMPA antagonist GYKI 52466 was tested on alcohol relapse behavior by Sanchis-Segura et al. [31]. This compound dose-dependently reduced cue-induced reinstatement and the alcohol deprivation effect (Figs 1, 2). Similar to the observations in GluR-A and GluR-C knockouts, GYKI 52466 did almost not affect baseline alcohol consumption in the home cage nor did it disrupt operant self-administration of ethanol in rats [32]. In conclusion, the GluR-C subunit seems to be involved in the neuroplastic changes underlying the ability of cues to promote alcohol-seeking behavior and therefore GluR-C containing AMPA receptors could be a suitable therapeutic target to prevent relapse.

Another antagonist of the AMPA/Kainate receptor, NBQX (2,3-dihydroxy-6-nitro-7-sulfamoylbenzo(f)quinoxaline), was only effective in reducing operant responding for ethanol or sucrose at doses (3 and 6 mg/kg) which also led to a significant disruption of the animals´ locomotor activity [32]. However, the AMPA/kainite antagonist CNQX (6-cyano-7-nitroquinoxaline-2,3-dione disodium) attenuated reinstatement of cue-induced ethanol-seeking behavior significantly ([21]; Fig. 2) supporting the role of AMPA/kainite receptor in alcohol craving and relapse.

Substances acting at metabotropic glutamate receptors (mGluR)

mGluRs comprise several subtypes (mGluR1-8), divided into three subgroups (I-III), that are based on sequence similarities, intracellular second messengers, and agonist activities [33]. Subgroup I (mGluR1,5) stimulates phosphatidyl inositol (PI) hydrolysis/Ca^{2+} signal transduction, whereas subgroups II (mGluR2,3) and III (mGluR4,6,7,8) are negatively coupled to cyclic adenosine monophosphate formation through adenylyl cyclase.

mGluRs are known to be involved in different CNS disorders [34]. In particular, using the mGluR5 antagonist 2-methyl-6-(phenylethynyl)-pyridine (MPEP) in various studies on anxiety and depression [35, 36], fear conditioning [37], or drug self-administration and conditioned place preference [38–40], the importance of mGluR5 in different CNS disorders including addiction has been demonstrated. MPEP is a highly potent, selective, non-competitive mGluR5 antagonist with minimal activity at other mGluRs or ionotropic glutamate receptors.

Sharko et al. [41] were the first to describe that MPEP interferes with alcohol self-administration during periods of peak consumption in C57BL/6J mice. MPEP dose-dependently and selectively decreased operant lever pressing for 10% ethanol, but not for water. When administered alone, the compound did not produce conditioned place preference, indicating that it does not have rewarding effects itself. In the same study administration of mGluR1 and mGluR2/3 antagonists had no effects on ethanol or water intake in these animals. As a potentially promising agent in the treatment of relapse, MPEP was further investigated by Bäckström et al. [42] for assessing its anti-relapse/anti-craving potential. In a combination of the alcohol deprivation effect model and the reinstatement model of ethanol-seeking behavior by drug-associated cues, valuable results were obtained. Subchronic treatment of Wistar rats with MPEP (1, 3, and 10 mg/kg) attenuated both relapse to ethanol consumption and stimulus-induced ethanol-seeking significantly and in a dose-related manner (Fig. 1, 2). Administration of the same doses of MPEP also influenced baseline drinking of ethanol, even though this was not as pronounced as on the ADE. This study confirmed that pharmacological targeting of mGlu5 receptors may be a promising approach for the treatment of alcoholism.

The mechanism by which mGluR5 is involved in ethanol craving and relapse behavior remains unclear. There seems to be functional coupling between the NMDA receptor complex and the mGlu5 receptors so that simultaneous activation of mGluR5 could enhance and modulate synaptic transmission at NMDA receptors [43–46]. In summary, the functional coupling of the mGluR5 and NMDA receptor suggests that the blockade of mGluR5 by MPEP reduces glutamatergic signalling through NMDA receptors and thereby interacts with ethanol-seeking and relapse behavior.

Conclusions

In summary, the glutamate synapse, including signal transduction pathways, offers a variety of targets for pharmacological intervention in influencing alcohol craving and relapse behavior. Competitive and uncompetitive NMDA receptor antagonists should be further tested in clinical trials, as they may act as substitution drugs for ethanol. However, in order to achieve substitution for some of the subjective effects of ethanol, high doses should be used which might also produce some side-effects. Convincing preclinical evidence comes from studies on mGlu5 receptors showing that mGluR5 antagonists could be promising compounds for the treatment of alcoholism. Thus, MPEP – a selective mGluR5 antagonist – proved to be effective in all standard models of preclinical alcohol research and should therefore be tested in a clinical phase I/II study.

Preclinical studies, however, should not only focus on glutamate receptors as potential targets for pharmacological intervention. Thus, it has been shown that glutamate release mechanisms mediate several neurobehavioral effects of alcohol [22] and in a preliminary study, the glutamate release inhibitor lamotrigine was able to reduce relapse drinking behavior. Furthermore, in genetically modified mice that exhibit a down-regulation of the glutamate transporter EAAT-1, which leads to augmented glutamate levels in the brain, enhanced alcohol consumption is observed, indicating that glutamate uptake mechanisms could also present an important site of action in modulating alcohol consumption and subsequently addictive behavior [47]. Furthermore, the NO/cGMP signalling pathway that is coupled to NMDA receptors is also critically involved in alcohol drinking behavior [48, 49]. Thus a variety of genes and their protein products of the glutamatergic system are involved in the action of alcohol in the CNS and promising compounds are in the pipeline of several pharmaceutical companies, waiting to be tested on alcoholic subjects.

References

1 Tsai G, Gastfriend DR, Coyle JT (1995) The glutamatergic basis of human alcoholism. *Am J Psychiatry* 152(3): 332–340
2 Tsai G, Coyle JT (1998) The role of glutamatergic neurotransmission in the pathophysiology of alcoholism. *Annu Rev Med* 49: 173–184
3 Krystal JH, Petrakis IL, Mason G, Trevisan L, D´Souza DC (2003) N-methyl-D-aspartate glutamate receptors and alcoholism: reward, dependence, treatment, and vulnerability. *Pharmacol Ther* 99: 79–94
4 Johnson TD (1996) Modulation of channel function by polyamines. *Trends Pharmacol Sci* 17(1): 22–27
5 Parsons CG, Danysz W, Lodge D (2002) Introduction to glutamate receptors, their function and pharmacology. In: D Lodge, W Danysz, CG Parsons (eds): *Ionotropic glutamate receptors as therapeutic targets*. FP Graham Publishing Co., Johnson City, TN, 1–27
6 Spanagel R, Bienkowski P (2002) Glutamatergic mechanisms in alcohol dependence and addiction. In: D Lodge, W Danysz, CG Parsons (eds): *Therapeutic Potential of Ionotropic Glutamate Receptor Antagonists and Modulators*. Graham Publishing Co., Johnson City, TN, USA, 375–403
7 Lovinger DM, White G, Weight FF (1989) Ethanol inhibits NMDA-activated ion current in hippocampal neurons. *Science* 243(4899): 1721–1724
8 Ren H, Honse Y, Peoples RW (2003) A site of alcohol-action in the fourth membrane-associated domain of the n-methyl-d-aspartate receptor. *J Biol Chem* 278(49): 48815–48820
9 Rassnick S, Pulvirenti L, Koob GF (1992) Oral ethanol self-administration in rats is reduced by the administration of dopamine and glutamate receptor antagonists into the nucleus accumbens. *Psychopharmacology* 109(1–2): 92–98
10 Shelton KL, Balster RL (1997) Effects of gamma-aminobutyric acid agonists and N-methyl-D-aspartate antagonists on a multiple schedule of ethanol and saccharin self-administration in rats. *J Pharmacol Exp Ther* 280(3): 1250–1260
11 Danysz W, Dyr W, Jankowska E, Glazewski S, Kostowski W (1992) The involvement of NMDA receptors in acute and chronic effects of alcohol. *Alcohol Clin Exp Res* 16: 499–504
12 Parsons CG, Danysz W, Quack G (1999) Memantine is a clinically well tolerated N-methyl-D-aspartate (NMDA) receptor antagonist – a review of preclinical data. *Neuropharmacology* 38: 735–767
13 Hölter SM, Danysz W, Spanagel R (1996) Evidence for alcohol anti-craving properties of memantine. *Eur J Pharmacol* 314: R1–R2
14 Piasecki J, Koros E, Dyr W, Kostowski W, Danysz W, Bienkowski P (1998) Ethanol-reinforced

behavior in the rat: effects of uncompetitive NMDA receptor antagonist, memantine. *Eur J Pharmacol* 354(2–3): 135–143
15 Parsons CG, Danysz W, Quack G (2000) Memantine and the amino-alkyl-cyclohexane MRZ 2/579 are moderate affinity uncompetitive NMDA receptor antagonists – *in vitro* characterisation. *Amino Acids* 19: 157–166
16 Bienkowski P, Krzascik P, Koros E, Kostowski W, Scinska A, Danysz W (2001) Effects of a novel uncompetitive NMDA receptor antagonist, MRZ 2/579 on ethanol self-administration and ethanol withdrawal seizures in rat. *Eur J Pharmacol* 413: 81–89
17 Bienkowski P, Koros E, Kostowski W, Danysz W (1999) Effects of N-methyl-D-aspartate receptor antagonists on reinforced and non-reinforced responding for ethanol in rats. *Alcohol* 18: 131–137
18 Hölter SM, Danysz W, Spanagel R (2000) Novel uncompetitive N-methyl-D-aspartate (NMDA)-receptor antagonist MRZ 2/579 suppresses ethanol intake in long-term ethanol-experienced rats and generalizes to ethanol cue in drug discrimination procedure. *J Pharmacol Exp Ther* 292(2): 545–552
19 Vengeliene V, Bachteler D, Danysz W, Spanagel R (2005) The role of the NMDA receptor complex in alcohol relapse: a pharmacological mapping study using the alcohol deprivation effect. *J Neuropharmacology; in press*
20 Bachteler D, Economidou D, Danysz W, Ciccocioppo R, Spanagel R (2005) The effects of acamprosate and neramexane on cue-induced reinstatement of ethanol-seeking behavior in rat. *Neuropsychopharmacology; in press*
21 Bäckström P, Hyytiä P (2004) Ionotropic glutamate receptor antagonists modulate cue-induced reinstatement of ethanol-seeking behavior. *Alcohol Clin Exp Res* 28(4): 558–565
22 Hundt W, Danysz W, Hölter SM, Spanagel R (1998) Ethanol and N-methyl-D-aspartate receptor complex interactions: a detailed drug discrimination study in the rat. *Psychopharmacology* 135: 44–51
23 Colombo G, Grant KA (1992) NMDA receptor complex antagonists have ethanol-like discriminative stimulus effects. *Ann N Y Acad Sci* 654: 421–423
24 Malenka RC (2003) Synaptic plasticity and AMPA receptor trafficking. *Ann N Y Acad Sci* 1003: 1–11
25 Luscher C, Xia H, Beattie EC, Carroll RC, von Zastrow M, Malenka RC, Nicoll RA (1999) Role of AMPA receptor cycling in synaptic transmission and plasticity. *Neuron* 24: 649–658
26 Zamanillo D, Sprengel R, Hvalby O, Jensen V, Burnashev N, Rozov A, Kaiser KM, Koster HJ, Borchardt T, Worley P et al. (1999) Importance of AMPA receptors for hippocampal synaptic plasticity but not for spatial learning. *Science* 284(5421): 1805–1811
27 Lu L, Grimm JW, Shaham Y, Hope BT (2003) Molecular neuroadaptations in the accumbens and ventral tegmental area during the first 90 days of forced abstinence from cocaine self-administration in rats. *J Neurochem* 85: 1604–1613
28 Sutton MA, Schmidt EF, Choi KH, Schad CA, Whisler K, Simmons D, Karanian DA, Monteggia LM, Neve RL, Self DW (2003) Extinction-induced upregulation in AMPA receptors reduces cocaine-seeking behaviour. *Nature* 421(6918): 70–75
29 Brückner MK, Rossner S, Arendt T (1997) Differential changes in the expression of AMPA receptors genes in rat brain after chronic exposure to ethanol: an *in situ* hybridization study. *J Hirnforsch* 38(3): 369–376
30 Cowen MS, Schroff KC, Gass P, Sprengel R, Spanagel R (2003) Neurobehavioral effects of alcohol in AMPA receptor subunit (GluR1) deficient mice. *Neuropharmacology* 45: 325–333
31 Sanchis-Segura C, Vengeliene V, Zghoul T, Bachteler D, Gass P, Sprengel R, Spanagel R (2004) Involvement of AMPA receptor GluR-C subunit in alcohol relapse. *Alcohol Clin Exp Res* 285: 54A
32 Stephens DN, Brown G (1999) Disruption of operant oral selfadministration of ethanol, sucrose, and saccharin by the AMPA/kainate antagonist, NBQX, but not the AMPA antagonist, GYKI66. *Alcohol Clin Exp Res* 23: 1914–1920
33 Conn P, Pin JP (1997) Pharmacology and functions of metabotropic glutamate receptors. *Annu Rev Pharmacol Toxicol* 37: 205–237
34 Spooren W, Ballard T, Gasparini F, Amalric M, Mutel V, Schreiber R (2003) Insight into the function of group I and group II metabotropic glutamate (mGlu) receptors: behavioral characterization and implications for the treatment of CNS disorders. *Behav Pharmacol* 14: 257–277
35 Spooren W, Vassout A, Neijt HC, Kuhn R, Gasparini F, Roux S, Porsolt RD, Gentsch C (2000) Anxiolytic-like effects of the prototypical metabotropic glutamate receptor 5 antagonist

2-methyl-6-(phenylethynyl)-pyridine in rodents. *J Pharmacol Exp Ther* 295: 1267–1275

36 Tatarczynska E, Klodzinska A, Chojnacka-Wojcik E, Palucha A, Gasparini F, Kuhn R, Pilc A (2001) Potential anxiolytic- and anti-depressant-like effects of MPEP, a potent, selective and systemically active mGlu5 receptor antagonist. *Br J Pharmacol* 132: 1423–1430

37 Schulz B, Fendt M, Gasparini F, Lingenhohl K, Kuhn R, Koch M (2001) The metabotropic glutamate receptor antagonist 2-methyl-6-(phenylethynyl)-pyridine (MPEP) blocks fear conditioning in rats. *Neuropharmacology* 41: 1–7

38 Chiamulera C, Epping-Jordan MP, Zocchi A, Marcon C, Cottiny C, Tacconi S, Corsi M, Orzi F, Conquet F (2001) Reinforcing and locomotor stimulant effects of cocaine are absent in mGluR5 null mutant mice. *Nature Neurosci* 4: 873–874

39 Paterson NE, Semenova S, Gasparini F, Markou A (2003) The mGluR5 antagonist MPEP decreased nicotine self-administration in rats and mice. *Psychopharmacology* 167: 257–264

40 McGeehan AJ, Olive MF (2003) The anti-relapse compound acamprosate inhibits the development of a conditioned preference to ethanol and cocaine but not morphine. *Br J Pharmacol* 138: 9–12

41 Sharko AC, Kimberly I, Koenig H, Lau K, Ou CJ, Olive MF, Camarini R, Hodge CW (2002) Involvement of metabotropic glutamate receptor subtype 5 (mGluR5) in alcohol self-administration. *Alcohol Clin Exp Res* 26: 112A

42 Bäckström P, Bachteler D, Koch S, Hyytiä P, Spanagel R (2004) mGluR5 antagonist MPEP reduces ethanol-seeking and relapse behavior. *Neuropsychopharmacology* 29: 921–928

43 Sorensen SD, Conn PJ (2003) G protein-coupled receptor kinases regulate metabotropic glutamate receptor 5 function and expression. *Neuropharmacology* 44: 699–706

44 Kotecha SA, Jackson MF, Al-Mahrouki AA, Roder JC, Orser BA, MacDonald JF (2003) Co-stimulation of mGluR5 and NMDA receptors is required for potentiation of excitatory synaptic transmission in hippocampal neurons. *J Biol Chem* 278: 27742–27749

45 Hermans E, Challiss RA (2001) Structural, signaling and regulatory properties of the group I metabotropic glutamate receptors: prototypic family C G-protein-coupled receptors. *Biochem J* 359: 465–484

46 Schoepp DD, Conn PJ (1993) Metabotropic glutamate receptors in brain function and pathology. *Trends Pharmacol Sci* 14: 13–20

47 Spanagel R, Pendyala G, Abarca C, Zghoul T, Sanchis-Segura, Magnone MC, Lascorz J, Depner M, Holzberg D, Soyka M et al. (2004) The circadian clock gene *Period2* alters the glutamatergic system and thereby modulates alcohol consumption. *Nature Med*; *in press*

48 Spanagel R, Siegmund S, Cowen M, Schroff KC, Schumann G, Fiserova M, Sillaber I, Wellek S, Singer M, Putzke J (2002) The neuronal nitric oxide synthase gene is critically involved in neurobehavioral effects of alcohol. *J Neurosci* 22(19): 8676–8683

49 Werner C, Raivich G, Cowen M, Strekalova T, Sillaber I, Spanagel R, Hofmann F (2004) Importance of NO/cGMP signaling via cGMP-dependent protein kinase II for mediating emotionality and neurobehavioral effects of alcohol. *Eur J Neurosci*; *in press*

Drugs for Relapse Prevention of Alcoholism
Edited by R. Spanagel and K. Mann

Future perspectives on relapse prevention

Rainer Spanagel and Karl F. Mann

Department of Psychopharmacology and Department of Addictive Behavior and Addiction Medicine, Central Institute of Mental Health, University of Heidelberg, J5, 68159 Mannheim, Germany

In 1751, William Hogarth successfully launched what in present times would be known as a public awareness campaign: in his famous work, he contrasted the chaotic "gin lane" with a neat and well-organized "beer street" (Figs 1 and 2). His "intervention" was most influential in a debate which in its conclusion was to witness a sharp rise in community tax on gin, with the consequence of having a dramatic drop in gin consumption in Old England. While these circumstances are well known, they are providing some of the first evidence that taxes are a main regulator of alcohol intake rate per inhabitant. Another aspect of his campaign merits equal attention, namely, the fact that he suggested something like a substitution therapy. Thus, in his view, gin drinking ought to be substituted and replaced by the less harmless consumption of beer. From today's perspective, Hogarth's provocative idea recently became reality, with the novel concept of harm reduction in conjunction with substitution therapy. Thus, a new generation of compounds that substitute for the discriminative stimulus properties of alcohol – namely, uncompetitive NMDA receptor antagonists – might have the ability of altering current day-to-day treatment programs for alcoholic patients.

Harm reduction in combination with substitution therapy

In line with the results from drug discrimination studies in animals, clinical experiments have indicated that NMDA receptors mediate, at least in part, the subjective effects of ethanol. Krystal et al. [1] have shown that ketamine – an uncompetitive NMDA receptor antagonist – produced dose-related ethanol-like effects in detoxified alcoholics. The effects of ketamine were more similar to the sedative than to the stimulant effects of ethanol and the subjects rated ketamine as more similar to ethanol than to marijuana or cocaine. Importantly, ketamine did not increase craving for ethanol. In a more recent study, Soyka and co-workers [2] have also found that the low-affinity uncompetitive NMDA receptor antagonist dextromethorphan can produce some ethanol-like subjective effects in both alcoholics and controls. These findings prompted the idea

Fig. 1. Gin Lane, engraving by William Hogarth, 1751. (The British Museum, London, UK.)

that these compounds have the ability to substitute for acute alcohol intake. Due to the lack of producing craving by themselves, they could be used in relapse prevention. This idea is outlined in several chapters of the present book and constitutes one important pillar of a major theory in the alcohol field – the glutamatergic theory of alcoholism. It derives from the findings that high intoxicating doses of ethanol decrease the activity of glutamate signaling in the brain. As a consequence of chronic alcohol intake, adaptive changes in the glutamate system are observed, which result in a hyper-glutamatergic state [3, 4]. Considering the fact that at least 70–80% of excitatory neurons work via glu-

Fig. 2. Beer Street, engraving by William Hogarth, 1751. (The British Museum, London, UK.)

tamatergic neurotransmission, augmented activity of the glutamate system due to chronic alcohol intake puts our brain into a hyper-excitatory state. In such a situation alcohol can be regarded as a cure, which has the ability to normalize general brain activity. Thus, it is suggested that a hyper-glutamatergic state is a trigger for alcoholic patients to further drink alcohol in a withdrawal state or to relapse after a period of abstinence. That this most likely represents a pharmacologically driven behavior has recently been demonstrated in the alcohol-deprivation-effect model of relapse drinking behavior [5]. Thus, following alcohol deprivation, an increase in alcohol consumption is observed after having regained access to alcohol. This behavior, however, is completely abolished by experimentally administered ethanol using the intraperitoneal route. In conclusion, the acute pharmacological effect of ethanol may fully substitute for such a complex behavior as relapse. It is therefore only a logical step further that "substitution drugs" such as uncompetitive NMDA receptor antagonists should dampen hyper-glutamatergic activity and also abolish or reduce relapse drinking behavior, a hypothesis which has been successfully and repeatedly demonstrated in laboratory animals [5–7].

Recent clinical findings, however, have not been in line with this hypothesis. In a currently still unpublished study, the efficacy of neramexane, a low-affinity NMDA receptor antagonist, was tested in a multi-center randomized, double-blind, placebo-controlled study. 249 alcoholic patients and the same number of control subjects were included in this study. One dose of neramexane *versus* placebo was tested. The outcome was negative: the verum group did not differ from the placebo group in terms of number of relapses and total abstinence days during the 6-month treatment period. It is important to note, however, that a low dose of neramexane was used, which is unfavorable in a substitution study, as preclinical studies indicate that only high doses would be helpful in preventing relapse behavior [5–7]. Furthermore, compliance was a problem in this study. Although the use of a low dose of neramexane, in addition to the compliance problem, might have influenced the outcome of this study, the negative results obtained are a threat to the glutamatergic theory of alcoholism and forces researchers to reassess their interpretations and conclusions in this regard. Thus, in a more critical perspective, one has to state that ethanol's interactions with the glutamatergic system are still far from clear. In fact, many important questions remain unanswered. For example, it is of crucial importance to examine whether long-lasting changes in the glutamatergic system occur in alcohol-dependent animals as well as in alcoholic patients. Thus, further preclinical and clinical studies including molecular, genetic, neuroimaging and behavioral techniques should address the following points: (I) Identification of adaptations within the glutamatergic system and its signalling pathways associated with chronic alcohol use as well as the time course of these adaptive processes. In particular, it should be examined whether long-term or even persistent changes within the glutamatergic system can account for craving and relapse in laboratory animals. With the aid of recently improved *in vivo* spectroscopic measurements of glutamate [8], those examinations may now also be conducted in abstinent alcoholic patients. (II) Further assessment of interactions of various NMDA receptor antagonists with specific alcohol-induced behavioral effects including aversive, reinforcing and subjective actions. (III) Evaluation of different classes of NMDA receptor antagonists as potential anti-craving/anti-relapse drugs in clinical studies. (IV) Systematic studies on the genetic basis for the susceptibility of alcohol dependence and relapse behavior within the glutamatergic system.

All these studies are necessary to ultimately provide substantial support for the concept of substitution therapy. However, even if the scientific basis of a substitution therapy would be of substantial and sound nature, it cannot be foreseen whether the majority of professionals in the addiction field would accept such a treatment concept. In particular, psychologists, social workers, members of self-help groups and counselors are quite hesitant in promoting the use of any kind of medication. Thus, it will be an issue of enhancing educational activities so as to increase the acceptance of these medications for the benefit of our patients. Another way to encourage professionals who are still reluctant in prescribing anti-relapse drugs, is to combine pharmacotherapy with behavioral and/or psychosocial therapy.

Pharmacotherapy in combination with behavioral and psychosocial therapy (COMBINE)

This book concentrates on pharmacological relapse prevention only. The whole field of behavioral and psychosocial interventions for relapse prevention was deliberately excluded. The authors agree with the vast majority of colleagues who, on the basis of empirical evidence, acknowledge the need for combining both treatment strategies [9]. However, this would certainly merit a book in its own right, especially since the individual "cocktail" for a given patient, in other words a targeted approach, is far from being clear at this point. However, some progress can be expected from project COMBINE. In this large-scale study, more than 1300 patients are treated with either acamprosate, naltrexone, placebo or with a combination of acamprosate and naltrexone. Whilst half the patients receive a low dose standard supportive therapy (Medical Management), the other half receives more intense psychotherapy, cognitive behavioral intervention [10]. Additionally, one group is treated by cognitive behavioral intervention without medication. The results of the study will be available in 2005/2006 and should help us better understand the contribution of each drug alone, their interaction and especially the interaction with two forms of psychotherapy. One single-center study by Kiefer et al. [11] already suggests that a combination of naltrexone and acamprosate may be more effective than either drug alone. It is hoped that the results of the COMBINE study will further support this finding and may finally end in an algorithm which may tell us more about an evidence-based stepped care approach for the individual patient.

Individual adapted pharmacotherapy (PREDICT): improving clinical efficacy

A German study that has been closely planned with the Steering Committee of project COMBINE aims to identify responders to either acamprosate or to naltrexone treatment *a priori*. In order to predict treatment response, different craving types are assessed in this study. There are at least two different pathways which can induce alcohol craving and relapse [12] – in other words, it is suggested that there are at least two different craving types. In the project PREDICT, the goal is to see whether naltrexone is specifically effective among individuals who experience positive motivational effects upon alcohol consumption and whether acamprosate is most effective among patients who consume alcohol to counteract conditioned withdrawal and negative mood states [13]. Animal studies are performed in close parallel to the clinical project PREDICT. Thus, changes in reinstatement of alcohol-seeking behavior in rats are analyzed in response to the same craving-inducing situations and the same pharmacological treatments that are used in humans. It has already been observed that reinstatement of alcohol-seeking behavior induced by alcohol-associated cues can be blocked by naltrexone and acamprosate [14]; however, it still remains an open question whether one of these drugs also interferes with withdrawal-induced reinstatement of alcohol-seeking behavior.

Another issue that is considered in project PREDICT is the genetic make-up of an individual. Thus, response to pharmacological treatment may be influenced by genetic polymorphisms of drug target genes. Developing a new approach which takes into account craving/relapse-specific stimuli and situations as well as genetic make-up may hold the potential for a break-through in the management and prevention of relapse. In fact, only recently, it was shown that genetic differences in the mu opioid receptor might predict naltrexone efficacy [15]. In summary, the effectiveness of acamprosate and naltrexone seems to be valid beyond doubt [16], although more research needs to be done in identifying potential responders. One major question remains, however, namely, why are these medications so rarely prescribed?

Involvement of the pharmaceutical industry in bridging the gap between the academic world and alcoholic patients

The chapters of this book have outlined the evidence for the pharmacological treatment of alcoholism. While the prevalence rates for alcohol dependence and harmful use (abuse) are on the order of more than 5% of the population in developed countries, the pharmaceutical industry is still hesitant in promoting their products. A recent survey in the US of 1,388 physicians specialized in substance abuse has revealed that only 13% prescribed naltrexone and a mere 9% prescribed disulfram (acamprosate was registered in the US in April 2004 and is, therefore, not included in this survey). In contrast, these addiction specialists prescribed anti-depressants for their primary alcohol-dependent patients in 46% of the cases and benzodiazepines in 11%, both being medications known for no detectable effectiveness in prevention of relapse to alcoholism. Interestingly, these physicians felt much more educated about anti-depressants or neuroleptics than about anti-craving substances [17]. The same article refers to 171,000 prescriptions for naltrexone and 246,000 prescriptions for disulfram in 1999. In the same year, 6,662,000 prescriptions were given for anti-psychotic and 23,138,000 for the leading anti-depressant drugs. To a certain extent, these numbers reflect the money being spent for marketing. For example, 7.8 million free samples of sertraline and 1.4 million samples of risperidone were distributed in 1999. In contrast, DuPont Pharma, the producer of naltrexone, never distributed free samples of Revia. The same holds true for journal advertisements, etc. In summary, $60 million for the promotion of sertraline and $20 million for risperidone were spent in 1999, in contrast to $0 for naltrexone. It seems to be this apparent lack of interest on the part of the pharmaceutical industry which contributes to the low use of anti-craving/relapse drugs by alcohol-dependent patients. In a recent debate with representatives of the pharmaceutical industry, the challenges of marketing strategies and further drug development in alcoholism have been outlined by an expert group [18].

A major challenge that was recognized by this expert group is that the development of marketing strategies and drugs by the pharmaceutical industry is hampered by their unenthusiastic and reluctant attitude to invest money in the field of alcoholism. In order to advance in this respect, industry's commitment in this field ought to be enhanced, especially by means of socio-political pressure. Until the very present, most of the advancements in alcohol research and drug development for relapse prevention are currently taking place in university-based laboratories, but this rarely progresses to higher levels of clinical trials. As demonstrated in this book, preclinical research has produced potentially interesting targets and drugs; however, a full-scale clinical trial of a candidate medication is always an expensive undertaking and rarely supported by funding bodies. One solution to this problem would be that before cost-intensive phase II/III studies are initiated on the basis of preclinical findings, small orientation studies using few patients and controls would aid in rapidly translating preclinical findings into clinical trials. Although small orientation studies always bear the danger of false positive and false negative results, they would certainly be a good help in guiding companies in the decision-making on cost-intensive endeavors. A convincing example of this approach is the medication development of acamprosate. Thus, the effectiveness of acamprosate to decrease alcohol consumption was first demonstrated in preclinical studies [19]. Then, in a small orientation study by Lhuintre et al. [20], it was shown that of the 70 patients who completed the study, 33 received acamprosate and 37 placebo. 20 patients on acamprosate did not relapse, compared with 12 on placebo. Following this study, several multi-center, double-blind, placebo-controlled studies were initiated and based on the successful results of these studies, acamprosate was approved in most of the European countries [21]. Again, small orientation studies can be easily performed at universities and this knowledge could finally be used by the pharmaceutical industry to perform phase II/III studies with cost-intensive monitoring.

However, so as to achieve approval of a novel anti-relapse drug, it is necessary to conduct clinical trials in a manner deemed acceptable to the regulatory authorities in different countries. In contrast to many other areas of medication development, there are no agreed guidelines from the regulatory authorities for the conduct of clinical trials in alcoholism. Consequently, there is a degree of variability in the conduct of trials amongst different centers and among countries as well. Clinical trials in Europe also tend to be more "naturalistic" and more prolonged than trials in the US, making it difficult to compare data across continents. Identified areas in which guidelines would be particularly valuable include minimizing the bias associated with patients leaving trials prematurely, and a preferred main end-point in assessing outcomes. However, it must be recognized that the preferred main end-point probably cannot be identical for all medications because of the need to match clinical end-point with the mechanism of action of the drug. For example, although anti-relapse drugs such as naltrexone and acamprosate might share an end-point such as "cumulative abstinent days" during the trial, this would not be appropriate for an agent aimed at reducing excessive drinking (e.g., galanthamine) without necessarily producing abstinence at all. Such an agent might be therapeutically very valuable but would not be identified as effective by a purely "abstinence-based" outcome measure. Overall, it is clear that pre-trial motivation for patients in reducing their alcohol consumption must be assessed and incorporated into study design and/or statistical analysis. Also, ideally, outcome measures should reflect clinical relevance, quality of life and economic cost/benefit analysis. Regardless of the complexity, it would certainly increase the industry's enthusiasm for the development of medications if some guidelines for acceptability were issued by the appropriate regulatory authorities.

Educational programs for improving the clinical and social "climate" for pharmacological relapse prevention

A major factor in the low level of clinical use of validated pharmacotherapy in the area of alcoholism may be the general public's view that alcoholism is a "behavior disorder", rather than a "medical condition". This notion is compounded by the important current role of psychosocial treatment in alcoholism and by the role of the social services in recovery and rehabilitation. This has led to some opposition (or at least antipathy) to medical treatment of alcoholism by the psychosocial community, who may feel that their position could be undermined by effective pharmacotherapy. Unless these barriers are overcome, medications that are effective treatments will have a hard time to reach the patients and therefore will not become profitable. This alone is sufficient in dissuading the pharmaceutical industry, or any other researcher interested in the practical value of his/her research, from becoming committed to medication development for alcoholism. This failure to penetrate a potential market is perhaps the major challenge to an effective drug development strategy. Better dissemination of information and more aggressive "educational marketing" (by government and professional organizations, as well as industry) are required if medication development in alcoholism is to have a brighter future. This is particularly important for primary healthcare practitioners. The major challenges appear to be in explaining and marketing drugs for alcoholism so that physicians and patients alike may genuinely accept them. Additionally, there is a compelling need to explain and integrate medical and psychosocial treatments so that there is less opposition to the introduction of a pharmacotherapy.

Conclusion

For relapse prevention, two effective drugs are currently on the market and many more drug targets and compounds have already been developed. The effectiveness of acamprosate and naltrexone may further be improved by either combining the two drugs or by combining the drug with behavioral and psychosocial therapy (project COMBINE). Yet another possibility to improve their clinical efficacy is to identify treatment responders *a priori*, either by matching with a specific "craving type" or by matching with the genetic make-up of a particular alcoholic patient (project PREDICT). Although both concepts may improve clinical efficacy of acamprosate and naltrexone, one has to emphasize that both compounds are clearly superior to a combined treatment of placebo and behavioral/psychosocial therapy. However, compared to other classes of psychopharmacological therapeutics, anti-relapse drugs are rarely prescribed. This is even more surprising, considering the fact that differences in the clinical efficacy among anti-depressants and active placebos are small [22]. Nevertheless, anti-depressants have for many years now conquered the world market. The question as to why anti-relapse drugs are so infrequently prescribed is best left addressed to the pharmaceutical industry. They are still hesitant in using appropriate marketing strategies, with only little interest in investing money in further drug development. Another problem, which adds to the low prescription rates, is the general public's view that alcoholism is a "behavioral disorder". This stigma is reinforced by the majority of professionals in the addiction field including psychologists, social workers, members of self-help groups and counselors, who are quite hesitant in promoting the use of any kind of medication. The only way out of this dilemma is a better dissemination of information and more aggressive educational marketing by government and professional organizations. The main player in the future in this regard, however, could be the pharmaceutical industry, but only if they take it upon themselves to conquer this potentially lucrative market, with the main beneficiaries being the patients themselves.

References

1 Krystal JH, Petrakis IL, Webb E, Cooney NL, Karper LP, Namanworth S, Stetson P, Trevisan LA, Charney DS (1998) Dose-related ethanol-like effects of the NMDA antagonist, ketamine, in recently detoxified alcoholics. *Arch Gen Psychiatry* 55: 354–360

2 Soyka M, Bondy B, Eisenburg B, Schutz CG (2000) NMDA receptor challenge with dextromethorphan – subjective response, neuroendocrinological findings and possible clinical implications. *J Neural Transm* 107: 701–714

3 Tsai G, Coyle JT (1998) The role of glutamatergic neurotransmission in the pathophysiology of alcoholism. *Annu Rev Med* 49: 173–184

4 Krystal JH, Petrakis IL, Mason G, Trevisan L, D`Souza DC (2003) N-methyl-D-aspartate glutamate receptors and alcoholism: reward, dependence, treatment, and vulnerability. *Pharmacol Ther* 99: 79–94

5 Vengeliene V, Bachteler D, Wojciech D, Spanagel R (2004) NMDA receptors and alcohol relapse: *Alcohol Clin Exp Res* 28: 18A

6 Hölter SM, Danysz W, Spanagel R (1996) Evidence for alcohol anti-craving properties of memantine. *Eur J Pharmacol* 314: R1–2

7 Hölter SM, Danysz W, Spanagel R (2000) Novel uncompetitive N-methyl-D-aspartate (NMDA)-receptor antagonist MRZ 2/579 suppresses ethanol intake in long-term ethanol-experienced rats and generalizes to ethanol cue in drug discrimination procedure. *J Pharmacol Exp Ther* 292: 545–552

8 Hurd R, Sailasuta N, Srinivasan R, Vigneron DB, Pelletier D, Nelson SJ (2004) Measurement of brain glutamate using TE-averaged PRESS at 3 T. *Magn Reson Med* 51: 435–440

9 Miller WR, Wilbourne PL (2002) Mesa Grande: a methodological analysis of clinical trials of treatments for alcohol use disorders. *Addiction* 97: 265–277

10 COMBINE Study Research Group (2003) Testing combined pharmacotherapies and behavioral interventions in alcohol dependence: rationale and methods. *Alcohol Clin Exp Res* 27: 1107–1122

11 Kiefer F, Jahn H, Tarnaske T, Helwig H, Briken P, Holzbach R, Kampf P, Stracke R, Baehr M, Naber D, Wiedemann K (2003) Comparing and combining naltrexone and acamprosate in relapse prevention of alcoholism: a double-blind, placebo-controlled study. *Arch Gen Psychiatry* 60: 92–99

12 Verheul R, Van den Brink W, Geerlings P (1999) A three-pathway psychobiological model of craving for alcohol. *Alcohol Alcoholism* 34: 197–222

13 Mann K, Smolka M, Spanagel R, Heinz A (2004) Searching for the acamprosate and naltrexone responders: PROJECT PREDICT. *Alcohol Clin Exp Res* 28(8): 86A

14 Bachteler D, Economidou D, Danysz W, Ciccocioppo R, Spanagel R. (2004) The effects of acamprosate and neramexane on cue-induced reinstatement of ethanol-seeking behaviour. *Neuropsychopharmacology, in press*

15 Oslin DW, Berrettini W, Kranzler HR, Pettinati H, Gelernter J, Volpicelli JR, O'Brien CP (2003) A functional polymorphism of the mu-opioid receptor gene is associated with naltrexone response in alcohol-dependent patients. *Neuropsychopharmacology* 28: 1546–1552

16 Mann K (2004) Pharmacotherapy of alcohol dependence. *CNS Drugs* 18: 485–504

17 Mark TL, Kranzler HR, Song X, Bransberger P, Poole VH, Crosse S (2003) Physicians' opinions about medications to treat alcoholism. *Addiction* 98: 617–626

18 Littleton J, de Witte P, Litten R, Gessa GL, Spanagel R, Kranzler H, Lehert P, Johnson B, Saunders J, Berglund M et al. (2004) Challenges to medications development in alcoholism: an international perspective. *Alcohol Alcoholism* 39: 271–275

19 Boismare F, Daoust M, Lhuintre JP, Saligaut C, Moore N, Chrétien P, Durlach J (1984) A homotaurinate derivative reduces the voluntary intake of ethanol by rats: are cerebral GABA receptors involved? *Pharmacol Biochem Behav* 21: 787–789

20 Lhuintre JP, Daoust M, Moore ND, Chretien P, Saligaut C, Tran G, Boismare F, Hillemand B (1985) Ability of calcium bis acetyl homotaurine, a GABA agonist, to prevent relapse in weaned alcoholics. *Lancet* 1: 1014–1016

21 Mann K, Lehert P, Morgan MY (2004) The efficacy of acamprosate in maintaining abstinence in alcohol-dependent individuals: results of a meta-analysis. *Alcohol Clin Exp Res* 28: 51–63

22 Moncrieff J, Wessely S, Hardy A (2004) Active placebos *versus* antidepressants for depression. *Cochrane Database Syst Rev* CD003012

Appendix – Chemical structures

Agent interfering with ethanol metabolism

Disulfiram (Antabuse®)	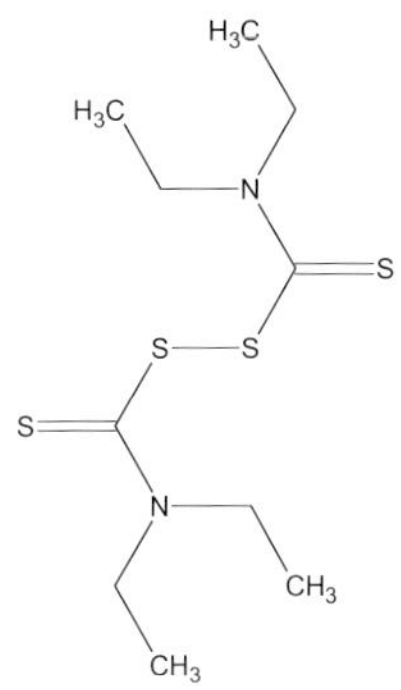	blocks the oxidation of EtOH at the acetaldehyde stage

Agents acting at **glutamate** receptors

Acamprosate	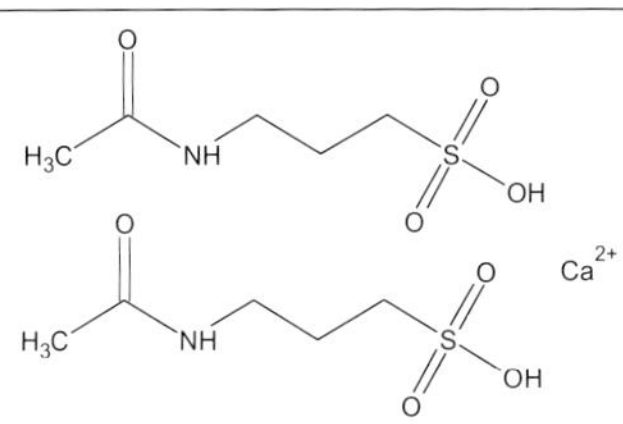	functional NMDA antagonist
AP-5	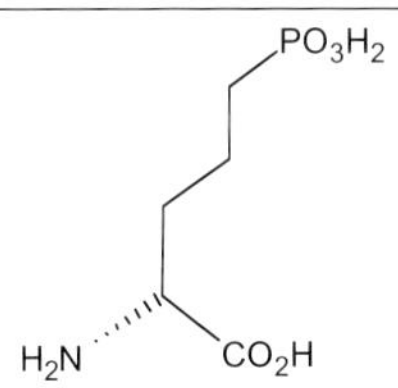	competitive NMDA antagonist
CGP37849	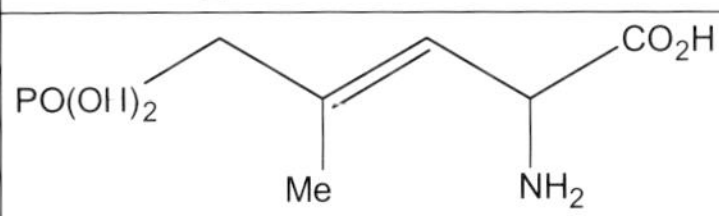	compctitive NMDA antagonist
CPPene	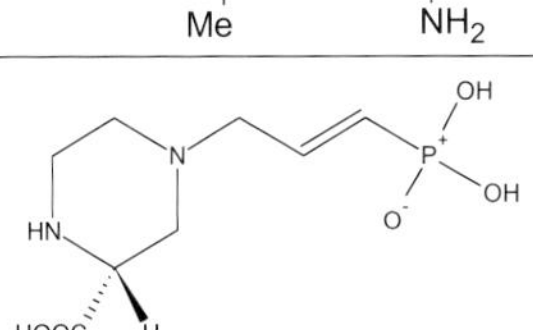	competitive NMDA antagonist
Ifenprodil	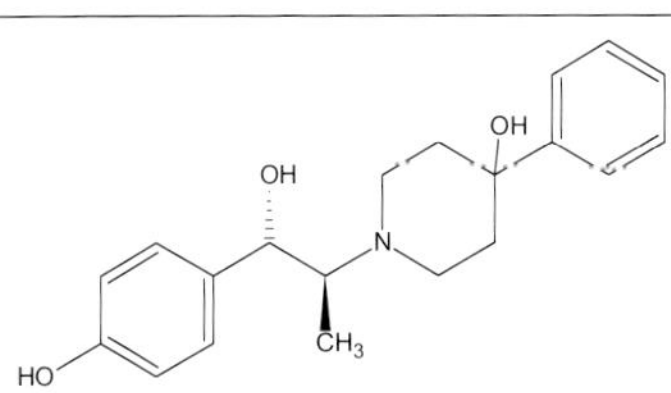	NMDA NR2B antagonist

Agents acting at **glutamate** receptors (continued)

L-701.324	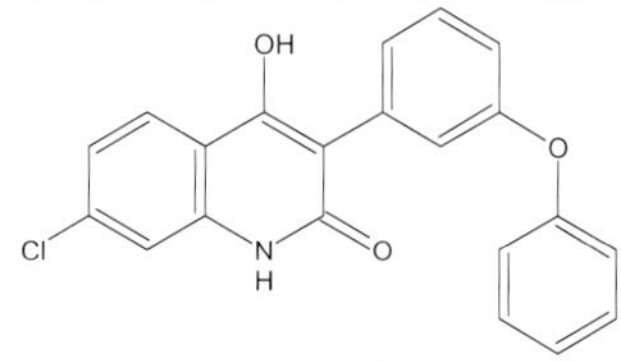	NMDA GlycineB antagonist
Memantine	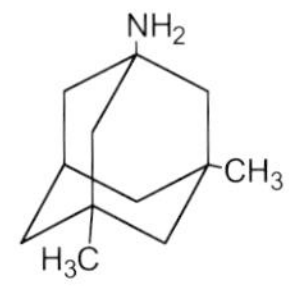	uncompetitive NMDA antagonist
MK-801 (dizocilpine)	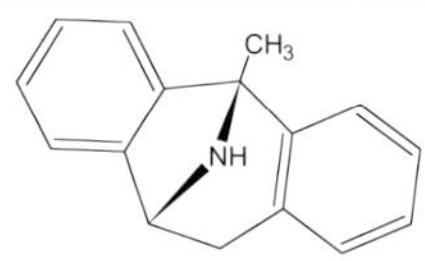	uncompetitive NMDA antagonist
Neramexane	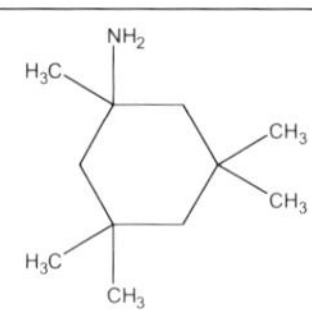	uncompetitive NMDA antagonist
PCP (phencyclidine)	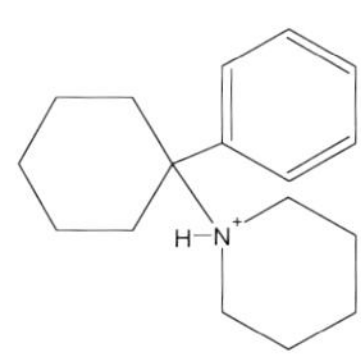	uncompetitive NMDA antagonist
GYKI 52466	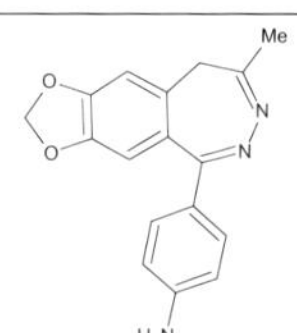	non-competitive AMPA antagonist
NBQX	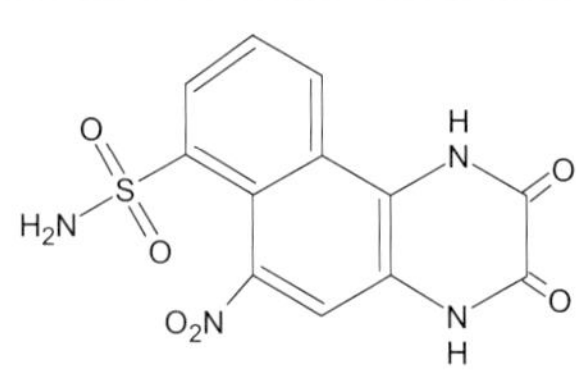	AMPA antagonist
MPEP	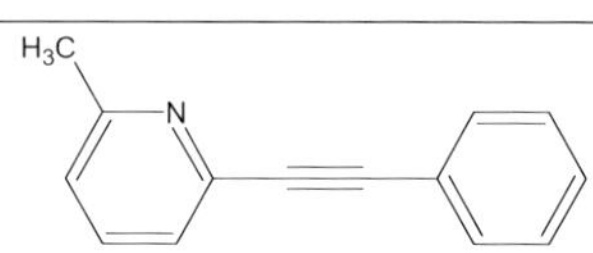	mGluR5 antagonist

Agents acting at **GABA** receptors

Baclofen	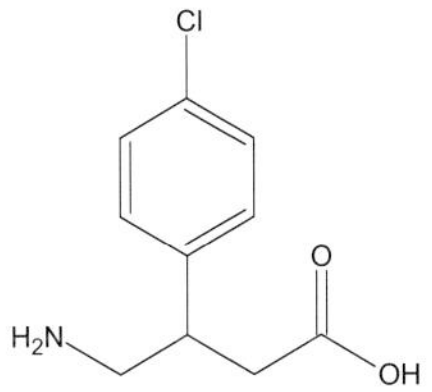	$GABA_B$ agonist
CGP 44532	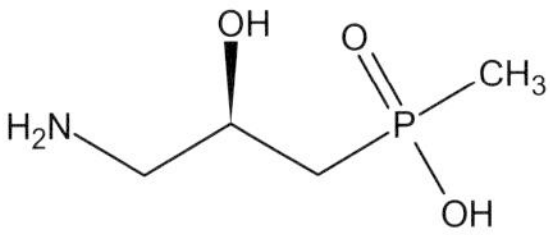	$GABA_B$ agonist

Agents acting at **serotonin** receptors and the serotonergic system

Amperozide	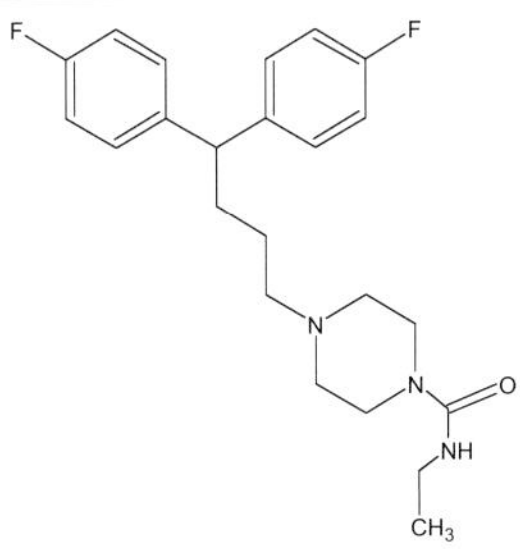	selective $5\text{-}HT_2$ antagonist
Buspirone	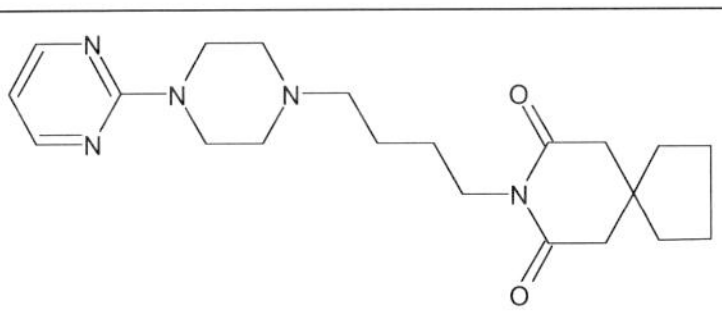	$5\text{-}HT_{1A}$ partial agonist
Citalopram	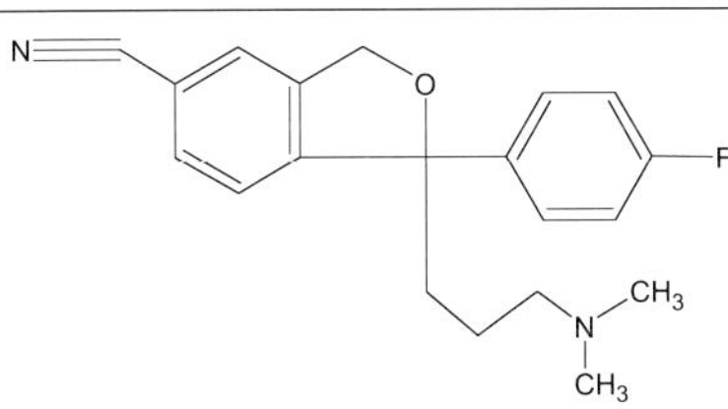	selective serotonin reuptake inhibitor
CGS 12066B (dimaleate)	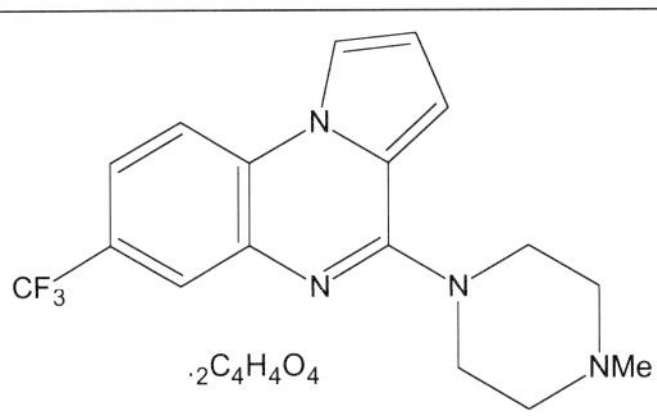	$5\text{-}HT_{1B/1A}$ agonist

Agents acting at **serotonin** receptors and the serotonergic system (continued)

CP-94,253 (hydrochloride)	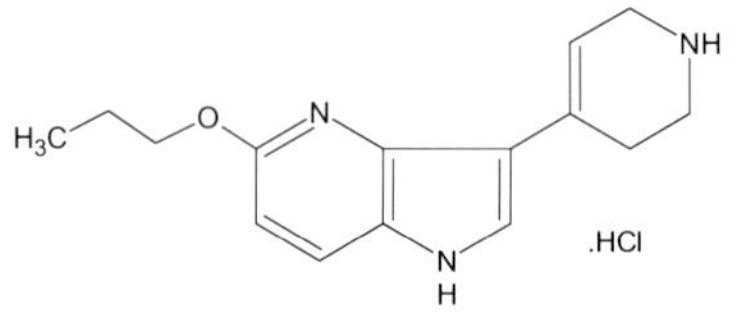	5-HT_{1B} agonist
DOI (hydrochloride)	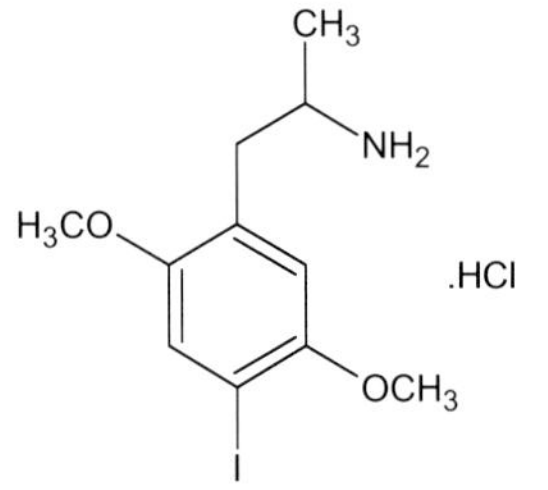	5-HT_{2A} agonist
Fluoxetine	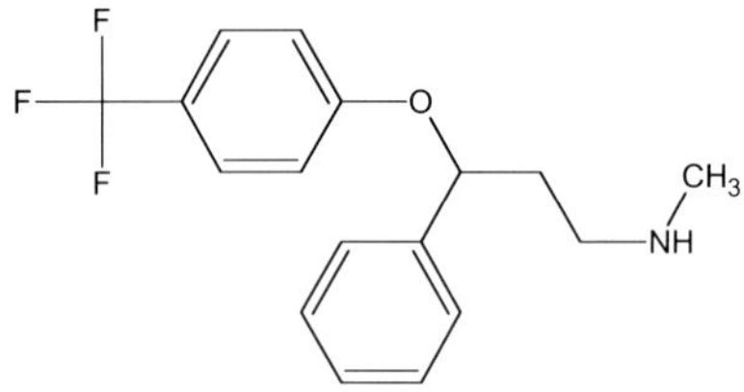	selective serotonin reuptake inhibitor
Gepirone	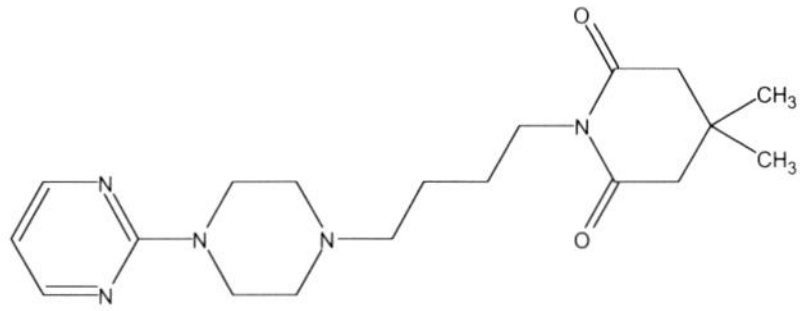	5-HT_{1A} agonist
GR127935	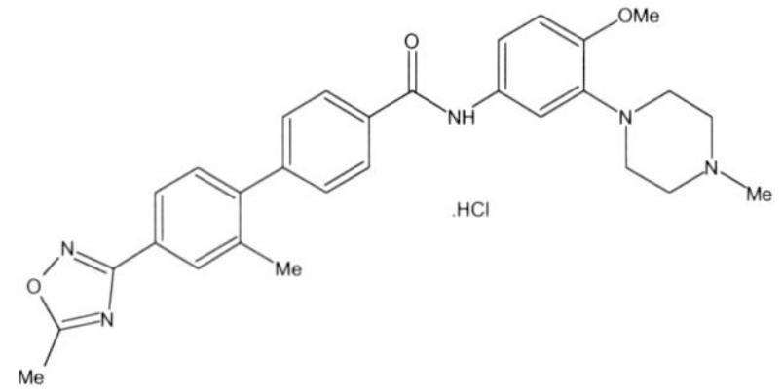	selective 5-HT_{1B} antagonist
Ipsapirone	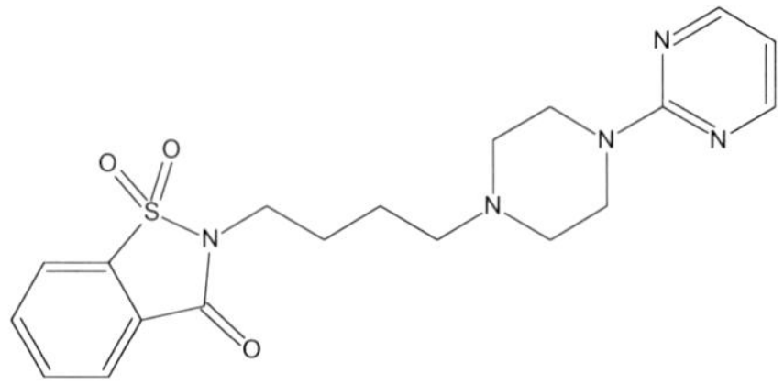	5-HT_{1A} agonist

Agents acting at **serotonin** receptors and the serotonergic system (continued)

mCPP	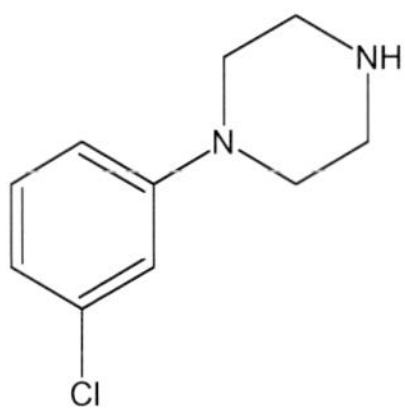	5-$HT_{1B/2C}$ agonist
MDL 100,907	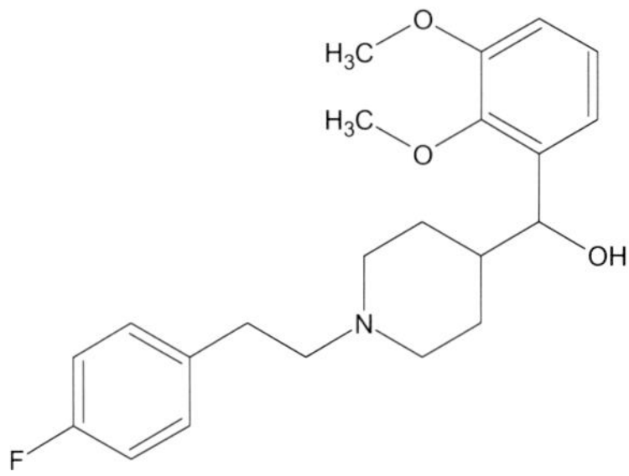	5-HT_{2A} antagonist
Mianserin	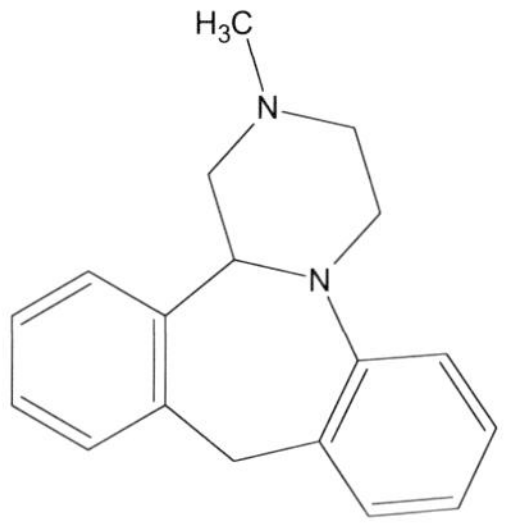	non-selective 5-HT_2 antagonist
8-OH-DPAT	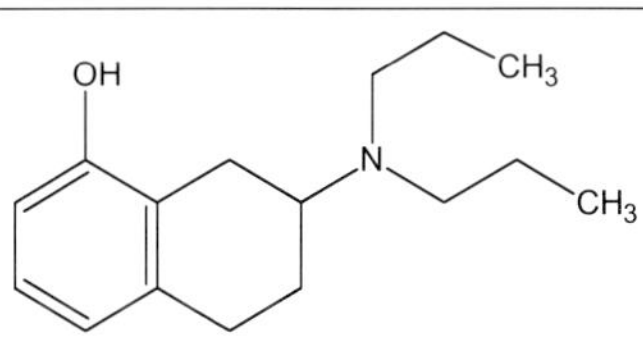	5-HT_{1A} agonist
Ondansetron	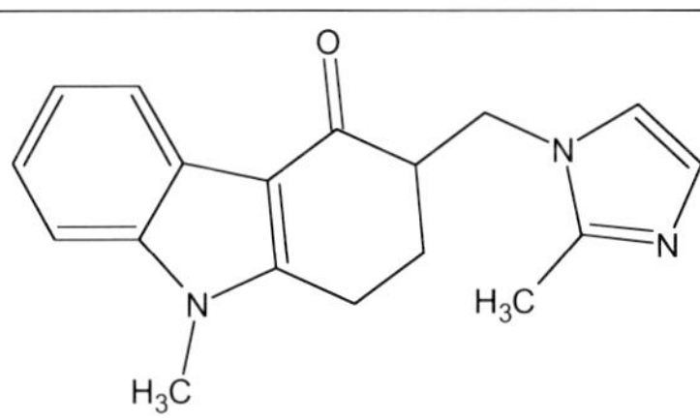	5-HT_3 antagonist
pCPA	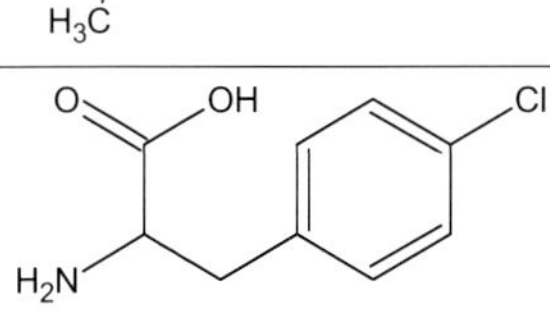	serotonin synthesis inhibitor

Agents acting at **serotonin** receptors and the serotonergic system (continued)

Ritanserin	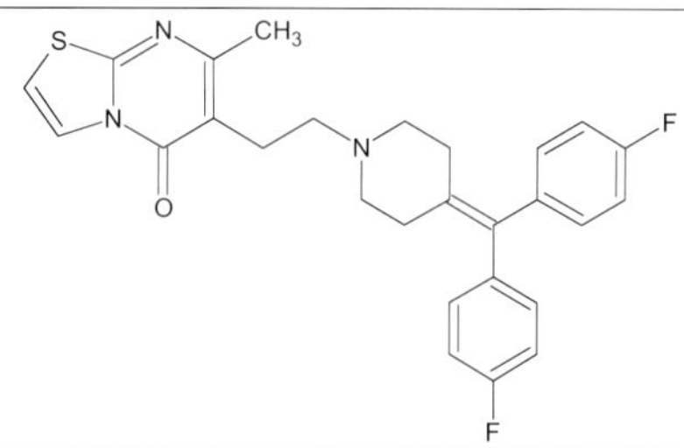	5-HT_2 antagonist
Ro 60-0175 (fumarate)	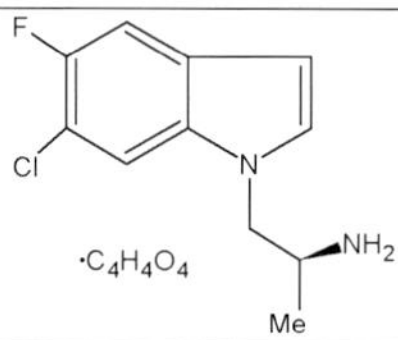	5-HT_{2C} agonist
RU24969 (hemisuccinate)	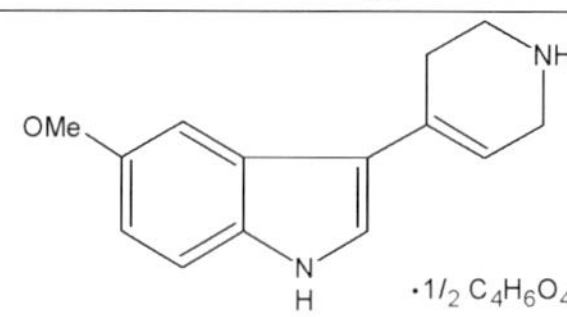	5-$HT_{1B/1A}$ agonist
Sertraline	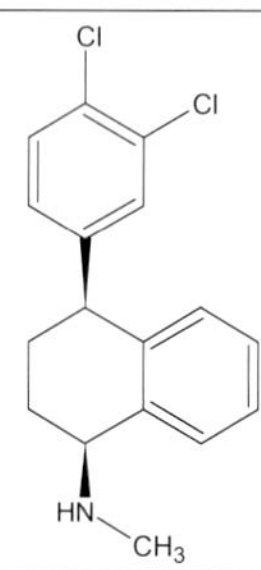	selective serotonin reuptake inhibitor
TFMPP	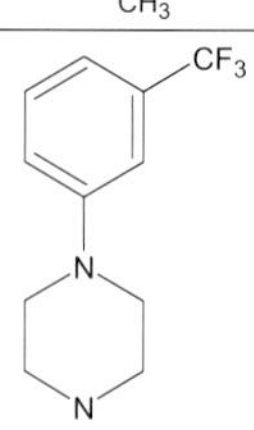	5-$HT_{1B/2C}$ agonist
WAY 100635	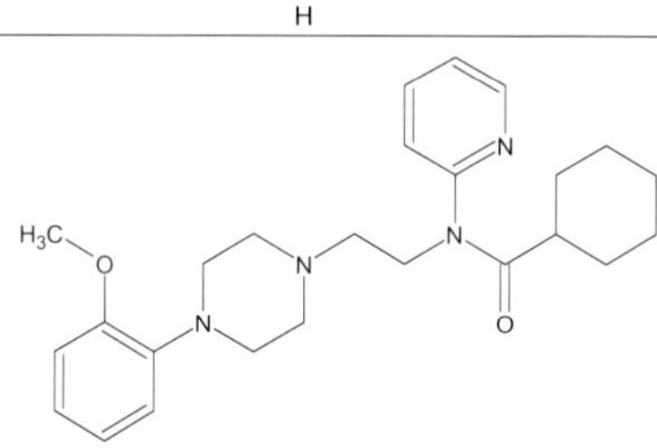	highly selective 5-HT_{1A} antagonist

Agents acting at **dopamine** receptors

Apomorphine	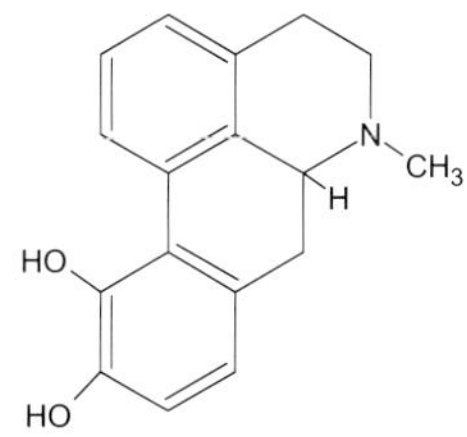	non-selective dopamine receptor agonist
Bromocriptine	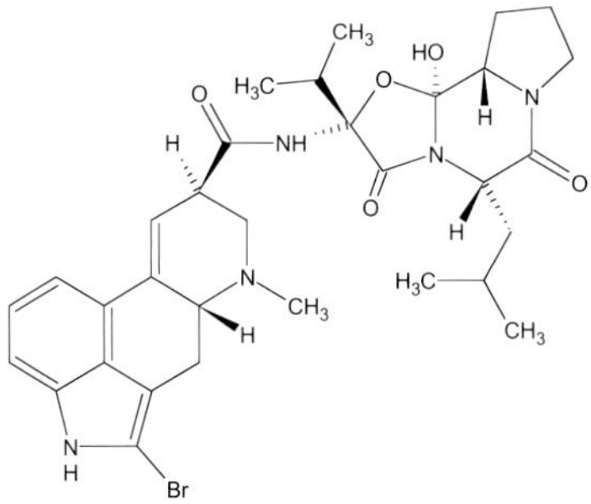	dopamine D2 agonist
Eticlopride	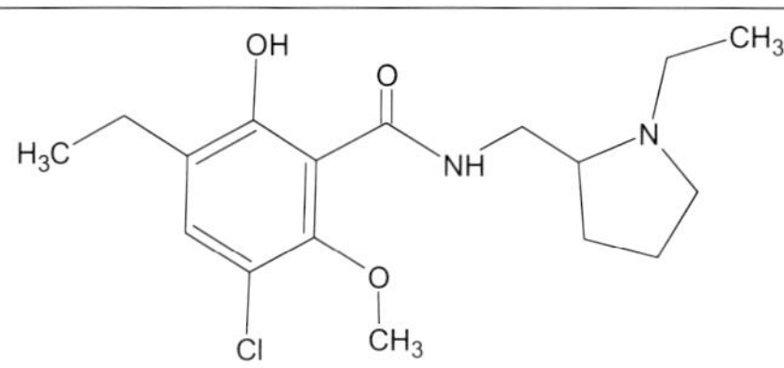	dopamine D2 antagonist
Flupenthixol	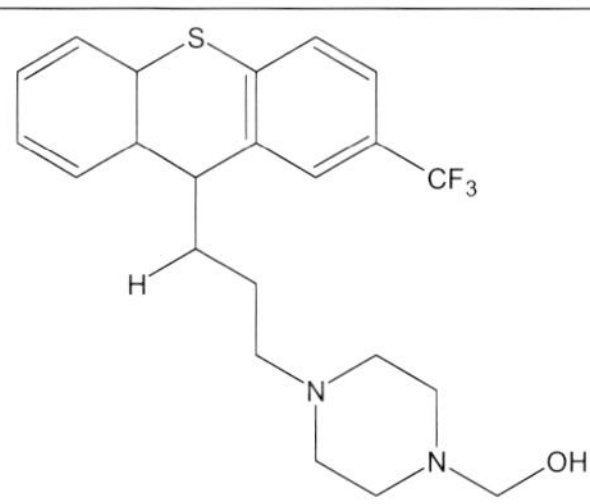	D1/2 antagonist
Haloperidol	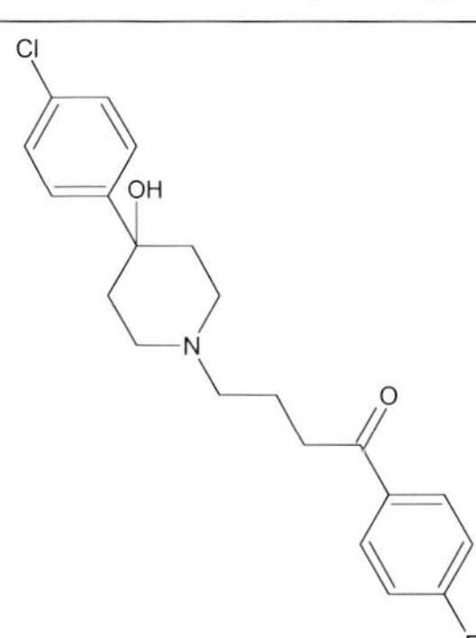	dopamine D2 antagonist

Agents acting at **dopamine** receptors (continued)

Lisuride (maleate)	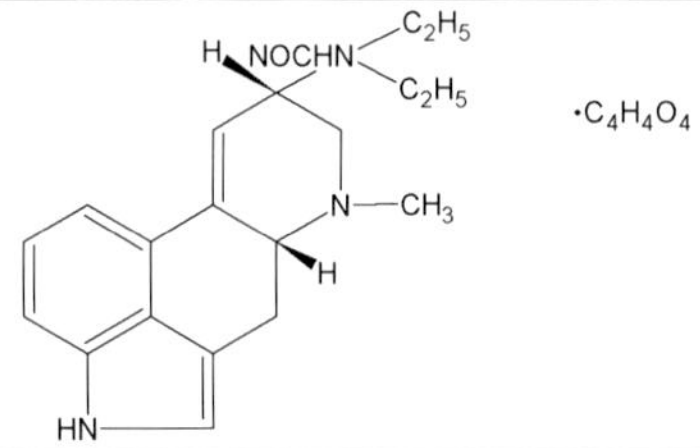	dopamine D2 agonist
Methoxybenzamide	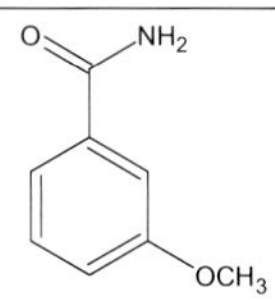	selective dopamine D2 antagonist
Raclopride	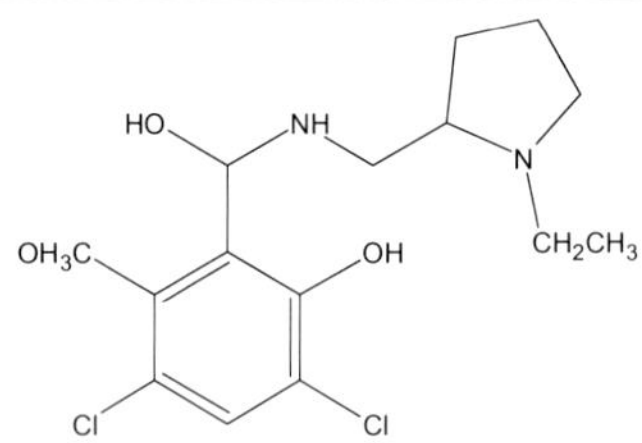	selective D2 antagonist
Remoxipride	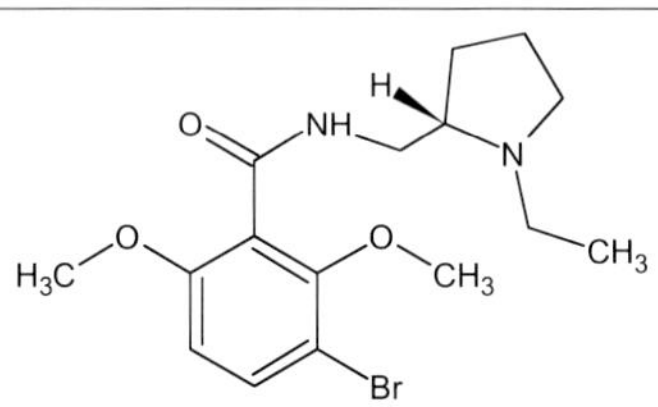	D2 antagonist
SCH 23390	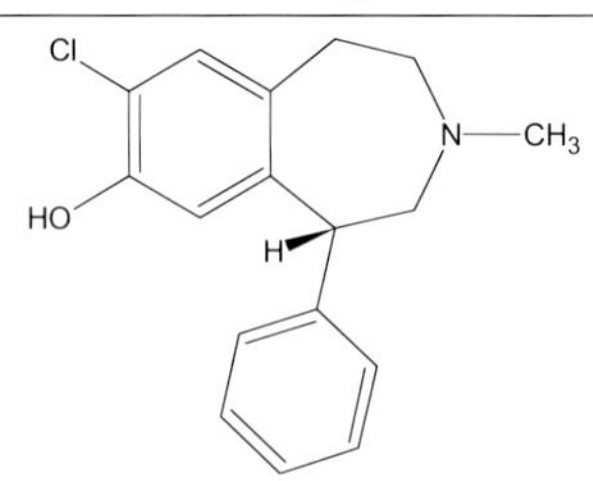	dopamine D1 antagonist
SDZ 208-911	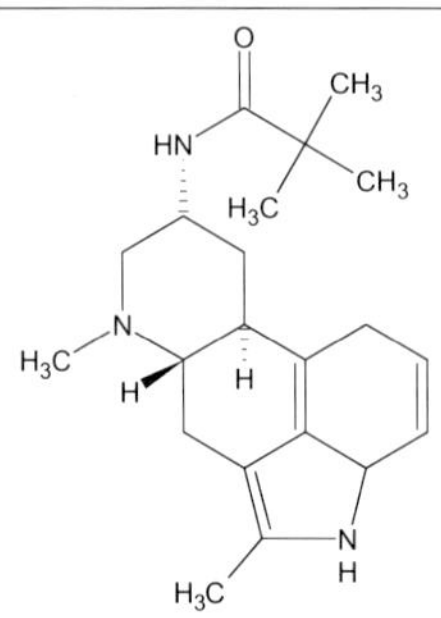	partial D2 agonist

Agents acting at **dopamine** receptors (continued)

Terguride	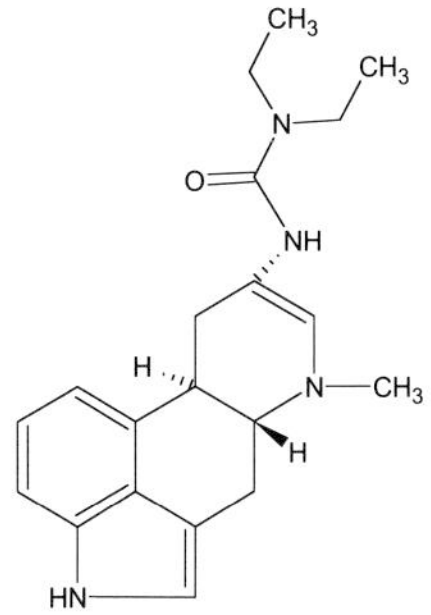	partial D2 agonist
Tiapride HCl	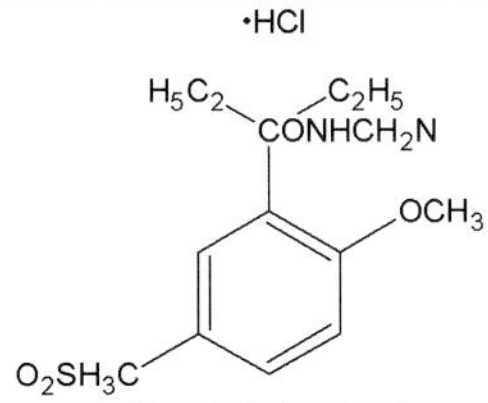	selective D2 antagonist
U99194A	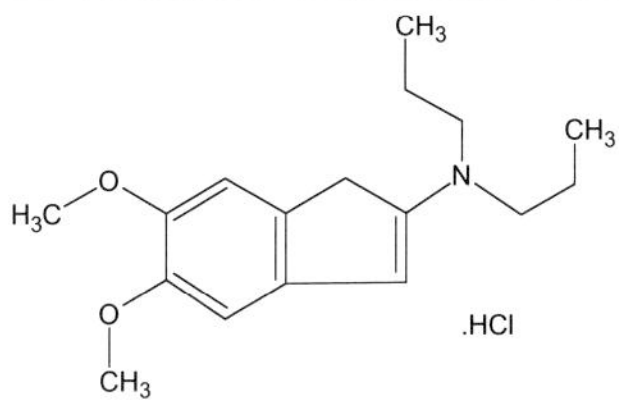	dopamine D3 antagonist

Agents acting at **neuropeptide Y** receptors

BIBP 3226	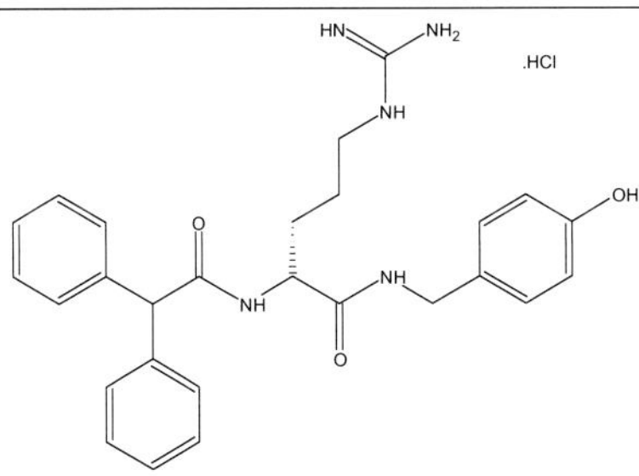	selective neuropeptide Y1 antagonist
BIIE 0246	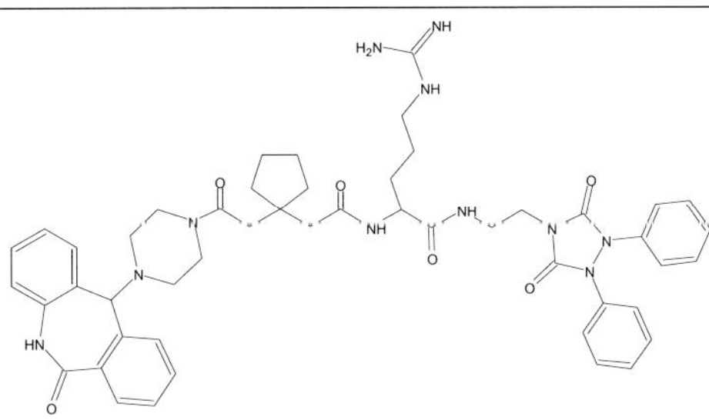	non-peptide neuropeptide Y2 antagonist

Agents acting at **opioid** receptors

_-funaltrexamine	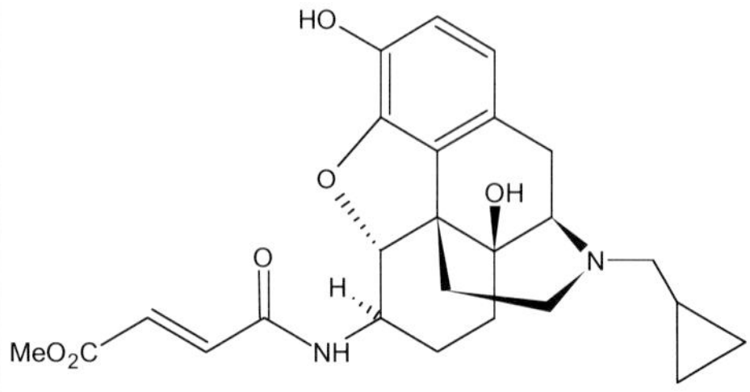	μ-antagonist
CTOP	H-D-Phe-Cys-Tyr-D-Trp-Orn-Thr-Pen-Thr-NH_2	μ-antagonist
ICI 174864	*N,N*-diallyl-Tyr-Aib-Aib-Phe-Leu	δ-antagonist
Nalmefene	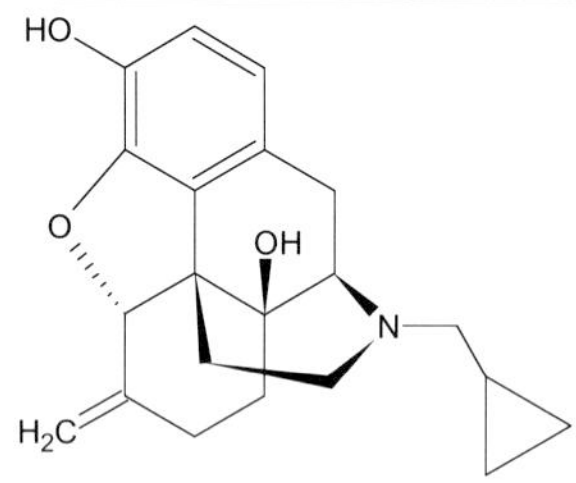	μ-antagonist
Naloxonazine	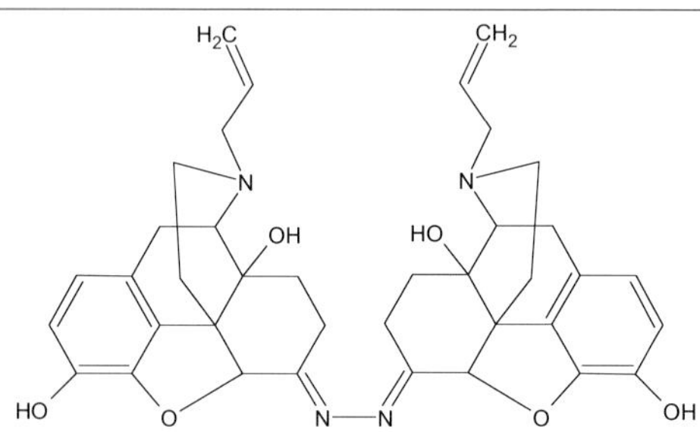	μ1-antagonist
Naloxone	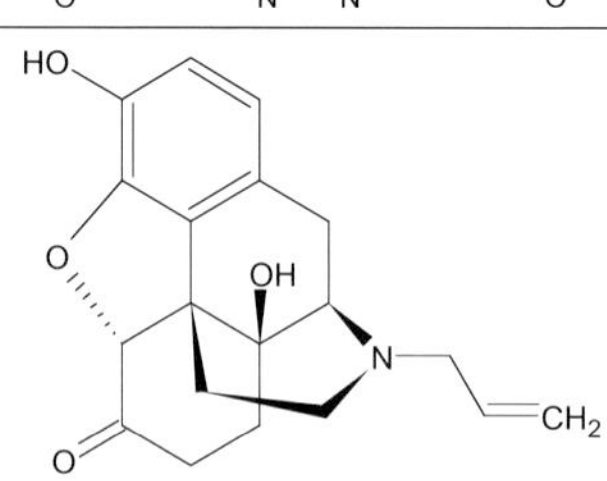	nonselective antagonist
Naltrexone	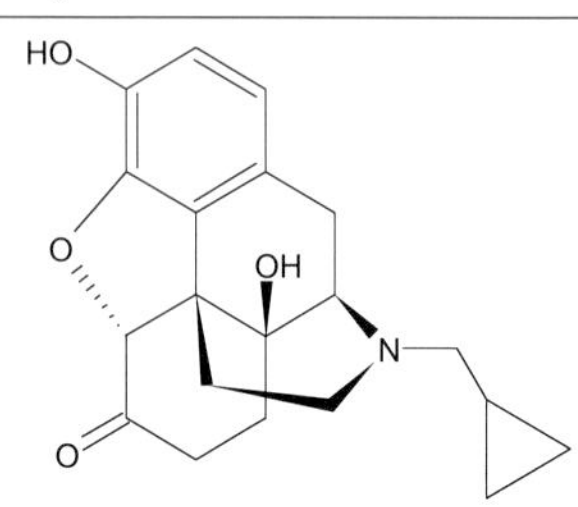	nonselectiv, but higher affinity μ-antagonist

Agents acting at **opioid** receptors (continued)

Naltriben	DPhe-Cys-Tyr-DTrp-Orn-Thr-Pen-Thr-NH_2	δ2 antagonist
Naltrindole	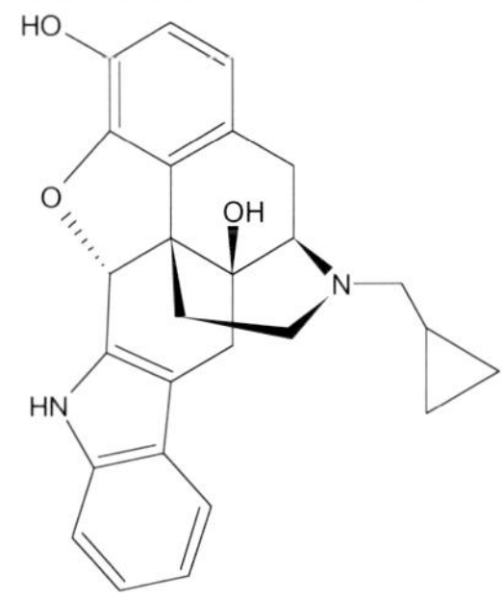	δ antagonist

Agents acting at **cannabinoid** receptors

CP55,940	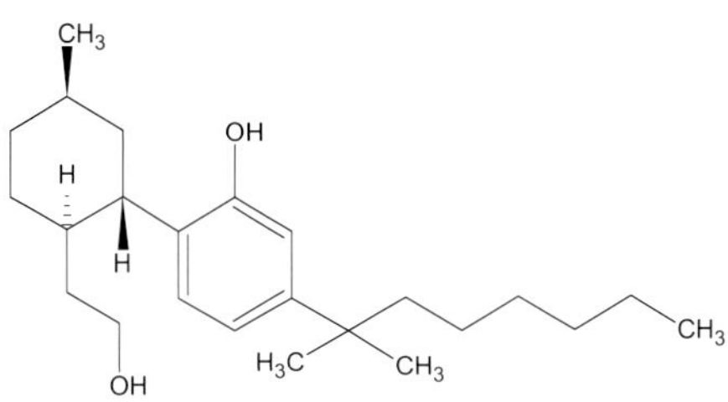	CB1/CB2 agonist
SR 141716 (A) (Rimonabant)	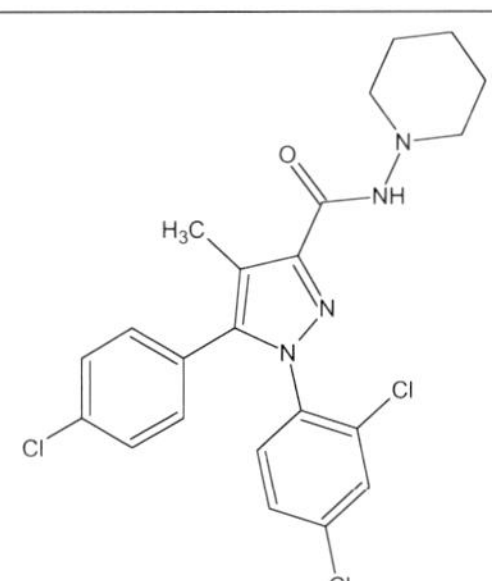	CB1 antagonist
WIN 55,212-2	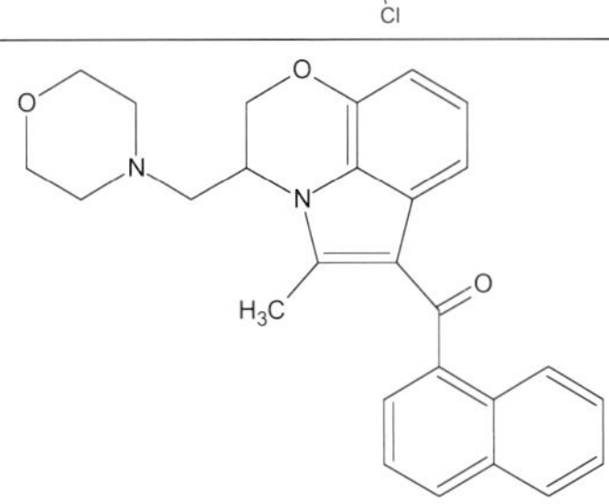	CB1 agonist

Index